VETERINARY TOXICOLOGY

(A Textbook as per VCI Syllabus)

VETERINARY TOXICOLOGY

(A Textbook as per VCI Syllabus)

Edited by

Dr. Satish K. Garg

BVSc & AH (Hons) MVSc PhD FST
Associate Professor
Department of Veterinary Pharmacology and Toxicology
College of Veterinary Science and Animal Husbandry
C.S. Azad University of Agriculture and Technology
Mathura Campus, Mathura - 281 001 (INDIA)

CBSPD

CBS Publishers & Distributors Pvt Ltd

New Delhi • Bengaluru • Chennai • Kochi • Kolkata • Lucknow • Mumbai
Hyderabad • Jharkhand • Nagpur • Patna • Pune • Uttarakhand

Notice

The Editor and the Contributors have made every possible effort to ensure that the line of treatment suggested in this book are as per the rationality and the recent information available in the literature. The choice of drug(s) and drug dosages are in accordance with the accepted standards and practices. However, Medicine is an ever and fast changing science and the research, clinical experience/evaluations and clinical standards broaden our knowledge and accordingly, the changes in the treatment are required. In view of the possible human error and also changes in medical science, the Editor, the Contributors and the Publishers warrant that though the information given in this book is accurate and complete but they should not be held responsible for any errors or omissions as a result of the use such information. The reader is encouraged to refer to the product information of each and every drug which he/she intends to use in the handling and management of cases of poisoning. This information is particularly important with the new and infrequently used drugs.

ISBN: 978-81-239-0705-5

First Edition: 2002
Reprint: 2003, 2004, 2006, 2007, 2009, 2010, 2011, 2012, 2014, 2018, 2023
Copyright © 2000, Satish K Garg

Published by **Satish Kumar Jain** and produced by **Varun Jain** for

CBS Publishers & Distributors Pvt Ltd
4819/XI Prahlad Street, 24 Ansari Road, Daryaganj, New Delhi 110 002, India.
Ph: 011-23289259, 23266861, 23266867   Website: www.cbspd.com
Fax: 011-23243014   e-mail: delhi@cbspd.com; cbspubs@airtelmail.in.

Corporate Office: 204 FIE, Industrial Area, Patparganj, Delhi 110 092

Ph: 011-4934 4934   Fax: 011-4934 4935; e-mail: publishing@cbspd.com; publicity@cbspd.com
Branches

- **Bengaluru:** Seema House 2975, 17th Cross, KR Road, Banasankari 2nd Stage, Bengaluru 560 070, Karnataka, India
  Ph: +91-80-26771678/79   Fax: +91-80-26771680   e-mail: bangalore@cbspd.com
- **Chennai:** 7, Subbaraya Street, Shenoy Nagar, Chennai 600 030, Tamil Nadu, India
  Ph: +91-44-26680620, 26681266   Fax: +91-44-42032115   e-mail: chennai@cbspd.com
- **Kochi:** 42/1325, 1326, Power House Road, Opp KSEB, Power House, Ernakulum Kochi 682 018, Kerala, India
  Ph: +91-484-4059061-65,67   Fax: +91-484-4059065   e-mail: kochi@cbspd.com
- **Kolkata:** 147, Hind Ceramics Compound, 1st Floor, Nilgunj Road, Belghoria, Kolkata-700056, West Bengal, India
  Ph: +033-25633055, 033-25633056   e-mail: kolkata@cbspd.com
- **Lucknow:** Basement, Khushnuma Complex, 7 Meerabai Marg (Behind Jawahar Bhawan),Lucknow-226001, UP, India
  Ph: +0522-4000032   e-mail: tiwari.lucknow@cbspd.com
- **Mumbai:** PWD Shed, Gala no 25/26, Ramchandra Bhatt Marg, Next to JJ Hospital Gate no. 2, Opp. Union Bank of India,
  Noorbaug, Mumbai-400009, Maharashtra, India
  Ph: 022-66661880/89   e-mail: mumbai@cbspd.com

Representatives

- Hyderabad   0-9885175004
- Patna   0-9334159340
- Jharkhand   0-9811541605
- Pune   0-9623451994
- Nagpur   0-9421945513
- Uttarakhand   0-9716462459

Printed at Glorious Printer, Dilshad Garden, Delhi

Foreword

Development of Veterinary Science Education in the country has been mainly responsible for usherring White Revolution and increasing livestock production. This has brought in economic prosperity to farming community. Our scientists have utilized developments in science and technology for bringing economic benefits to the farmers. Since the establishment of the first State Agricultural University at Pantnagar there has been tremendous expansion of Veterinary Education in recognition of the fact that this would lead to increasing production of animals and animal products. In India, the animals are taken as a part of family and owners have tremendous attachement to animals because they ensure for them livelihood, food and nutritional security.

For the healthy growth and higher productivity of animals, issues relating to toxicology become very critical and important. Our scientists have been taking advantage of developments in medical sciences for developing better and targeted drugs for improving animal health. During undergraduate and postgraduate education special emphasis is given on skill development relating to various issues of Pharmacology and Toxicology. One of the requirements for skill development in this area is the availability of good text books covering various aspects of Toxicology. Many of the text books developed by foreign authors may not be very relevant to India in view of the toxicity problems that exist in our country. Therefore there is need to develop excellent text-books covering all aspects of toxicology by Indian Scientists. Dr. Satish Kumar Garg, Department of Pharmacology and Toxicology, College of Veterinary Science and Animal Husbandry, C.S. Azad University of Agriculture and Technology, Mathura has undertaken this important task of developing the book of "Veterinary Toxicology". He has invited contributors who have specialized in different areas to contribute. The book covers all subjects from general toxicology to toxicology of metals, non-metal, plants, agrochemicals and drugs, environmental toxicology and drugs, chemicals and plants- induced carcinogenicity, genotoxicity and reproductive toxicity.

I would therefore like to commend the efforts of Dr. Satish K. Garg for bringing out an excellent publication. My congratulations to all the authors who have contributed for this text book. This would be extremely useful to students, staff, research scholars, field Veterinarians and all those interested in good health of animals.

(S.L. Mehta)

New Delhi
March 13, 2000

Deputy Director General (Education)
Indian Council of Agricultural Research
Krishi Anusandhan Bhawan
Pusa, New Delhi- 110012

Contributors

Dr. A.K. Srivastava, PhD
Professor and Head
Department of Pharmacology and
Toxicology
Punjab Agricultural University,
Ludhiana - 141 004

Dr. I.A. Siddiqui, PhD
Professor and Head
Department of Animal Nutrition
C.S.Azad Univ. of Agri. and Technology
Mathura Campus, Mathura - 281001

Dr. N.K. Mahajan, PhD
Senior Assistant Professor
Department of Veterinary Public Health
CCS Haryana Agricultural University
Hisar-125004

Dr. Om P. Sharma, PhD
Senior Scientist
Indian Veterinary Research Institute
Regional Research Station,
Palampur-176061

Dr. R.K. Dawra, PhD
Senior Scientist
Indian Veterinary Research Institute
Regional Research Station,
Palampur - 176061

Dr. Satish K. Garg, PhD FST
Associate Professor
Department of Pharmacology and
Toxicology
C.S.Azad Univ. of Agri. and Technology
Mathura Campus, Mathura - 281 001

Dr. S.P. Verma, PhD
Professor
Department of Pharmacology and
Toxicology
CCS Haryana Agricultural University,
Hisar - 125004

Dr. Vinod Kumar Dumka, MVSc
Assistant Professor
Department of Pharmacology and
Toxicology
Punjab Agricultural University,
Ludhiana - 141 004

Dr. H.S. Panwar, PhD
Professor and Head
Department of Pharmacology and
Toxicology
C.S.Azad Univ. of Agri. and Technology
Mathura Campus, Mathura - 281001

Dr. M.A. Ayub Shah, PhD
Senior Assistant Professor
Department of Pharmacology and
Toxicology
C.S.Azad Univ. of Agri. and Technology
Mathura Campus, Mathura - 281 001

Dr. Neeraj Sinha, DSc
Scientist, Division of Toxicology
Central Drug Research Institute
Lucknow - 226 001

Dr. Rakesh K. Chaudhary, PhD
Associate Professor
Department of Pharmacology and
Toxicology
Punjab Agricultural University,
Ludhiana - 141004

Dr. R.P. Uppal, PhD
Professor
Department of Pharmacology and
Toxicology
CCS Haryana Agricultural University,
Hisar - 125 004

Dr. S.P. Singh, PhD
Associate Professor
Department of Pharmac. & Toxicology
G.B. Pant Univ. of Agri.and Technology
Pantnagar - 263145

Dr. V.P. Vadlamudi, PhD
Professor and Head
Department of Pharmacology and
Toxicology
Marathwada Agricultural University,
Prabhani - 431 402

Dr. Yogeshwer Shukla, PhD
Assistant Director
Environmental Carcinogenesis Division
Industrial Toxicology Research Centre
Lucknow - 226 001

Preface

The overall standards of Undergraduate Veterinary Education in India are being governed by the Veterinary Council of India under Minimum Standards of Veterinary Education Degree Course -B.V.Sc.&A.H. Regulations 1993 and the Regulations for Minimum Standards of Post-graduate Education in Veterinary Sciences have already been drafted by the Veterinary Council of India and are almost in final stages of promulgation.

The Veterinary Toxicology, in recent years, has acquired greater importance than ever before because of the increasing and indiscriminate use of farm chemicals and drugs and the impact of environmental pollutants on the health of animals. In addition, domestic animals are also exposed to natural toxicants in the form of phytotoxins and mycotoxins due to their indiscriminate feeding behaviour. This fact, in turn, has given fillip to research on various aspects of Toxicology such as toxicoepidemiology, toxicokinetics, toxicodynamics, diagnosis including differential diagnosis and line of treatment in different species of animals. Unfortunately, most of the information on different aspects of Veterinary Toxicology are scattered in different journals, many of which are not easily accessible to the readers, veterinary clinicians and researchers. Further, there are limited Text Books on this subject which too are cost prohibitive for the students and teachers of developing countries. These constraints prompted the Editor to edit the multiauthored book of "Veterinary Toxicology", keeping in view primarily the Course Curricula prescribed by Veterinary Council of India and the requirements of the practicing Veterinarians and post-graduate students, with the coopration and contributions of eminent teachers and researchers having wide experience of teaching and research in different subdisciplines of Toxicology.

It is needless to say that the intent of the Editor and the contributing authors of this text book is primarily to provide a comprehensive information on the basic and applied principles of toxicology including Clinical Toxicology in one book. It is hoped that this book will be well received and be of great value to all those concerned with Veterinary Toxicology, be they students, teachers, investigators or clinicians. The Editor and the contributors sincerely seek critical feedback from the students, academicians, clinicians and others as to whether or not the book is meeting its goals and objectives.

As the Editor of the book, I have the pleasure of expressing my sincere gratitude and appreciation to all my colleagues who have made the first edition of the "Veterinary Toxicology" a reality. It is especially important to acknowledge and thank all the authors without whose help and efforts, it would not have been possible to bring out this edition of the book. I wish to put on record my sincere thanks to Dr. M. A. Ayub Shah and Dr. H.S.Panwar for their whole hearted support from the beginning not only for contributing certain chapters but also for supporting in proof reading of the manuscript. Dr. R.P. Pandey, Senior Assistant Professor of Veterinary Surgery and Radiology and Dr. Rajesh Katoch of HPKV, Palampur deserve special thanks for excellent photography of certain toxic plants found in and around Mathura and Palampur, respectively. I am also indebted to Dr. Jitendra Kumar for his kind help in bringing and providing lot of literature from the Library of Haryana Agricultural University, Hisar. I am also grateful to Mr. Suresh Saraswat of M/s Saraswat Computer Graphics, Jamuna Bagh Road, Mathura for meticulous type setting of the manuscript. Last but not the least, the cooperation and help and the interest of CBS Publishers, New Delhi in bringing out this book is highly appreciated.

Mathura   **Satish Kumar Garg**
May 01, 2000

Toxicology has been recognised as a separate discipline in Veterinary Sciences both at Under-graduate and Post-graduate levels for the last almost four decades and a lot of emphasis has been given on its teaching, research and practical utility from clinical point of view. On the contrary, toxicology has not been given that much importance in Medical Curricula as it is being covered to a liimited extent that too as a component/part of Forensic Medicine.

This book also aims to atleast give a very basic and some of the clinical toxicological aspects to those associated with the health care of human beings. The suggestions, criticisms and comments from the readers are welcome.

Contents

CHAPTER 1

GENERAL TOXICOLOGY

Satish K. Garg

1.1 INTRODUCTION, HISTORY AND SCOPE OF TOXICOLOGY

INTRODUCTION TO TOXICOLOGY

The word 'toxicology' is derived from the Greek word 'toxicon' which means 'poison'. Thus, toxicology literally means the 'study of poisons'. It can thus be defined as "the study of poisons and their harmful effects on living organisms". It includes the sources of poisons, their identification, physicochemical properties, toxicity including median lethal dose , factors affecting toxicity, toxicokinetics, toxicodynamics, clinical signs of toxicity, post mortem changes, histopathology, diagnosis including differential diagnosis, analytical procedures and principles of treatment of the conditions that they cause. It also includes study of special effects of toxicants such as carcinogenesis, teratogenesis, mutagenesis, immunotoxicity, neurotoxicity, reproductive toxicity and ecotoxicity etc.

'Poison' may be defined as any substance which when taken inwardly or applied in any kind of manner to body, depresses the health or entirely destroys life. Toxicant is the another term used for poison. The word poison has been defined in many more ways :

*A substance which by its physical or physico-chemical properties damages functions of the living organisms once it has entered the body (Starkenstein *et al.*, 1929).

*A substance that being in solution in the blood or acting chemically on the blood either destroys life or impairs seriously the function of one or more of the organs of body (*Blakiston's New Gould Medical Dictionary*, 1949).

*A substance or matter (solid, liquid or gas) which when applied to the body outwardly or in any way introduced into the body can destroy life by its own inherent qualities without acting mechanically and irrespective of the temperature (*Black's Veterinary Dictionary*, 1953).

'Xenobiotics' (*Xeno* is a Greek word which means "strange or alien") are the substances which are foreign to the body and are biologically active. These can not be broken down to generate energy or be assimilated into a biosynthetic pathway. It is a very wide class which includes structurally diverse agents both natural and man made chemicals such as drugs, industrial chemicals, pesticides, alkaloids, secondary plant metabolites and toxins of molds, plants and animals and environmental pollutants.

Poison is any substance which when taken inwardly in a very small dose or applied in any kind of manner to a living body depraves the health or entirely destroys life (M.J.B. Orfila, 1821). However, Paracelsus (1493-1541) recognised that all substances are poisons, there is none which is not a poison. The right dose differentiates a poison from a remedy and put forth the concept of dose-dependency of the toxic effects of substances. This principle holds true even today.

Toxicity not only depends on the toxicant or poison, but also the quantity, species involved and route through which it enters the body. Common man does not realize that sugar and common salt could be poisonous or deleterious for the health if taken in excess. Similarly, some of the agents which are too potent and lethal may not be toxic e.g. cobra venom if taken orally is not lethal if the gastrointestinal tract (GIT) is intact, however, it is lethal if administered parenterally.

Any definition of toxicology is inadequate as what is harmful under one set of conditions or in one species of animals may be harmless under other circumstances or in other species of animals.

Toxicity : It may be defined as the inherent capacity of a substance to produce toxic effects or detrimental changes on the organisms.

Toxicosis: It is the state resulting from exposure to a poison. This term is being used interchangeably with poisoning and intoxication.

Toxins : Toxins are the toxicants or poisonous substances liberated / produced by living organisms and are generally not well defined chemically. These are also termed as biotoxins. Depending on their origin, these are grouped differently - *phytotoxins* (plant toxins), *mycotoxins* (fungal toxins), *zootoxins* (toxins of lower animals e.g. bufotoxin, snake venom etc.) and *bacterial toxins* (liberated by bacterial cells) which are further of two types, namely- *endotoxins* (found within or as a part of the bacterial cells) and *exotoxins* (elaborated from bacterial cells).

Toxinology : It is the branch of toxicology that deals with the study of toxic effects of toxins.

Hazard or risk : It is the likelihood of poisoning of a living organism after exposure to a particular toxicant. A highly toxic chemical may not be that hazardous as the least toxic chemical as is evident from the example given in Table 1.1.

HISTORICAL DEVELOPMENTS

Poisonous properties of certain plants and animals and the problems associated with them are as old as mankind. Poisoning sometimes assumes the dimensions of environmental disaster e.g. about 40,000 people died of ergotism in 992 in France and Spain, Bhopal gas tragedy due to methylisocyanate in India in December 1984 and nuclear accidents at Three Mile Island near Harrisburg, Pennsylvania in 1979 and Chernobyl in Soviet Union in 1986.

Table 1.1: Comparison of the toxicity and hazard

Compound	LD_{50} (mg/kg)	Toxicity	Use level (mg/kg)	Risk ratio
A	300.00	Less	100.00	3:1
B	20.00	More	1.00	20:1

Use of poisons is as old as the human race. Some people of India, Africa and Asia, particularly the tribals living in jungles and remote areas have good knowledge of the poisonous plants and animals as some of the animal venoms/toxins and plant extracts are being used by them for catching fish, poisoning arrow heads for hunting as warfares and assassination of their foes.

Ebers papyros (1500 BC) is an ancient document reflecting that Egyptians had vast knowledge of many poisons, including hemlock (the state poison of Greeks), aconite (a Chinese arrow poison), opium (used both as a poison and an antidote). Greeks, Romans and Italians used poisons for execution and murder of their political opponents. Socrates was executed with an extract of hemlock (*Conium maculatum*) and it was recognised as the state poison of Greek. Knowledge of poisons has been expeditiously used for suicides. Demonsthenes (385-322 BC) committed suicide by consuming a poison hidden in his pen. King Mithridates VI of Pontus used war prisoners and criminals for acute toxicity experimentation. And, in such an attempt discovered antidotes for several venomous reptiles and poisonous substances. He himself was so much scared of poisons that he used to regularly ingest an antidotal mixture of 36 ingredients to save him of any assassination attempt by his foes. Once he was caught in war by his enemies and wanted to commit suicide by consuming some poison, but failed because he had already been taking an antidotal mixture regularly.

Poisonings attained epidemic proportions in Rome during the fourth century BC. During this period, women made use of poisons to kill people, particularly their husbands to get their property and wealth. Theophrastus (370-286 BC) included numerous poisonous plants and their actions in *De Historica Plantarum*.

Greek physician Discorides (50 AD) classified poisons for the first time into plants, minerals or/and animal poisons. Arsenic containing perfumes were prepared by lady named Toffana and such cosmetics termed as *Aqua toffana* were used to kill the foes. Such arsenic containing cosmetics have been reported to be responsible for deaths even during the 20th century.

During middle ages, the toxic concoctions, onset of their action, potency, specificity, site of action and clinical signs and symptoms of poisoning were recorded. In France, Catherine de Medici, a lady along with Marchioness de Brinvillers used the most effective poisons on sick and poor people in the name of treating them and killed several people and Catherine was given the title of *"La Voisine"*. Later she was convicted of many poisonings including over 2000 infants.

'*Poison and their antidotes*' describing the treatment of poisonings from insects, snakes and mad dogs was written by Moses ben Maimon (1135-1204 AD). Probably the alchemists in search for an universal antidote were successful in getting 60 per cent ethanol beverage while distilling fermenting products.

The age of enlightenment started with the realization put forth by the famous Swiss physician Paracelsus (1493-1541) that every substance is a poison and it is the dose which distinguishes a poison from a remedy. He gave the concept of dose- dependency of the toxic effects of substances which remain an integral part of toxicology, pharmacology and therapeutics. He also emphasized the importance and necessity of experimentation in evaluating the responses to chemicals, therapeutics and toxic properties, degree of specificity etc. And based on these parameters, Paracelsus introduced the use of mercury as the drug of choice for treatment of syphilis.

Metals are probably the oldest toxins known to mankind. Lead was used even before 2000 BC. Hippocrates (370 BC) referred to abdominal colic in persons extracting metals - arsenic and mercury as cited by Theophrastus (370 - 236 BC). The major documentation work on occupational hazards associated with the people working in mines was a treatise entitled, "*Miner's sickness and other diseases of miners*" by Paracelsus in 1567. In 1700, "*Discourse on the Diseases of Workers*" was published by Ramazzini. Incidence of scrotal cancer in chimney sweepers due to exposure to polyaromatic hydrocarbons was recognised in 1775 by Percival Pott.

Experimental toxicology was promoted by Magendie (1783-1885), Mattie Josesph Benaventura Orfila (1787-1853) and Bernard (1813-1878). M.J.B. Orfila was the first toxicologist to make use of the autopsy material as an evidence in a medico-legal case of poisoning and laid the foundation of *forensic toxicology* and described the correlation between persistence of chemicals in the body and their physiological effects. He is recognised as the father of modern toxicology. To detect the presence of toxicants in the body, certain analytical methods were developed by him. He also devised certain antidotal therapies.

Magendie discovered the mechanism of action of emetine, strychnine and arrow poisons while Claude Bernard discovered that of curare and carbon monoxide. Phosgene ($COCl_2$) and mustard gas were used as war gases in world war I. Organophosphate compounds (cholinesterase inhibitors) were discovered during the world war II. Thereafter, vast expansion, on all aspects of science of toxicology including publication of text books and journals, legislations and formation of societies etc. has taken place. With the introduction of stringent legislations for use of colouring agents and preservatives in food stuffs, *analytical toxicology* has become much more important, and the environmental protection agencies are also impressing upon the realization of thorough understanding of the risks involved and the essentials of adequate protection of man, animals and the ecosystem. Survival of humanity depends on the survival of other species (plants and animals) and availability of clean water, soil and energy. Survival of humans, animals and the ecosystem alike is the ultimate goal of the study of toxicology. And, therefore, application of the discipline of toxicology in the safety evaluation and risk assessment is of utmost importance in today's modern life.

SCOPE OF TOXICOLOGY

Toxicology is a multidisciplinary science and thus to study the toxic effects and behaviour of toxic substances, thorough understanding of the anatomy, histology, physiology, biochemistry, pathology and analytical chemistry is important. Integration of cell biology, immunology, molecular biology and organic chemistry is must for proper understanding of the underlying toxicological processes and mechanisms (Fig.1.1). With rapid advancements in various scientific disciplines, to consider toxicology as a whole is not justified. Therefore, over the years, various specialized subdisciplines of toxicology have been recognised such as clinical toxicology, toxicoepidemiology, immunotoxicology, neurotoxicology, genetic toxicology, molecular toxicology, nutritional toxicology, occupational toxicology, environmental toxicology, ecotoxicology, predictive toxicology, forensic toxicology, etc.

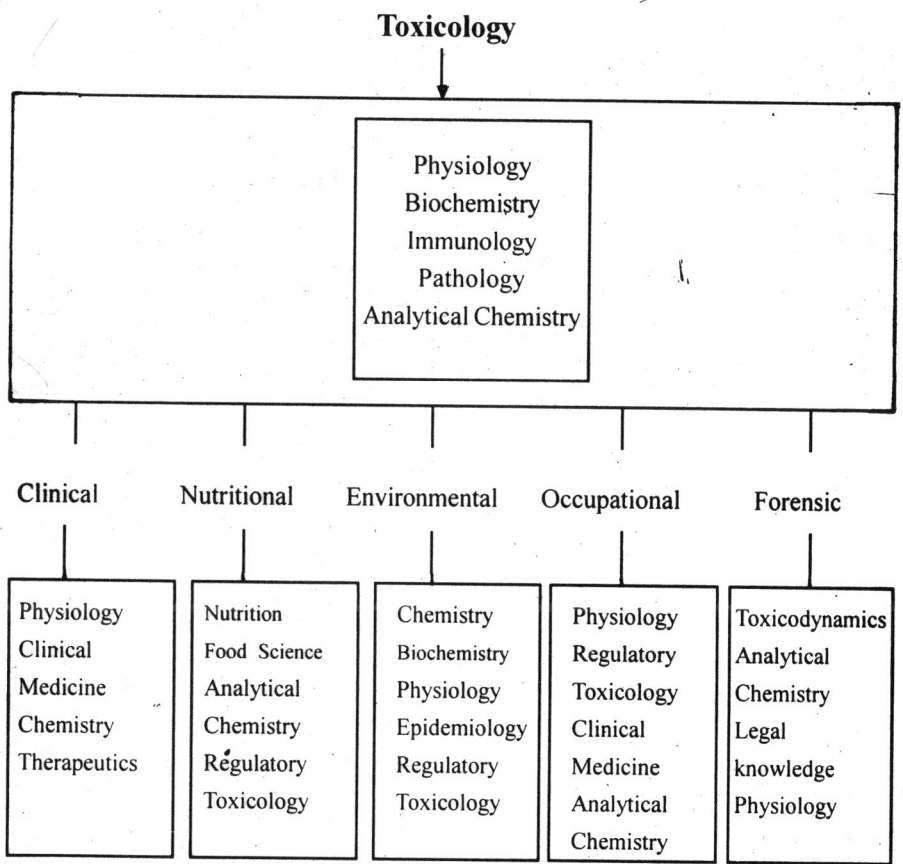

Fig. 1.1 : Toxicology, its emerging subdisciplines and their inter-relationship with biomedical and other sciences.

Clinical toxicology : It is the study of the effects of poisons/toxicants on human beings, animals and other living organisms, their diagnosis and treatment and methods for their detection etc.

Toxicoepidemiology: It refers to the study of quantitative analysis of the toxicity incidences in organisms, factors affecting toxicity, species involved and the use of such knowledge in planning of prevention and control strategies.

Nutritional toxicology : It is the study of toxicological aspects of food / feed stuffs and nutritional habits.

Environmental toxicology : It is the study of effects of toxicants, whether used/applied purposely (e.g. pesticides, herbicides) or as industrial effluents or pollutants/contaminants, on the health of organisms and environment.

Analytical toxicology : It is the application of analytical chemistry tools in the qualitative and quantitative estimation of the agents involved in the process of toxicity.

Occupational toxicology : It is the study of occupational hazards associated with individuals working in a particular industry/occupation and their correlation with the possible toxicants and also the possible remedial measures.

Ecotoxicology : It is the study of fate and effects of toxic substances on ecosystem.

Regulatory toxicology : It is the conduct of toxicological studies as per the content and characteristics of studies design prescribed by regulatory agencies.

Forensic toxicology : It is the study of unlawful use of toxic agents and their detection for judicial purposes.

Reproductive toxicology : It is the study of occurrence of adverse effects on the male or female reproductive system due to exposure to chemical or physical agents.

Developmental toxicology : It is the study of adverse effects on the developing organisms occurring any time during the life span of the organism due to exposure to chemical or physical agents before conception (either parent), during prenatal development or postnatally until the time of puberty.

Teratology : It is the study of malformations induced by the toxic agents during development between conception and birth.

Following exposure to a poison/toxicant, toxicity occurs as a result of the interaction between the toxicant and the target sites. Thus, understanding of the toxicokinetic and toxicodynamic concepts is of paramount importance.

Toxicokinetics : It refers to the process of absorption, distribution, biotransformation and excretion of toxicants in relation to time, i.e. what the body does to the poison with relation to time.

Toxicodynamics : It refers to the study of biochemical and physiological effects of toxicants and their mechanisms of action and comprises the sequence of events following interaction of the toxicants with target molecules.

1.2 SOURCES OF POISONING

Poisoning and the intensity of harmful effects depends on the form of exposure to toxic substances. Poisoning in human beings may be accidental, intentional, malicious or as an occupational hazard, however, poisoning in animals is generally either accidental or malicious. Intentional toxicity in humans may be through abuse (narcotics) or by way of therapy through inadvertent use of drugs (adverse effects of drugs) or for committing suicides.

Sources of poisoning are diverse. These may enter the body by way of contact, inhalation, ingestion etc. may be accidentally, unintentionally or intentionally. Depending on the sources, these are categorised as :

I. Natural sources : Most of the intoxicants of natural origin enter the body through food chain (food or water). Some of the rocks and soils are very rich in some of the toxic minerals e.g. fluorine, arsenic, selenium, lead etc. Thus, the plants, crops or grasses grown in these areas are rich sources of these intoxicants e.g. high quantity of nitrates in well water, copper in non-graminous plants, nitrate in spinach and selenium in seleniferous plants etc.

Certain areas are rich in poisonous plants depending on the agroclimatic conditions, geographical locations etc. Generally, the toxic plants have a repulsive odour and are usually refused by the animals, however, during the peak summer or winter months when there is scarcity of green fodder, some of the toxic/poisonous plants are consumed by the animals, especially when they are hungry e.g. *Lantana camara, Quercus incana, Acacia leucophloea,* Sudan grass etc. Some of the fodder crops are very rich in toxic principles such as cyanogenetic glycosides, nitrates, nitrites, oxalates, selenium etc.

Mycotoxins are another common natural source of poisoning in both, human beings and animals as stored feed stuffs, particularly the concentrates rich in maize, groundnut, rice bran, oil cakes etc. are very rich in mycotoxins. Several reports of mycotoxicosis in humans, animals and birds are there. Other natural sources of toxins are : bacterial toxins, venoms and other zootoxins e.g. scorpion toxin, tick toxins, bufotoxin etc.

II. Man made sources : Humans are one of the biggest, most common and threatening sources of poisoning. Humans induced poisoning may be accidental, malicious or occupational.

(i) **Accidental poisoning** : Poisoning through the contamination of food and water with toxicants, industrial effluents, improper and injudicious use of drugs and chemicals including food additives, food preservatives or colouring agents, rodenticides, pesticides, herbicides, fumigants, accidents in chemical processing units like Bhopal gas tragedy (Dec. 3-4, 1984), chemical and biological warfares, radiation hazards or accidents in atomic energy generating units (Chernobyl episode). Similarly, animals are also poisoned due to dumping of industrial wastes and pasture treatments. Careless handling and use of agrochemicals, disposal of containers of chemicals or agrochemicals and domestic materials add to man made sources of poisoning.

Industrial and petrochemical emissions and other environmental pollutants have over the years been recognised as one of the greatest threat not only to human and animal health but also the entire ecosystem. Environmental problems have assumed dimensions of global magnitude.

(ii) Malicious or intentional poisoning : It is the unlawful or criminal killing of human beings or animals by administering certain toxic/poisonous agents. Incidence of such poisonings is more prevalent in human beings and less in animals. Yet some of the reports are there where *Abrus*, strychnine, rodenticides, arsenic trioxide or agrochemicals are mixed in food or drinking water or administered through some other route to kill the animals.

(iii) Occupational sources : Working environment presents the risk of worker's over-exposure to various chemicals or hazardous agents. Occupational poisoning refers to poisoning of persons working with hazardous / toxic chemicals. Occupational hazards associated with people working in mines were recognised during the fifteenth century. Ellenbug (1480) warned of mercury and lead toxicity in people working in goldmining. Occupational workers in the chemicals, metals, tanneries, glass, paint, coal, asbestos industries or the persons involved in public health programmes are prone to the toxic effects. e.g. use of DDT or some other chemicals under the National Malaria Eradication Programme, people involved in the spray of agrochemicals on the field crops, painters etc. are worst sufferers. Therefore, a rational biological monitoring of the exposure in terms of air borne concentration of chemicals, evaluation of the intensity of exposure and early markers of adverse effects on health are stressed upon which are possible only when sufficient toxicological information has been gathered on the mechanism of action and/or metabolism of xenobiotics to which workers may be exposed. With increasing incidences of occupational hazards, the concept of industrial toxicology is coming up with the objective of prevention of health impairments in workers engaged in industrial units. And, this objective can only be achieved if exposure limits are well defined, e.g. maximum allowable concentration (MAC), threshold limit values (TLVs), threshold limit value-ceiling (TLV-C), occupational exposure limits (OELs) etc.

Types of Poisoning / toxicity : Some of the common terms used while describing toxicity are:

Toxic dose: It is the lowest dose of a toxicant which produces toxic effects in living organisms.

Lethal dose: It is the lowest dose of a toxicant which causes lethality or death in animal(s) during observation period.

Median lethal dose (LD_{50}): It is the dose of a toxicant which produces lethality in 50 per cent of the population/animals exposed to that particular agents/compound.

Depending on the agent used, dose, route, frequency and duration of exposure/administration, latency period, rapidity, severity of the symptoms and lethality etc., toxicity may generally be divided into three subtypes :

Acute toxicity : Acute toxicity results from exposure of the organisms to relatively high doses of a compound usually a single dose or exposure over a short period (a few days). Symptoms of toxicity are more severe and onset of toxicity and/or death are rapid and some of the animals may even die without showing any symptoms. Acute toxicity data give an idea of the dose - response relationship, a quantitative estimate of LD_{50} (median lethal dose) values for comparison with other intoxicants, target organs and associated clinical manifestations. It also includes therapeutic management of the toxic effects. The LD_{50} value and other acute toxic effects should be determined following one or more routes of administration (one route being the intended route of application) in one or more species of animals, generally in rats, mice, rabbits, guinea pigs or dogs etc.

Comparative acute toxicity of different compounds/poisons is determined in terms of the LD_{50} values. Depending on the toxic effects of poisons and oral median lethal dose values (mg/kg), poisons may be grouped into : extremely toxic , highly toxic, moderately toxic, slightly toxic, practically non-toxic or relatively harmless. The general guidelines for classification of relative toxicities are presented in Table 1.2.

Table 1.2 : Classification of poisons based on the oral median lethal dose values

Classification	Median lethal dose
Extremely toxic	< 1 mg / kg
Highly toxic	1 - 50 mg/kg
Moderately toxic	50 - 500 mg/kg
Slightly toxic	0.5 - 5 g/kg
Practically non toxic	5 - 15 g/kg
Relatively harmless	> 15 g/kg

Subacute toxicity: It refers to the appearance of a mild form of toxicity symptoms following repeated administration/exposure to relatively low doses or small amounts of poisons for atleast 90 days. The toxicity symptoms develop gradually. These tests are performed to establish a "no observed adverse effect level" (NOAEL) or no observed effect level (NOEL) and to identify the specific organ(s) affected by the test compound after repeated administration.

NOEL : It is defined as the largest dose that will produce no deleterious effects when administered over a given period of time.

It is generally conducted in two species of animals (rat and dog) following administration at three dose levels by intended route of exposure. These studies not only provide data on dose - response relationship of a substance/toxicant after repeated administration, but also to predict and establish appropriate doses for chronic exposure studies.

Chronic Toxicity : It is a very mild form of toxicity syndrome following long term repeated exposure to relatively low doses of poisons over prolonged period of time, generally six months or more. Onset of toxicity is gradual, often appearing first as a decline in feed intake or weight gain etc. The protocol for chronic studies is similar to that for the subacute toxicity studies except the duration of exposure e.g. duration of exposure for rodents is six months to two years while usually one year for non-rodent species.

Chronic toxicity studies are conducted to assess the cumulative toxicity of intoxicants at maximum tolerable dose (MTD) levels. The MTD is defined as the dose that slightly suppresses body weight gain (10%) in a 90 day subacute toxicity study. Compounds that have almost negligible tendency to accumulate and are rapidly metabolised have the same chronic LD_{50} values as the acute LD_{50} values. However, the toxicants to which the subjects develop tolerance, the dose required to produce lethality on repeated exposure increases. For such agents, chronicity factor (C.F.) is determined. Chronicity factor gives an indication of the cumulative effects of poisons. It is the ratio of acute LD_{50} to chronic LD_{50} values :

Chronicity Factor (CF) = Acute LD_{50} / 90 day LD_{50}

A compound though may have low acute toxicity, but if has the tendency to accumulate in body tissues can cause subacute or chronic toxicity. Such toxicants are termed as cumulative poisons (e.g. DDT, lead, fluoride etc.). However, the toxicants which are rapidly detoxified and excreted are known as noncumulative poisons. Such agents have low potential to exhibit subacute or chronic toxicity.

In addition to the general type of toxicity described above, toxicological investigation should also include special toxicity evaluations such as mutagenicity, carcinogenicity, neurotoxicity, reproductive toxicity and teratogenicity in addition to general allergic or hypersensitivity and idiosyncrasy reactions induced by toxicants.

Mutagenicity : It is the ability of an agent to induce mutation or changes through a change in the genotype or genetic material of a cell by covalent modification of the bases in DNA particularly guanine of DNA which passes on when the cell divides. Such agents are termed as mutagens e.g. formaldehyde, hexamethylphosphoramide, cresols etc.

Carcinogenicity : It is the ability of a compound/toxicon to transform normal cells into progressively and uncontrollably proliferating ones and resulting into neoplasms or cancer. Generally, any substance shown to induce cancer / or neoplasm is termed as carcinogen e.g. soot, cadmium, methylchloranthrene.

Teratogenicity : It is the ability of an agent to induce gross structural (anatomical) or/and physiological malformations in a developing foetus during gestation.

The word 'teratology' or the study of congenital defects, is derived from the Greek word, "teras" which means "monster". Such agents are termed as teratogens e.g. thalidomide, methotrexate, colchicine, cortisone and acute principles of certain plants e.g. *Veratrum californicum*, *Lupinus caudatus*, *Oxytropis muricata* etc.

Allergy/hypersensitivity : It is an immunologically mediated reaction of the subject to a previous sensitization by the same or structurally similar chemical/poison/allergen e.g. penicillin, tubocurarine, salicylates etc. Compared to human beings, incidence of allergic reactions is far less in animals as the data are scanty.

Idiosyncrasy : It is a genetically determined qualitatively abnormal reaction of a subject to a poison/drug. In this case, out of a large segment of population, very few of the subjects may respond differently from rest of the individuals. e.g. the individuals deficient in glucose-6-phosphate dehydrogenase suffer with haemolytic crisis following administration of antimalarial drugs, barbiturates, griseofulvin, chlorates and glycosides from broad beans etc.

1.3 CLASSIFICATION OF POISONS

No single classification of poisons is applicable to the wide spectra of toxic agents. Poisons have been classified differently by different people. Different classifications have been put forth based on the physical state, chemical characteristics, interests, needs, uses, target organs, sources, biological effects, toxicity potential, mechanisms of action, environmental and human health considerations etc.

I.   **Based on the source :**

   (a)   **Naturally occurring poisons :** Plant poisons, animal toxins, microbial and fungal toxins, toxic minerals as ground water contaminants, e.g. fluoride, arsenic, lead etc.

   (b)   **Synthetic Poisons :** e.g. agrochemicals, drugs, food additives etc.

II.   **Based on physical state :** Solids, liquids, gases, dust, aerosol etc.

III.   **Physical characteristics :** Inflammable, non-inflammable, explosive, non-explosive etc.

IV.   **Physical effects :** Irritants, corrosives, non-corrosives or non-irritant, etc.

V.   **Based on chemical characteristics :** Inorganic, nitrogenous organic, non-nitrogenous organic, halogenated or non-halogenated hydrocarbons etc.

VI.   **Based on the use :** Food additives, insecticides, rodenticides, herbicides, acaricides, drugs, solvents etc.

VII. **Based on relative toxicity potential :** Extremely toxic, highly toxic, moderately toxic, slightly toxic, practically non toxic or relatively harmless.

VIII. **Based on the molecular mechanism of action :** Sulphydryl (-SH) inhibitor (e.g. arsenic, mercury), methaemoglobin producer (nitrite), cytochrome oxidase inhibitor (cyanide), cholinesterase inhibitor (organophosphate compounds), causing extensive haemorrhages (sweet clover) etc.

IX.   **Based on environmental and human health considerations :** Air pollutants, water pollutants, radiation hazards, occupational hazards etc.

X.   **Based on type of toxicity :** Acute, subacute or chronic toxicity.

XI.   **Based on target organs :**

(a) Respiratory system :	Nitrites, carbon monoxide, chlorate, *alpha* naphthyl thiourea (ANTU), cyanide etc.
(b) Nervous system :	Barbiturates, chloral hydrate, strychnine, chlorinated hydrocarbons, common salt etc.
(c) Liver :	Carbon tetrachloride, mycotoxins, lantadenes etc.
(d) Kidneys :	Mercury (Hg), aminoglycosides, mycotoxins etc.
(e) GIT :	Insecticides, mercury, arsenic, common salt etc.
(f) Heart :	Cardiac glycosides.
(g) Skeletal system :	Lead, fluorine, selenium, mycotoxins etc.

1.4 TARGETS AND MODE OF ACTION OF POISONS

Toxic effects primarily depend on the concentration and the persistence of toxic agents at their target sites. Target site / tissue is the site where a toxicant produces its deleterious effect. Some of the toxic agents directly injure or destroy the biological system e.g. strong acids and alkalies, heavy metals, carbon monoxide, hydrocyanic acid etc. while others interfere with the membrane permeability e.g. carbon tetrachloride (CCl_4), *alpha* naphthyl thiourea (ANTU), histamine, cholera toxin etc. In certain cases, parent compound is not toxic, rather the metabolite having structural similarity with the parent compound competes with the substrate for the target site e.g. parathion, fluoroacetate etc. and sometimes even generate highly reactive groups such as free radicals, electrophiles, nucleophiles or redox/active reactants and the biotransformation results in more efficient interaction with receptors or enzymes or other endogenous ligands.

The phenomenon of conversion of a non toxic or less toxic agent to a more toxic product by way of biotransformation is termed as bioactivation/toxication or lethal synthesis. e.g. organophosphate insecticide parathion is metabolized to paraoxon which is a strong cholinesterase inhibitor. Similarly rodenticide fluoroacetate is converted to fluorocitrate which enters the tricarboxylic acid (TCA) cycle and inhibits the aconitase enzyme.

Depending on the mode of action of poisons, following targets of toxicants action have been identified:

(a) Enzymes : e.g. organophosphate compounds produce toxicity through inhibition of cholinesterase, hydrocyanic acid by cytochrome oxidase, lead via inhibition of membrane bound Na^+- K^+, -ATPase, δ-aminolevulinic acid synthetase and ferrochelate etc.

(b) Ion-channels : e.g. chlorinated hydrocarbon insecticides (DDT, aldrin) and pyrethroids (fluvalinate, deltamethrin), tetrodotoxin, saxitoxin etc. through Na^+ channels.

(c) Blood Constituents : e.g. haemoglobin for nitrates, nitrites and chlorates and erythrocytes for carbon monoxide.

(d) Nucleic acid : e.g. aflatoxins, mutagenic and carcinogenic agents, radiation etc.

(e) Bone marrow : e.g. chloramphenicol, radiations etc.

(f) Bone cartilages and joints: e.g. fluoroquinolones.

(g) Liver : e.g. carbontetrachloride, chloroform, paracetamol (high doses), lantadenes, pyrrolizidine alkaloids etc.

(h) Kidney : e.g. lead, mercurial salts, aminoglycosides etc.,

(i) Haematopoietic system : e.g. benzene, lead, some snake venoms etc.

(j) Central nervous system: e.g. lead, mercury, barbiturates, strychnine etc.

1.5 TOXICOKINETICS

Toxicokinetics refers to the study of absorption, distribution, biotransformation and excretion of toxicants or xenobiotics in relation to time. An important parameter in toxicokinetics is the change in blood concentration of the toxicant with time. Therefore, understanding of the basic processes of absorption, distribution, biotransformation and excretion is essentially required.

Absorption : Toxic effects are evident only after absorption. It is important to know by which route the xenobiotic has entered the body and what are the processes involved in the passage of xenobiotics across the membranes i.e. translocation of xenobiotics.

To arrive at the receptor or target site, the xenobiotics must be translocated across the cell membrane. Translocation takes place by the following processes:

Simple diffusion : It is the most common route by which xenobiotics pass through membranes. It is not substrate specific. Small molecular weight (600) hydrophilic toxicants cross membranes by simple diffusion through the aqueous pores. However, high molecular weight hydrophobic molecules diffuse across the lipid domain of membranes and the rate of translocation depends on their lipid : water partition coefficient value, concentration gradient and degree of ionization which further depends on pka value of the toxicant and pH of the environment. Higher lipid : water partition coefficient value and lower degree of ionization favour absorption of toxicants by this process. It is characterised by movement of xenobiotics down a concentration gradient.

Filtration : Passage of small molecules of xenobiotics through the pores or aqueous channels in cell-membrane is restricted to hydrophilic compounds and low molecular mass. It takes place either by hydrostatic or osmotic force. Kidney glomeruli have relatively large sized pores (70 nm) and allow passage of water soluble molecules upto a molecular weight of 60,000, however, channels in most cells are much smaller (< 4 nm) and allow passage of xenobiotics of small molecular weight upto a few hundred only.

Facilitated diffusion : It is a carrier-mediated process and moves the toxicants across the concentration gradient. The transport process does not require the input of energy. It is more rapid than simple diffusion. But translocation depends on limitation of the carrier molecules e.g. transportation of glucose across basolateral membrane of the intestinal epithelium, from plasma to erythrocytes and blood into the CNS takes place by this process.

Active transport : It is a carrier - and energy -dependent process and translocates the xenobiotics against the concentration gradient. It is a saturable process and is blocked by metabolic inhibitors e.g. excretion of strongly acidic or basic drugs/xenobiotics and metabolites into bile or urine. Removal of penicillins from CNS at the choroid plexus also takes place by active transport process.

Pinocytosis : It is a process in which there is an invagination of the cell membrane around the xenobiotic moiety. It is particularly important in the removal of large molecules e.g. particles from the alveoli in the lungs.

Routes of entry of xenobiotics into the body : Three principle routes of entry of the xenobiotics into body are: percutaneous, respiratory and oral.

1. Percutaneous : Skin is relatively impermeable and makes it reasonably good barrier. Lipophilic xenobiotics of low molecular weight penetrate the body through intact skin by diffusion. Nonpolar substances diffuse at a rate which is directly proportional to their lipid : water partition coefficient and inversely proportional to molecular weight. Polar substances appear to enter via water of hydration associated with proteins. The extent and rate of percutaneous absorption depends on physico-chemical properties of the toxicants, blood perfusion, ambient temperature and pH of the toxicant. Absorption takes place more rapidly from oily solutions and emulsions. It is a time- dependent process as the passage through striatum corneum is rate limiting. It varies from species to species and even within in species depending on the diffusibility and thickness of striatum corneum. Application of solvents e.g.

dimethylsulfoxide (DMSO), methanol, ethanol, hexane, acetone etc. increase the permeability of skin and thus increase the percutaneous absorption of many agents/ toxicants.

2. Respiratory tract/lungs : Lungs constitute an important route of entry of the toxicants dispersed in air, volatile gases and organic compounds. Surface area of the alveoli is very large, alveolar epithelium is extremely thin, pulmonary capillaries have a large surface area and pulmonary blood perfusion is very high. All these factors make lungs as one of the most effective absorptive areas in the body. Low molecular weight and highly lipophilic agents are rapidly absorbed e.g. nitrous oxide, hydrocyanic acid, chloroform etc.

Readily water soluble gases are removed to a certain extent in the nasopharyngeal and tracheo-bronchial region while poorly water soluble gases reach the alveoli. The amount of toxicant reaching lungs in gaseous, aerosol or liquid form depends on concentration of the toxicants in air, minute volume and solubility of gas in blood and particle size e.g. larger particles (75 µm) are deposited in the nasopharyngeal region, 2-5 µm in the tracheobronchial region and may be cleared by mucocilliary escalator and finally come out in sputum or enter the gastrointestinal tract. However, small particles of 1 µm or less than 1 µm may be absorbed or deposited in the alveoli, may be phagocytised and thrown out or enter the lymphatics or may dissolve in alveolar fluid and enter the circulation. Air pollutants e.g. metallic oxides, poisonous gases, volatile liquids, herbicides, insecticides and fumigants etc. enter the body through inhalation.

3. Oral : It is the most usual route of entry of toxicants. Oral absorption of toxicants begins in the mouth and oesophagus, however, it is insignificant in most of the cases. Presence of food in the stomach, gastric acid, gastric enzymes, bacteria etc. alter the toxicity of xenobiotics by influencing the absorption or modifying the compounds.

Most of the absorption takes place in small intestine in almost all species of animals as it has the largest absorptive surface area due to the presence of villi and microvilli. However, absorption also takes place from the stomach in dogs and cats, rumen and reticulum in ruminants and large intestines in non-ruminant herbivores.

Majority of the lipid soluble acid and base xenobiotics are absorbed by passive diffusion. Absorption is dependent on lipid : water partition coefficient, degree of ionization and the concentration in and outside the cell. The degree of ionization depends on pKa value of the toxicants and pH of the environment. Degree of ionization can be determined by Henderson-Hasselbalch equation :

For acids : pKa - pH = log [unionized] / [Ionized]; and

For bases : pKa - pH = log [ionized] /[unionized]

when pKa value of a toxicant is equal to pH of the absorption medium, 50% of the toxicant is in ionized and 50 % in nonionized form.

Gastrointestinal tract possesses specialized carrier systems for certain nutrients. Some xenobiotics use these carrrier systems for their passage through the cells.

Certain xenobiotics absorbed from the gastrointestinal tract are first transported to liver via venous blood and thus, in many cases, a large proportion of the toxicant is either removed from blood and excreted into bile with or without prior biotransformation. The phenomenon of removal of the chemicals before entrance to the systemic circulation is referred as *first pass effect* or *pre-systemic elimination*. However, pre-systemic elimination does not always result in reduced toxicity by decreasing concentration of the toxicant in general circulation but may also result in accumulation of certain xenobiotics in the liver. Hepatic lesions are common to most of the xenobiotics e.g. carbontetrachloride, paracetamol, *Lantana camara* etc.

Distribution : Following absorption, xenobiotics are distributed throughout the body via blood stream. Kinetics of xenobiotics distribution in blood, from cell to cell or organ to organ takes place with the help of transport processes discussed earlier. The rate of distribution in blood, organs or tissues is influenced by several factors, such as dose, route of administration, lipid solubility, extent of binding to plasma proteins and extravascular tissue constituents and blood flow rate through organs and tissues and redistribution. The concentration of a xenobiotic in the tissues/organs is directly proportional to the free xenobiotic concentration in plasma which further depends on its binding to plasma proteins. Some of the extravascular binding sites for certain xenobiotics are listed below (Table 1.3):

Table 1.3 : Some of the xenobiotics and their extravascular binding sites in the body

Xenobiotic(s)	Binding sites in the body
Fluoride and lead	Bones
Arsenic	Hair and nails
Iodine	Thyroid gland
Halogenated hydrocarbons	Adipose tissues
Primaquine	Liver
Mercurials	Kidneys
Paraquat	Lungs

Protein binding : Plasma proteins not only bind with some of the physiological constituents of the body and help in their transportation, e.g. transferrin for transport of iron, ceruloplasmin for copper, *alpha* and *beta* lipoproteins for lipid soluble components such as vitamins, steroids, cholesterol. Different xenobiotics differ in their extent of plasma protein binding e.g. antipyrine does not bind to proteins, binding of secobarbital is 50%, gentamicin 10-20%, phenylbutazone and warfarin is 98-99%.

Protein bound xenobiotics serve as depot as bound toxicants can not cross the capillary wall and thus the phenomenon is important in toxicology as it reduces the concentration of toxicant at the target site thereby lowering the intensity of toxic effects.

Severe toxic reactions can occur if a toxicant is displaced from plasma proteins by another agent. e.g. high doses of warfarin can be tolerated by animals without appreciable toxic manifestations; however, if some other agent/drug having high protein binding affinity is coadministered e.g. sulfonamides, phenylbutazone or dieldrin etc., these will displace warfarin from plasma proteins and result in increased free concentration of warfarin in blood, extensive haemorrhages and death. Thus, plasma proteins act as buffer and protect the tissues against toxic effects. However, it also has its own drawbacks i.e. prolongs the duration of action/effect and slows elimination.

Nonspecific and irreversible binding of xenobiotics to a carrier protein affects the transport of metabolites e.g. long-term administration of tetrasule (2, 4, 5, 4- tetrachlorodiphenyl sulfide) to rats causes development of goiter as thyroxine (T_4) is displaced from its carrier binding proteins (T_4-binding globulin and T_4 -binding prealbumin) which results in accelerated degradation of thyroxine.

Protein binding affinity and capacity for xenobiotics vary between species e.g. pilocarpine binds weakly with proteins in pigs while strongly in cows, horses, rabbits, sheep and other species of animals.

Blood brain barrier : The entry of toxins into brain or spinal cord is frequently more difficult than into other tissues. Blood brain barrier does not allow penetration and distribution of certain xenobiotics into the central nervous system (CNS). Capillary endothelial cells in CNS have tight junctions having a few or no intercellular pores and a sheat of glial cells or astrocytes lining the capillaries to a large extent which lack aqueous pores and it is termed as blood brain barrier. It only allows the passage of non ionized lipid soluble toxicants not the water soluble or ionized drugs.

Blood brain barrier is not fully developed at birth. Therefore, some xenobiotics are more toxic to new borns than to adults e.g. morphine is 3-10 times more toxic to new borns compared to adult rats. Similarly, lead induces encephalomyopathy in neonatal rats but not in adult rats.

Placental barrier : Placenta not only provides nutrition to the conceptus, but also helps in exchange of maternal and foetal blood gases, disposes foetal excretory wastes and maintains pregnancy through hormones. It also offers a conduit for passage and distribution of noxious agents from mother to the foetus.

Placental barrier is composed of a number of cell layers interposed between the foetal and maternal circulations. The number of layers varies with the species and stage of gestation. More lipid soluble xenobiotics rapidly cross the placenta by passive diffusion. The role of placenta in preventing the transfer of noxious agents from mother to foetus is not certain, but placenta has the capability of biotransforming some of the xenobiotics before reaching the foetus.

Redistribution : Different tissues in the body have different blood perfusion rates. Hence equilibrium of distribution occurs quickly in some while slowly in other tissues/organs e.g. following intravenous thiopentone administration, anaesthesia is induced very rapidly as it is a highly lipophilic compound and the peak thiopentone concentrations are build up in the brain. When equilibrium in drug concentration between blood and brain is disturbed, then in order to restore the equilibrium, the drug diffuses back to blood from brain and the process continues and decreases the concentration of thiopentone in brain leading to recovery from thiopentone anaesthesia. This phenomenon is known as redistribution.

A well perfused organ may initially attain higher concentrations of a xenobiotic, however, the affinity of a less perfused organ/tissue may be higher for a particular xenobiotic, thus redistribution of a xenobiotic may take place from the former to the latter site with time e.g. two hours following administration of lead, 50% of lead is present in liver, however, one month later, 90% of the lead present in body is found in the bones.

Elimination : It includes all the processes by which there is decrease in the amount of a xenobiotic in the body which eventually reduces the quantity of toxicant at its site of action. The main elimination processes are biotransformation and excretion.

Biotransformation (Metabolism) : It primarily takes place in the liver, but extra hepatic metabolism in the lungs and kidneys is also important. Hydrophilic compounds having high ionization properties are excreted unchanged (aminoglycosides, organic acids, quaternary ammonium compounds), however, lipophilic compounds are not excreted until these are converted to hydrophilic and more polar metabolites which are more suitable for carrier mediated excretion processes.

Majority of the xenobiotics undergo chemical and enzymatic changes and may result in bioinactivation (*detoxification*) or bioactivation (*lethal synthesis*) of xenobiotics. Some compounds require metabolic activation before they exert their toxic effects. The phenomenon of converting an inactive compound/agent into a toxic agent is termed as *bioactivation or lethal synthesis* e.g. conversion of parathion to paraoxon, harmless fluoroacetate to harmful/toxic agent fluorocitrate.

Bioinactivation or detoxification decreases intensity and duration of toxic effects e.g. LD_{50} of aminopyrine in mice is 0.24 mg/kg while that of its metabolite 4 amino antipyrine is 1.2 mg/kg.

Biotransformation of xenobiotics takes place in two phases i.e. phase I and phase II reactions. Phase I reactions involve oxidation, reduction or hydrolysis whereas phase II involves conjugation with some highly water soluble endogenous moieties. Products of biotransformation are occasionally unstable and decompose to release highly reactive compounds such as free radicals, strong electrophiles or epoxides etc. Epoxides have a strong electrophilic character and are generally unstable and react with nucleophilic groups in macromolecules such as proteins, RNA and DNA. As a result, faulty replication, transcription and synthesis of abnormal proteins may occur. These reactions lead to mutations and carcinogenic changes.

(a) **Phase I reactions :** Phase I biotransformation reactions involve the processes of oxidation, reduction or hydrolysis by which there is unmasking or introduction of a reactive polar group, such as -OH, SH, -NH$_2$ or -COOH. These reactions are catalysed by different enzyme systems (Table 1.4). These functional groups enable the compound(s) to undergo conjugation with endogenous substances such as glucuronic acid, acetate, sulfate and amino acids etc. The xenobiotic-conjugates, so formed, are water soluble and invariably inactive.

Table 1.4 : Types of reactions and enzymes that participate in biotransformation reactions.

Phase 1 Reactions	Phase II Reactions
Oxidation	**Conjugation**
Cytochrome P450-dependent monooxygenase	Glucuronyl transferase
Dioxygenase	Sulfotransferases
Xanthine oxidase	Glutathione-S-transferase
Peroxidase	Glucosyltransferase
Amine oxidase	Thiol transferase
Monoamine oxidase	Amide synthesis (transcyclase)
Dehalogenase	
	Methylation
Reduction	O-methyl transferase
Cytrochrome P 450-dependent reductase	N-methyl transferase
Aldehyde reductase	S-methyl transferase
Keto reductase	
Glutathione peroxidases	**Acetylation**
N-oxide reductase	N-acetyl transferase
	Acyl transferases
Hydrolysis	
Epoxide hydrolase	
Carboxylesterase	
Glucosidase	
Amidases	
Alcohol dehydrogenases	
Aldehyde dehydrogenases	
Aryl sulfatase	
Superoxide dismutase	

Phase I metabolic reactions usually yield products with decreased activity, some may give rise to products with similar or even greater activity as shown in table 1.5.

1. Oxidation : It is the most common reaction and may take place in a number of ways such as hydroxylation, deamination, desulfurization, dealkylation or sulfoxide formation (Table 1.6). In the biotransformation of lipophilic xenobiotics, microsomal oxidation is the most prominent reaction where microsomal enzymes associated with smooth endoplasmic reticulum of hepatocytes are involved and the enzyme cytochrome

P-450, a haem- protein, which is a part of an enzyme system termed as mixed function oxidase (MFO) system, plays an important role. The hepatic microsomal P-450's role is very important in determining the intensity and duration of action of drugs and detoxification of toxic chemicals. The cytochrome P-450 are a family of haemoprotein oxidoreductases. The haeme ion in cytochrome P-450 is usually in ferric (Fe^{3+}) state. When reduced to ferrous (Fe^{2+}) state, it can bind ligands such as O_2 and carbon monoxide (CO). This CO derivative of the reduced form shows absorption maximum at 450 nm from which cytochrome P-450 derives its name.

Table 1.5 : Phase I reactions and the activity of metabolite

Process	Xenobiotics	Metabolite
Detoxification	Pentobarbital	p-hydroxypentobarbital
	Procaine	p-aminobenzoate
Active xenobiotic to	Phenylbutazone	Oxyphenbutazone
active metabolite	Aspirin	Salicylic acid
	Chloroform	Phosgene
	Aniline	p-amino phenol
Lethal synthesis	Parathion	Paraoxon
	Fluoroacetate	Fluorocitrate
	Prontosil	Sulfanilamide

This enzyme has a specific requirement for reduced nicotinamide adenine dinucleotide phosphate (NADPH) and molecular oxygen. The basic reaction catalysed by cytochrome P-450 is mono-oxygenation in which one atom of oxygen is incorporated into a substrate, RH, and the other is reduced to water with reducing equivalents derived from NADPH, as shown below:

$$RH + O_2 + NADPH + H^+ \longrightarrow ROH + H_2O + NADP^+$$

The other enzyme systems of Phase I biotransformation (Table 1.4) are involved in metabolism when the appropriate functional groups are available e.g. alcohol dehydrogenase is involved in the biotransformation of alcohols and aldehydes, monoamine oxidase is a flavine adenine dinucleotide (FAD) -containing enzyme that catalyses the oxidative deamination. Epoxide hydrolases are enzymes that add water across epoxide bonds to form diols. A number of carbroxyl esterases are responsible for biotransformation of certain compounds including organophosphates. The extent to which these metabolic reactions take place appear to vary with the species.

2. Reduction : Reduction reactions are far less common. These are carried out by xenobiotic-reducing enzymes which carry out reduction of carbonyl-, nitro- and azo groups. Reductive reactions involve anaerobic conditions, require NADPH and are almost certainly mediated by FAD-containing enzymes. Aldehydes and ketones may be reduced to corresponding alcohol and nitro compounds to amines as shown below:

Chloral hydrate $\longrightarrow$ Trichloroethanol
Prontosil (azo dye) $\longrightarrow$ Sulfanilamide
Halothane $\longrightarrow$ 1, 1, 1- trifluoroethane

Table 1.6: Some examples of the oxidation of xenobiotics

Xenobiotic		Metabolite
Acetanilide	hydroxylation $\longrightarrow$	paracetamol
Codeine	dealkylation $\longrightarrow$	Morphine or norcodeine
Amphetamine	hydroxylation $\longrightarrow$	p-hydroxyamphetamine
Amphetamine	oxidation and deamination $\longrightarrow$	Benzoic acid
Ethanol	Non-microsomal oxidation $\longrightarrow$	Acetaldehyde
Histamine	Deamination $\longrightarrow$	Imidazoleacetaldehyde
Adrenaline	Monoamine oxidase - $\longrightarrow$	3,4- dihydroxymandelic acid

3. Hydrolysis : It is the process of cleaving of a foreign compound by the addition of water. It occurs both in the cytoplasm and smooth endoplasmic reticulum. It is an important metabolic pathway for compounds with an ester linkage (-CO. O-) or an amide (-CO. NH-) bond. The cleavage of esters or amides generates nucleophilic components which undergo conjugation. Esterases hydrolyse esters and amides to corresponding carboxylic acids, alcohols or amines as depicted below:

Pethidine $\longrightarrow$ Pethidinic acid
Procaine $\longrightarrow$ Para-aminobenzoic acid
Acetylcholine $\longrightarrow$ Choline + Acetic acid

(b) Phase II reactions or conjugation/synthetic reactions :

Synthetic reactions may take place when a xenobiotic or a polar metabolite of phase I metabolism containing -OH, -COOH, $-NH_2$ or -SH group undergoes further transformation to generate non-toxic products of high polarity which are highly water soluble and readily excretable by combining with some hydrophilic endogenous moieties. Conjugating agents are glucuronic acid, acetyl, sulfate, glycine, cysteine, methionine and glutathione which conjugate with different functional groups of xenobiotics as shown in Table 1.7. Most of the phase-II biotransforming enzymes (Table 1.4) are located in the cytosol with the exception of uridine diphosphate glucuronyl transferases (UDPGT) which is a microsomal enzyme. Phase II reactions proceed faster than the Phase-I reactions. Therefore, the rate of elimination of a compound ,whose excretion depends on biotransformation by cytochrome P450s followed by phase II conjugation is generally determined by first order reaction.

Table 1.7: Conjugation reactions and different functional groups of xenobiotics

Conjugation reaction	Functional groups of Xenobiotics	Conjugate
Glucuronide conjugation	--OH, --COOH, -NH$_2$, -SH	Glucuronic acid
Sulfate ester formation	-OH, -NH$_2$	Sulfate
Methylation	-OH, -NH$_2$, -SH	Methyl group
Acetylation	-NH$_2$, -SO$_2$, -NH$_2$	Acetyl group
Glutathione conjugation	-F, -Cl, -Br	Glutathione
Amino acid conjugation	-COOH	Glycine, taurine

Conjugation reactions may be divided into electrophilic conjugations involving glucuronide, sulfate, acetate, glycine, glutathione and methyl transfer or nucleophilic conjugations involving only glutathione.

The conjugating moieties do not react directly with the xenobiotics or the metabolites of phase I reactions, but do so either in an activated form (usually nucleotides) or with an activated form of the xenobiotic. The reaction between the nucleotide and the xenobiotic is catalysed by an enzyme. The conjugation reaction requires a conjugating agent, a nucleotide containing either the conjugating agent or the xenobiotic and a transferring enzyme. Depending on the occurrence of conjugating agents or the amount of the transferring enzyme, variations in conjugation reactions are there in different species of animals. However, phase I reactions are ubiquitous throughout mammalian species. Hence certain synthetic reactions are either slow or absent in some species of animals e.g. glucuronide synthesis takes place at slow rate in cats, acetylation is absent in dogs and foxes and sulfate conjugation is low in pigs.

1. Glucuronidation : Formation of glucuronides is quantitatively most important. It is carried out by the smooth endoplasmic reticulum-bound glucuronyl transferase. The glucuronic acid is donated by uridine diphosphate glucuronic acid (UDPGA) which serves as an endogenous substrate for the enzyme glucuronyl transferase. Substrates for glucuronide formation are phenols, alcohols, carboxylic acids, amines, hydroxylamines and mercaptans. Glucuronides are mainly excreted via bile, hydrolysed in gut and then may be reabsorbed and delivered again to the liver where conjugation may occur again. Glucuronides synthesis is slow in cats due to low level of glucuronyl transferase while it is absent in certain breeds of fish due to deficiency of the nucleotide UDPGA.

2. Sulfation : Another common phase II reaction is sulfate conjugation. The enzymes involved are cytoplasmic sulfotransferases, a group of soluble enzymes located primarily in the liver, kidneys, intestines, lungs and brain. 3' phosphoadenosine 5' phosphosulfate (PAP) serves to transfer the sulfuryl group to a nucleophilic position (O, N or S). This process yields ethereal sulfates of various aromatic and aliphatic hydroxyl compounds e.g. phenols, alcohols, chloramphenicol, steroids (androgens and estrogens). The n-hydroxy compounds are substrates for conjugation with sulfate. Sulfate conjugation involves the transfer of SO_3^- (not SO_4^-) from PAPs to the xenobiotic and conjugates are mainly excreted in urine. Sulfate conjugation capacity is limited in pigs due to deficiency of the enzyme sulfotransferase.

3. Acetylation : N-acetylation is the major route of biotransformation for xenobiotics containing an aromatic amine ($R-NH_2$) or a hydrazine group ($R-NH-NH_2$) e.g. isoniazid, sulfonamides etc. It is carried out by cytoplasmic enzymes N-acetyltransferases found in liver and other tissues. The acetyl donor is acetyl coenzyme A. Acetylation decreases water solubililty as well as lipid solubility. Dogs and foxes do not acetylate the aromatic amino group as they lack the enzyme arylamine acetyl transferase due to the presence of a factor in blood which decreases the arylamine acetyl transferase.

4. Glautathione conjugation : It is an important detoxification mechanism. Glutathione (GSH) is a tripeptide found in most of the tissues, especially in high concentrations in liver and plays an extremely important role in protecting hepatocytes, erythrocytes and other cells against toxic injury. It is involved both in enzymatic and nonenzymatic reactions. Non enzymatically, it acts as a low molecular weight scavenger of reactive electrophilic xenobiotics and competes with DNA, RNA and proteins in capturing electrophiles.

Enzymatic reactions involving glutathione are catalysed by the enzyme glutathione-S-transferase. It catalyses the reaction between glutathione and aliphatic or aromatic epoxides and halides. Glutathione conjugates formed in the liver are excreted intact in bile or they are converted to mercapturic acids in the kidneys, which is highly water soluble and excreted in urine.

Glutathione-S-transferase also catalyses reactions of organic nitrates with glutathione. Nitrates are reduced to nitrite which interact with amines and result in the formation of carcinogenic nitrosamines. Depletion of glutathione predisposes to hepato-toxicity and mutagenicity.

5. Methylation : It is generally a minor pathway of biotransformation. The cofactor for methylation is S-adenosyl methionine (SAM). Methylation reactions are catalysed by cytoplasmic enzyme methyl transferase. Substrates are phenols, catechols, aliphatic and aromatic amines and sulfhydryl-containing compounds.

6. Conjugation with amino acids or amino acid conjugation : Conjugation with amino acids is carried out by mitochondrial enzymes and N-acetyl transferases. Substrates for such conjugation are carboxylic acids. Xenobiotics containing carboxylic acid group conjugate with $-NH_2$ group of amino acids such as glycine, glutamine, taurine etc. However, the xenobiotics containing an aromatic hydroxylamine group conjugate with serine and proline. Other acceptor amino acids for xenobiotic conjugation are ornithine, arginine, histidine, serine and aspartic acid.

Bile acids are endogenous substrates for glycine and taurine conjugation. Amino conjugates are primarily eliminated in urine. Conjugation with amino acids varies with species and also with xenobiotics e.g. bile acids conjugate with glycine and taurine in most species except cats and dogs where these conjugate with taurine only. Benzoic acid conjugates with glycine in most species except in birds and reptiles where it conjugates with ornithine.

Benzoic acid + Glycine $\longrightarrow$ Hippuric acid

Excretion : Most of the xenobiotics are eliminated by a combination of biotransformation and excretion processes. Biotransformation decreases lipid solubility, the metabolites are readily excretable polar compounds with low lipid solubility and are mainly eliminated through kidneys. The other important route of excretion is via faeces that may be due to excretion through bile, saliva and pancreatic secretions. The gaseous metabolites are mainly excreted through lungs which is an important route of excretion of lipophilic substances as the capillary and the alveolar membranes are thin and in close proximity. Certain toxicants are also excreted via various body secretions such as sweat, milk and tears. Excretion of xenobiotics through milk poses risk of fatal poisoning to young suckling animals and human beings and thus a cause of public health concern.

1.6 FACTORS AFFECTING TOXICITY

Although toxicity of any agent is mainly determined by the dose, but for toxic effects, the toxic agent or the metabolite must reach the target site in sufficiently high concentrations and for sufficient time to initiate toxic changes. Considerable differences in sensitivity to toxicants between individuals and between species are observed. These differences in sensitivity between species are mainly attributed to the differences in anatomy, body size and physiology etc. while between individuals within a species generally may be due to variations in heredity, age, body weight and size, dietary habits, disease and simultaneous exposure to other toxicants/agents etc. All the factors, either singly or in combination, influence the outcome of toxicity of a toxicant. Some of the important factors are :

1. Dose : The occurrence and magnitude of effects depends on the amount of substance absorbed into the blood stream which further depends on the dose administered. Thus, higher the level of exposure, more intense is the toxicity. Depending on the dose and the agent used, it may produce therapeutic, toxic or lethal effects. Even the normal body or essential dietary constituents, if present in higher concentrations, become toxic e.g. copper, iodine, sodium chloride etc.

2. Physico-chemical properties/characteristics : Toxicity depends on physical state of the compound (gas, liquid or solid - coarse or finely powdered), solubility (soluble in water, oil or some other solvent). A compound in solid state is less toxic than its powdered form. Finely powdered forms are more toxic compared to coarse powders and gaseous forms are more toxic than liquid or solid forms. Nature of solvent/vehicle, if present in solution form, is an important determinant of toxicity e.g. dermal application of insecticides in aqueous medium is less toxic than in oily medium as dermal absorption in the former is less while much greater when sprayed in the oily medium. Toxicity also depends on chemical nature of the compounds e.g. red phosphorus is nontoxic as is insoluble, nonvolatile and excreted from body in unchanged form while yellow phosphorus is toxic. Trivalent arsenic is more toxic than the pentavalent arsenic, barium carbonate is toxic while barium sulfate is non toxic, carbon monoxide is more toxic than CO_2, nitrates are reduced to nitrites by ruminal and intestinal microflora and toxicity is produced by nitrites not nitrates and copper is absorbed in cupric (Cu^{2+}) not cuperous (Cu^+) form from the gastrointestinal tract.

3. Nature of exposure / exposure mode : Intensity of poisoning depends on whether the animal is exposed to a single or multiple doses. Generally, several low doses of poison will be more effective in producing toxicity than a single moderate dose, particularly in cumulative type of poisons e.g. lead, fluorine, DDT, copper etc. In cumulative toxicity, toxic manifestations are not immediate rather quite delayed e.g. carcinogenicity, mutagenicity, radiation hazards, fluorosis etc. Poisoning may be categorised as acute, subacute, subchronic and chronic depending on exposure to a chemical for less than or upto 24 hours, repeated exposure for less than or one month, between 1-3 months and more than 3 months, respectively. Sometimes resistance or tolerance develops when the subjects are repeatedly exposed to certain compounds e.g. barbiturates, opiates, sedatives etc.

4. Route of exposure : It determines the rate and extent of absorption following extravascular route of exposure e.g. DDT is about equally toxic to insects and mammals when given parenterally, but when applied externally it is considerably more toxic to insects as the chitinous exoskeleton of insects is more permeable to DDT than mammalian skin.

5. Age : Very young and very old subjects are more susceptible to the effects of poisons as age is an important determinant of biotransformation of xenobiotics. The mechanism(s) of detoxification and excretion are not fully developed in young ones while older subjects have less efficient detoxifying and excreting mechanisms. The biochemical background of age dependent biotransformation differences are cellular changes in the development of endoplasmic reticulum and components of the cytoplasmic P450 system. Some of the drugs induce toxicity· in neonates e.g. chloramphenicol induced grey-baby syndrome in human infants is due to inadequate conjugation with glucuronic acid as infants are deficient in glucuronyl transferase.

6. Sex : Generally females are more susceptible to the action of poisons. Sex related differences in biotransformation of xenobiotics are due to a combination of factors like hormones, enzymes etc. Some of the cytochrome P450 isozymes have been found to be sex-dependent. Compared to male rats, female rats are more sensitive to red squill and parathion toxicity. Some of the agents having teratogenic potential are not given to pregnant animals or women e.g. thalidomide, plants of *Veratrum* sp. and *Lupinus* sp. Castration increases the susceptibility of males to poisons and administration of testosterone in females increase their resistance to poisons. Therefore, testosterone appears to play an important role in making the males less susceptible to the action of poisons. It has recently been suggested that pituitary releases a so called feminising factor which causes the liver in female mammals to act as a female liver, the release of this factor is controlled by hypothalamus. In males, function of hypothalamus is inhibited by hormones and thus does not release feminising factor. Sex - dependent variations in effects of poisons are more pronounced in laboratory animals (rats and mice), however, these appear to be less significant in humans and other species of animals.

7. **Species :** Toxicological effects of poisons vary quantitatively and qualitatively from species to species and suggest that data from one species should not be extrapolated for another species e.g. mammals and birds can synthesize vitamin C (ascorbic acid) while primates, guinea pigs, fruit eating bats and birds need exogenous vitamin C as these species lack the biosynthetic pathway for ascorbic acid. Methotrexate, an anticancer drug, is very toxic to human beings, mice, rats and dogs but is not toxic to guinea pigs and rabbits. The LD_{50} value of permethrin in rats is 1400 times higher compared to that for the desert locust. Similarly, the LD_{50} value for the highly toxic dioxin 2, 3, 7, 8- tetrachlorobenzo-p-dioxin differs more than 1000 fold between guinea pigs and hamsters. *Belladona* is toxic for all species of animals except rabbits, red squill is toxic for rats not dogs, sheep are more susceptible to chronic copper poisoning etc. All the above mentioned examples suggest that it should be mandatory that the toxicological data should be generated in different target species of animals.

Some of the species dependent differences in toxicity have been attributed to differences in biotransformation reactions e.g. dogs and foxes do not acetylate aromatic amines as these are deficient in N-acetyl transferase while cats are deficient in glucuronyl transferase. Guinea pigs do not form marcapturic acid conjugates and pigs do not form ethereal sulfates. These examples amply support species-dependent differences in toxicity and suggest the importance of choice of an appropriate animal model to conduct toxicity studies.

The phenomenon of selective toxicity and species variation has been extensively exploited for therapeutic purposes e.g. sulfonamides inhibit bacterial growth while host cells are not affected due to differences in metabolic pathways. Malathion is oxidised by cytochrome P-450 enzyme system to malaoxon which is a potent inhibitor of acetylcholinesterase. It is about 38 times less toxic when given orally to rats than when applied topically to houseflies because mammals possess very active esterases which inactivate malaoxon by hydrolysing the ester groups. Insects also contain esterases, but these act much slower compared to that of mammalian esterases. Therefore, malathion is more toxic against insects compared to mammals.

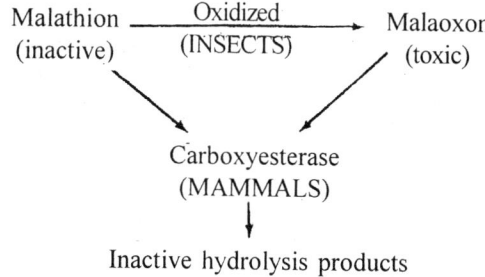

Inactive hydrolysis products

Though biotransformation of xenobiotics also takes place in non-mammals (fish, birds, reptiles, insects and even microorganisms too) similar to that in mammals, however, fish generally have a lower metabolic capacity for xenobiotics than birds and the birds are less efficient compared to mammals.

8. Genetic differences : Differences between strains within one species have been observed in man, rats and mice due to certain enzymatic anomalies. Genetic differences in biotransformation of xenobiotics have been observed in human beings as a single gene has been found to regulate the oxidative metabolism of a large number of compounds. Acetylation of drugs/poisons is regulated by hepatic N-acetyl transferase. Deficiency of N-acetyl transferase is a genetic trait which is highest among blacks and Caucasians, lessor among Japanese and Chinese and lowest among Eskimos. Three phenotypes called as fast, slow and very slow acetylators have been found in human beings. A genetic deficiency of the acetylating enzyme is encountered among certain groups in the population, both in human beings and animals. Ninetysix per cent of Eskimos, eight per cent of Asians and forty per cent of the Europeans have been detected to be fast acetylators. Therefore, depending on the differences in acetylation, isoniazid, an antitubercular drug is expected to produce neuropathy in slow acetylators if doses are not adjusted. Although the deficient enzyme is not essential for survival and such individuals can lead a normal life, but toxicity may occur when they are challenged with a drug or a xenobiotic. However, the subjects lacking diaphorase I suffer with intense cyanosis and those deficient in glucose-6-phosphate dehydrogenase suffer with severe haemolytic crisis on exposure to certain drugs or toxicants.

9. Pathological condition : Biotransformation and excretory capacity of the body is altered during disease state due to change in the physiological and functional ability of liver, kidneys, lungs, heart, intestinal wall, intestinal flora etc. Thus, the weak and debilitated animals are more vulnerable to the action of poisons. Constipation increases the toxicity of orally administered/ingested substances due to enhanced absorption as passage of toxicants is delayed. Healthy individuals can tolerate even large doses of those agents which have a high protein binding affinity e.g. dicoumarol, phenylbutazone, sulfonamides etc. without any apparent serious toxic manifestations. However, if the subject is hypoproteinemic particularly hypoalbuminemia, intensity of toxicity is increased much more for such poisons/agents.

10. Concurrent exposure : If the animals or human beings are being simultaneously exposed to some other drugs or xenobiotics (pesticides, heavy metals or environmental pollutants (present in soil, air, water or food), hepatomicrosomal enzyme activity may be increased or decreased and accordingly, the detoxifying mechanisms get induced or suppressed and influence the toxicity of xenobiotics. Metabolism of the offending xenobiotics may be accelerated or retarded depending on induction or inhibition/suppression of hepatomicrosomal conjugation enzymes. Thus on concurrent exposure, the phenomenon of potentiation, synergism or antagonism may be observed.

Enzyme induction : It may be defined as the qualitative and quantitative change in the metabolism of xenobiotics generally as a result of repeated exposure to the same or some other xenobiotics due to increase in the biosynthesis of an enzyme rather than increased activity. The phenomenon was first observed in studies involving N-demethylation of amino azo dyes in rat livers. Some of the enzyme inducers are phenobarbital, phenylbutazone, 3-methylcholanthrene, pregnenolone-16- α-carbonitrile, polycyclic aromatic hydrocarbons, DDT, aldrin, dieldrin, hexachlorobenzene, lindane, polychlorinated and polybrominated biphenyls etc.

Enzyme inhibition / suppression : Certain xenobiotics and drugs may bind to active site of the enzyme and retard the metabolic processing of other xenobiotics. This phenomenon is known as enzyme suppression or inhibition and such agents are termed as inhibitors. Inhibition of enzyme may be reversible or irreversible. Some of the common examples of reversible hepatomicrosomal enzyme inhibitors are SKF-525 A (2-diethylamino ethyl 2, 2-diphenylvalerate), *alpha*-naphthoflavone while carbon-tetrachloride is an irreversible inhibitor of cytochrome P-450. Other inhibitors of cytochrome P-450 are cimetidine, chloramphenicol, ketoconazole, macrolide antibiotics (erythromycin, trolandomycin etc.).

11. Dietary factors : Presence or absence of certain constituents in feed, fodder or diet alters the toxicity or adverse effects of drugs, chemicals or poisons. Certain dietary components of plants e.g. phytic acid (inositol hexaphosphate) chelates metals and minerals and reduces their absorption. Similarly, dietary calcium and zinc compete with lead and other toxic metals and dietary tannins and proteins form complex with toxicants and reduce their absorption from the gastrointestinal tract. Animals on low crude fiber diet are more prone to toxicity. A high protein diet usually increases drug biotransformation capacity and a low protein diet reduces it. Toxicity of carbon tetrachloride is more in animals on high fat diet. Low protein diet reduces aflatoxins-induced hepatotoxicity as during protein deficiency cytochrome P-450 and NADPH-cytochrome P-450 reductase activity drops and UDP-glucuronyl-transferase activity increases. Deficiency of vitamin C and E result in increased damage from free radicals.

12. Source of poison : Humans and animals may take up the poison through food/feed, water or air as the intoxicants may be present in feeds and forages, drinking water or air as pollutants. Chances of poisoning may be enhanced or reduced based on its nature and source e.g. poisonous plants may not necessarily cause severe toxicity during a particular season or stage of growth of the plants. Toxic principle(s) are present in highest concentration in certain parts of the plants like root, stems, shoot, leaves, flowers, fruits or seeds.

Improperly stored feeds particularly rich in maize and groundnut are richest source of mycotoxins. Other common sources of poisoning are agrochemicals, paints, heavy metals etc. Sometimes even the drinking water rich in nitrates, arsenic, fluorine etc also cause serious health problems both in humans and animals.

1.7 DIAGNOSIS OF POISONING

Diagnosis of poisoning in animals and birds is not an easy task. Therefore, it should never be made in haste based on any single observation. Under field conditions, any ailments, if not diagnosed properly, is suspected to be a case of some sort of poisoning, particularly if there is mortality. Attempts should be made to reach at a confirmatory diagnosis. To establish the specific cause of poisoning, if possible, abnormality of the structure and function(s) produced by the causative agent should be keenly observed. Rational treatment and effective control measures can only be adopted if proper diagnosis is made. However, first step should be rapid assessment of the vital body functions, particularly the respiratory and cardiovascular functions. Once the vital signs are stable and adequate, further evaluation of poisoning may be taken up. Diagnosis in case of poisoning may be of three types, namely- tentative, presumptive and confirmative based on the history, physical examination of the affected animals, circumstantial evidences, pathological and laboratory observations and investigations and analytical evidences.

1. History : Though history is of limited value, yet detailed history of poisoning should be extracted from the owner or attendant of the animal. Sometimes it may not be possible to get all the details, or at times it may be inaccurate and misleading. While taking history, identity of the subject, age, toxic agent, its vehicle, how it has entered the body-through ingestion, spray on the skin/dermal application, inhalation or parenteral route i.e. mode of exposure, location of exposure, amount of exposure, poisoning was accidental, advertent, inadvertent or malicious and how much time has elapsed between exposure and the onset of toxicity symptoms (latent period) and typical signs of toxicity etc.

2. Physical examination and clinical evidences : Physical examination of the subject suspected for poisoning should be made keenly for apparent clinical symptoms. The affected animals should be systematically examined for all the body organs/ systems and functions irrespective of the route of exposure. Observations of the vital clinical parameters, namely- temperature, pulse, respiration, blood pressure etc. are of primary importance e.g. organophosphate compounds induce hypothermia while chlorinated hydrocarbon insecticides and belladona induce hyperthermia, bradycardia in barbiturates and mushroom poisoning while tachycardia with belladona and cocaine, respiratory rate is increased in carbonmonoxide, hydrocyanic acid and fluoroacetate poisoning while decreased in narcotic analgesics and sedatives poisoning. Similarly, observations like miosis in opium poisoning, mydriasis in belladona poisoning, dryness of mouth in belladona, moist mouth alongwith increased salivary and bronchial secretions in OP compounds poisoning, bitter almond smell of breath in cyanide poisoning, garlic odour of expired air in phosphorus and arsenic and mice smell in hemlock poisoning. Similarly, colour of the urine also gives an indication of the possible toxicant, e.g. dark green urine with phenols and cresols, red with phenothiazine or compounds inducing haematuria, brown black in acorns and deep yellow after phenacetin, picric acid, furazolidone or metronidazole etc. Other symptoms like diarrhoea, constipation, vomiting, abdominal pain, colic, CNS symptoms like ataxia, convulsions, paralysis, coma, death etc. alongwith history, pathological observations and knowledge of access to the suspected toxicant may aid atleast in making some presumptive diagnosis.

3. Circumstantial evidence : Thorough investigation of the owner regarding sudden change in the feed, fodder, feed additives, pastures, administration of any drug(s), pasture treatment with fertilizers or herbicides, spray of agrochemicals/pesticides on the animals or the premises, painting of the buildings with lead based paints, repair of the roof or animals had access to any other thing like containers of agrochemicals, petroleum products, fertilizers, disinfectants, rodenticides, baits or discarded equipments containing some chemicals etc. The presence and previous use of poisons on the premises should also be determined. Existence of poisonous/toxic plants, fungi, blue green algae in the surroundings or the movement of snakes around the premises may also be taken into consideration while arriving at any diagnosis in cases of poisoning. Similarly, drinking water should not be ruled out as the possible source of poisoning. Higher levels of nitrite, fluorine, arsenic in ground water or the contamination of water with salt, lead, industrial effluents etc. may be responsible for poisoning of animals or birds. Circumstantial evidences should not necessarily and always be considered responsible for poisoning unless supported by other tests and evidences.

4. Pathological evidence : Gross and microscopic examination of the tissues/ organs aids in diagnosis of some of the poisoning cases. Sometimes, lesions of poisoning are characteristic and may be of pathognomonic importance, however, generally these are non-characteristic and of generalized type. Pathological investigations including necropsy findings provide definite clue to the nature of poison(s) and / or the system involved and affected e.g. jaundice indicates hepatic damage, may be due to carbontetrachloride, plant toxins, mushrooms, drugs or any other agent, cyanotic mucous membranes indicate carbon monoxide, nitrite, nitrate or chlorates poisoning while bright red colour of blood and mucous membranes indicates hydrocyanic acid poisoning. Similarly, odour of the abdominal contents e.g. bitter almond in cyanide and rotten eggs in hydrogen sulfide or the colour of ruminal/ abdominal contents give indication of possible toxicants involved e.g. greenish blue in copper, yellow to orange or green with chromic compounds, yellow with picric and nitric acid and black due to corrosive acids. Presence of leaves, twigs, some other things like flakes of paint or polythenes also give clue of the poisoning. Signs of generalized haemorrhage may indicate sweet clover poisoning. Similarly, hepatic and renal lesions also help in arriving at the possible toxicant(s) coupled with other evidences..

5. Laboratory investigations : Diagnosis of poisoning may be confirmed by routine haematological and biochemical tests, the results of which further depend on the appropriate biological samples, preservatives used and transportation of the samples. Laboratory examinations help the physician/veterinarian to assess the level of dysfunction of various organs/systems and to know which system and to what extent has been affected. It not only helps in diagnosis but also in formulating the line of treatment and other resuscitative measures.

6. Analytical evidence : Confirmatory diagnosis of poisoning depends on the qualitative and quantitative detection/estimation of a significant amount of the toxic agent in the biological samples e.g. blood, faeces, vomitus, urine, adipose tissue, hair or skin and environmental samples, such as- food, feed, forage, water, baits, chemicals, solvents, pesticides, drugs or medicaments. Sometimes, poisoning can not be confirmed by laboratory examination; either there are no suitable or possible analytical methods or tests or the true toxicant was not suspected or the samples collected or the preservative(s) used or the transportation of the samples was not proper. All these factors limit the usefulness of analytical evidence in arriving at a confirmatory diagnosis of poisoning.

To sum up, a tentative diagnosis can be made based on the case history, clinical examination of the affected animals and gross necropsy examination of the dead animals. The presumptive diagnosis can be made on the basis on the history of poisoning cases, clinical examination, necropsy findings, circumstantial evidences of the access of animals to the suspected poisons and the positive response to antidotal treatment. However, confirmatory diagnosis can only be made on the basis of qualitative and quantitative detection of the toxicant in feed, water, ruminal contents/ gastrointestinal contents, animal blood or other biological samples.

1.8 GENERAL PRINCIPLES OF TREATMENT OF POISONING

Poisoning is always an emergency and need to be managed immediately with appropriate measures using specific antidotes whenever available. However, in majority of poisoning cases treatment with an antidote is not possible due to lack of confirmative diagnosis. In acute poisonings, first of all truely emergency phases should be managed to improve condition of the animal by providing appropriate supportive care to ensure its survival. For this, focus on prompt removal or the neutralization of poison whilst maintaining vital functions of the body by restoring respiration by giving artificial respiration and or drugs acting on cardiovascular system, CNS stimulants or depressants, emetics etc. depending on the clinical state of the patient/animal until slow acting and specific treatment is instituted.

Intensity of toxicity depends on the dose of poison and concentration of the same at the receptor/target site. Translocation of the drug/poison to the target site depends on the absorption, distribution, metabolism and excretion characteristics of the drug/chemical/poison. Following exposure through oral or some other route, concentration of the toxicant increases at the target site/ or in the blood as a function of time. Toxic manifestations are observed beyond certain levels of toxicants in the body as shown in Fig. 1.2.

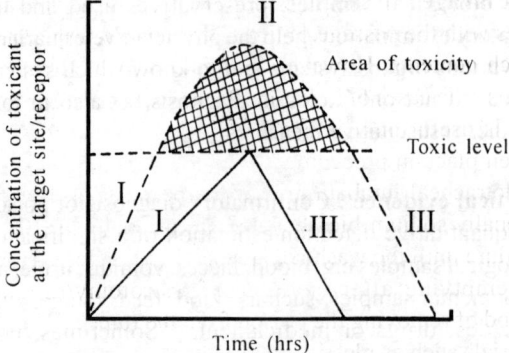

Fig. 1.2: Time-concentration profile of toxicant indicating the toxic level and different phases of increase and decrease in the concentration of toxicant at target site.

Phase I and phase III of the figure indicate the increase and decrease in the concentration of toxic substance in the body, respectively. Shaded area in the figure (Phase II) represents the time of onset and disappearance of toxic effects. From the figure, it is evident that toxic manifestations can be avoided/reduced by manipulating the phase I, II and III. Treatment should be aimed at decreasing the slope of phase I and increasing the slope of phase III so that phase II is not reached or the duration of phase II is reduced.

In order to treat the poisoning cases effectively, a clear understanding of some of the basic principles of toxicokinetics, toxicodynamics and specific therapeutic goals is essential e.g. absorption of the toxic substances be minimised, antagonise the effects of absorbed toxicants, metabolic biotransformation of poisons with reduced toxicity be enhanced while biotransformation into toxic substances be inhibited/reduced and elimination from the body be enhanced. These goals can be achieved in the following ways:

1. Decrease the slope of ascending portion of the curve (Phase I) : It can be achieved by removing the source of poison or to remove the animals from the area in which poison is believed to be present. It is also important to limit and delay further absorption of the poison so that toxic levels are not built up in the body.

Gastrointestinal tract is an important site where from maximum absorption of toxicants takes place. Thus, prevention of gastrointestinal absorption is an important aspect in initial treatment of acute poisoning. It can be achieved by :

    a)   Removal of the poison from the stomach;

    b)   Hastening the passage through bowel;

    c)   Neutralization of the poison within the gastrointestinal tract.

a)   **Removal of the poison from the stomach :**

(i) **Gastric lavage :** It should be done as soon as possible after ingestion and its efficacy decreases with the passage of time. It is a rapid way of removing the poisons alongwith stomach contents. However, it should not be attempted in unconcious or convulsing animals or in cases of ingestion of caustic substances or petroleum distilllate hydrocarbons. It is useful only in small animals and is extensively used in dogs if ingestion has taken place in preceding 2-4 hours. It is indicated on the anaesthetized animals with endotracheal intubation. For the purpose, an isotonic sodium chloride solution (occasionally sodium bicarbonate) occasionally @ 10 ml /kg is indicated. Repeat the procedure until the washout stomach fluid is clear. However, in ruminants, manual gastric emptying after emergency gastrotomy/ruminotomy is the only satisfactory method of removing the poison from the rumen e.g. plant poisoning or for indigestible materials such as plastic, polyurethane foam etc.

(ii) **Emetics :** In dogs, cats and swines, vomition may be induced to empty the stomach. Emetics may be used if ingestion has taken place within the preceding 2-3 hours. But vomition should not be induced if there are seizures (unless controlled) or the animal is in comatose state or there is severe respiratory distress or has ingested caustic substances (strong acid, alkali or cationic detergent) or petroleum distillate (gasoline, kerosene, paint thinner etc.). Emesis may be induced chemically by making use of hypertonic solutions of copper sulfate or sodium chloride, but not reliable as these themselves may induce toxicity. Therefore, drugs like ipecac or apomorphine hydrochloride are used to induce vomition. However, there is a greater risk of aspiration pneumonitis due to aspiration of gastric contents into the trachea and lungs.

(a) Apomorphine hydrochloride is administered to dogs at the dose rate of 0.04 - 0.07 mg/kg by I/V, I/M or S/C or subconjunctival routes. Effect is prompt and emesis generally occurs within 2-10 min following S/C or conjunctival administration. It is contraindicated in the presence of existing central depression. Do not give emetics to large animals. Apomorphine is contraindicated in cats and pigs.

(b) Xylazine is used in dogs and cats. Emesis is produced within 10-20 min. Emesis in cats is more consistent compared to dogs. Dose for cats is 1 mg/kg by IM route. However, slightly higher dose is recommended for dogs.

(c) Syrup of ipecac (10%) is administered orally; 10-20 ml (1-2 ml/kg) for dogs and 2-5 ml (3.3 ml/kg) for cats. Emesis is produced within 20-30 minutes.

If none of these drugs are available, common salt (sodium chloride) 1-3 teaspoons in warm water or hydrogen peroxide (3% solution) 1 ml/kg may be administered orally. However, copper sulfate should not be used as it is an irritant and may hasten the absorption of poisons from the stomach.

(b). **Hastening the passage through bowel :** Cathartics may be of some value particularly if some slow acting poison is involved. But use of cathartics is only of minor importance and no longer recommended in the management of orally ingested poisons as these may produce dehydration, hypernatremia, hypermagnesemia, hyperkalemia and hyperphosphatemia. Irritant purgatives or oil-based purgatives should never be used as these promote absorption of poisons. However saline purgatives such as sodium sulphate, magnesium sulphate or magnesium citrate may be used either orally or as enema. Dose of sodium sulphate for dogs and cats is 1.0 g/kg (2-25 g) while 100-200 gm for large animals. Orally liquid paraffin is also recommended for inducing evacuation of the bowel. Doses for dogs are 5-15 ml, cats 2-16 ml, adult cattle and horses 0.5-2.0 litres, calves and foals 60-120 ml and sheep and goats 250-500 ml.

(c). Neutralization of the poison within the gastrointestinal tract : Use of adsorbing agents like activated charcoal or neutralizing agents like acids, alkalies, chelating agents e.g. dimercaprol (British antilewisite, BAL), calcium EDTA or tannic acid, magnesium oxide, aluminium hydroxide gel etc. is recommended for various poisons, however, there is lot of controversy about their use and efficacy.

Activated vegetable charcoal is inert, non-absorbable, odourless and tasteless and binds or adsorbs organic materials to form stable complex(es) which is not absorbed. Currently, it is the single most effective and useful agent employed for prevention of absorption of ingested poisons. It is administered at the dose rate of 250-500 g in large animals and 5-50 gm in small animals, preferably as a suspension in water several minutes to 24 hrs after ingestion and before emetics, if possible. At one time, it used to be an important ingredient of the "universal antidote" in combination with kaolin, tannic acid and magnesium oxide. The universal antidote contains vegetable charcoal 10 g, magnesium oxide 5 g, kaolin 5 g, tannic acid 5 g and water added upto 200 ml. However, activated charcoal is not so much preferred now. Other adsorbant cholestyramine has also been tried, however, because of its efficacy against only a limited number of toxicants, it is generally not recommended.

Certain compounds can be used as local antidotes as these change the ionic form and absorption and thus modify toxicity e.g. sodium bicarbonate in iron poisoning, the resulting ferric carbonate is less irritant and less absorbed and calcium reduces toxicity of fluorine. Weak acids and alkalies should not be used to neutralize strong alkalies or acids because temperature of upto 100°C occurring during neutralization will damage the tissues. Flushing with large quantities of water is the best remedy in such cases.

Other substances which form a complex or insoluble precipitate with poisons and/or neutralize them in the gastrointestinal tract are water containing albumins which form insoluble complexes with heavy metals and also neutralize acids and bases. Tannins and ferric hydrate precipitate metals and alkaloids.

2. Increase the slope of descending portion of the curve (Phase III) : It can be achieved by limiting distribution of the toxicant by changing pH of the gastrointestinal tract or facilitating elimination of the poison from the site of exposure or absorption e.g. washing of the skin or eyes after topical application and gastric lavage, emesis or purgation after oral ingestion. Similarly, urinary excretion of poisons may be hastened by increasing the flow of urine (diuresis).

(a). **Increase glomerular filtration** : Use of potent diuretics is contraindicated. Administer 10% glucose or 10% mannitol or sterile saline solution by slow IV infusion.

* Large animals : 1-2 ml/kg/24 h

* Small animals : 1-2 ml/kg /24 h

b). **Reduce tubular reabsorption** : Changing pH of the urine by urinary acidifiers (ammonium chloride, ascorbic acid, sodium acid phosphate) or alkalizers (sodium bicarbonate, sodium acetate, potassium acetate) in an attempt to ionize the offending toxic substance will promote diuresis and urinary excretion of poisons - weak bases and weak acids, respectively. Commonly used urinary acidifiers and alkalizers are listed in Table 1.8.

Table 1.8: Some of the commonly used urinary acidifiers and alkalizers in domestic animals

Compound	Animals species	Dose	Route
Urinary acidifier			
Ammonium chloride	Cattle, goat, sheep	200 mg/kg	Oral
	Horse	4-15 gm	Oral
	Cat	20 mg/kg	Oral
	Dogs	40 mg/kg	Oral
Ascorbic acid	All species	40 mg/kg	IV
Sodium acid phosphate	Small animals	150-300 mg (three times a day)	oral
Urinary alkalizer :			
Sodium bicarbonate (1.4 % solution)	All species *	2-4 ml/24 hours	IV
Sodium acetate	Dogs	0.3-1.0 g	oral
Sodium citrate	Dogs	0.3-1.0 g	oral
Ringer lactate	All species	5-10 ml/kg/h	IV
Trometamol (3.66% solution)	Dogs	1-4 mg/kg/24 h	IV

* Monitoring of acid base balance is must.

(c). **Dialysis** : Haemo- or peritoneal dialysis with an appropriate dialysate can be attempted in dogs and cats, but its cost is prohibitive.

3. Elevating the threshold of the individuals to toxicity : Reduce the sensitivity of individual to toxicity by directly antagonising the effect of poison or correct the offending symptoms based on the principles of physiological or pharmacological antagonism using nonspecific or specific treatments. Nonspecific treatment is aimed at immediately mitigating the symptoms of poisoning to give relief and stabilize the condition of the animal e.g. use of analeptics in barbiturates poisoning, anticonvulsants for convulsive seizures, vasopressors for hypotensive crisis, bronchodilators, *beta* adrenoceptor agonists and antihistamines in anaphylactic shock and parasympatholytics to check excessive respiratory and salivary secretion etc.

Specific antidotal treatment of poisoning is highly desirable and most effective. Unfortunately, very few antidotes are available for a limited number of toxicants. Thus , these are being employed rather rare under such circumstances. An antidote may be defined as a specific remedy used for countering the action of a particular poison.

Depending on the mechanism of action, the antidotes may be classified as under:

      (i) Competitive antagonism : e.g. nalorphine for morphine poisoning.
      (ii) Non-competitive antagonism : e.g. atropine for carbamate poisoning.
      (iii) Chemical neutralization : e.g. sodium nitrite and sodium thiosulfate for cyanide poisoning.
      (iv) Metabolic inhibition : e.g. ethanol for methanol poisoning.
      (v) Oxidative reduction : e.g. methylene blue for nitrite poisoning.
      (vi) Chelation : e.g. dimercaprol (BAL) for arsenic, CaEDTA for lead, deferoxamine for iron poisoning etc.

Details of antidotal treatment, wherever available are given in respective chapters, however, some of the specific antidotes for offending chemicals/poisons are summarised in Table 1.9 for the benefit of readers.

Table 1.9: Some of the common poisons and their antidotes

Poison/Chemical/Drug/Toxin/	Antidotal treatment
Lead	BAL, Calcium disodium EDTA
Mercury	BAL
Arsenic	BAL (Dimercaprol), d-penicillamine
Iron	Deferoxamine
Zinc	Calcium disodium EDTA
Cyanide	Sodium nitrite, sodium thiosulfate
Nitrite, nitrate, chlorate	Methylene blue
Opiates	Nalorphine
Alkaloids	Tannic acids
Fluorides, oxalates	Calcium gluconate
Barbiturates	Bemegride
Copper	Penicillamine molybdenum salts
Molybdenum	Copper
Organophosphate compounds	Cholinesterase reactivators (oximes)
Coumarin anticoagulants	Vit. K$_1$
Ethylene glycol, methyl alcohol	Ethanol

CHPATER 2

TOXICITY OF METALS

R.P.Uppal

2.1 ARSENIC

Arsenic constitutes one of the most important toxicological hazards to farm animals. However, poisoning is relatively less frequent due to decreased use of arsenicals as pesticides, ant baits and wood preservatives. Toxicity varies with factors such as oxidation state of arsenic, solubility, species of animal involved and duration of exposure. Arsenic poisoning is caused by different types of inorganic and organic arsenical compounds.

Inorganic Arsenicals

Sources of poisoning : Poisoning occurs due to arsenic trioxide, arsenic pentoxide, sodium and potassium arsenate, sodium and potassium arsinite and lead or calcium arsenate. Trivalent arsenicals, also known as arsenites, are more soluble, and therefore, more toxic (5-10 times) than pentavalent or arsenate compounds.

Drinking water containing more than 0.25% arsenic is considered potentially toxic, especially to large animals. Herbivores are commonly poisoned because they eat contaminated forage. The inadvertent use of arsenicals (lead arsenate) owing to its resemblance with other preparations (lime) causes toxicity in dairy cattle. Deaths in sheep due to dipping in arsenical preparations has been on record. Lead arsenate is sometimes used as taenicide in sheep. Cats are poisoned because they ingest syrup baits intended for insects. Dogs are occasionally poisoned either maliciously or after being injected intravenously for therapeutic purposes due to overdosing. Swine and fowl are rarely poisoned.

Toxicokinetics : After absorption, arsenic is distributed throughout the body but tends to accumulate in the liver and kidneys. Pentavalent arsenic (arsenate) may be metabolized to trivalent (arsenite) but this does not occur with organic arsenicals. In domestic animals, arsenic does not stay in the soft tissues for a long period. It is rapidly excreted in bile, milk, saliva, sweat, urine and faeces.

After continuous intake, arsenic tends to accumulate in the bones, skin, and keratinised tissues such as hair and hoof. Arsenic stored at these sites may be found there for a long time. Arsenic probably crosses the placental barrier but the paucity of central nervous signs suggests that it does not cross the blood-brain barrier (BBB) very well. Milk from poisoned animals can be toxic to human beings but flesh of the surviving animals is said to be safe for human consumption.

Toxicodynamics : The lethal dose of sodium arsenite in most species is between 1 and 25 mg /kg. Cats may be more sensitive.

Arsenicals, particularly the arsenites are reported to act at molecular level primarily by combining with two sulfhydryl groups of *alpha*-lipoic acid (6, 8, dithioctic acid). Lipoic acid is an essential cofactor for the enzymatic decarboxylation of keto acids such as pyruvate, ketoglutarate and keto-butyrate. By inactivating lipoic acid, arsenic inhibits the formation of acetyl, succinyl and propionyl co-enzyme A. Probably other oxidative decarboxylation processes that utilize lipoic acid are also inhibited. The major consequences of lipoic acid inhibition would appear to be inhibition/slowing of glycolysis and TCA cycle.

$$
\begin{array}{ccc}
\underset{\text{α lipoic acid}}{\begin{array}{l} CH_2-SH \\ | \\ CH_2 \\ | \\ CH-SH \\ | \\ (CH_2)_4 \\ | \\ COOH \end{array}} &
\underset{\text{Trivalent arsenic}}{+\ R-As{=}O} \longrightarrow &
\underset{\text{Arsenic lipoic acid complex}}{\begin{array}{l} CH_2 \\ \quad \searrow S \\ CH_2 \quad \quad As-R\ +\ H_2O \\ \quad \nearrow S \\ CH \\ | \\ (CH_2)_4 \\ | \\ COOH \end{array}}
\end{array}
$$

Secondary molecular action of trivalent arsenic may include inactivation of sulfhydryl groups of oxidative enzymes and inactivation of -SH group of glutathione or other essential monothiols and dithiols.

Pentavalent arsenic can uncouple oxidative phosphorylation by forming unstable arsenate esters instead of phosphate esters that are required to provide high energy.

The effect of arsenic on gut, in part, can be attributed to its local corrosive effect though gastrointestinal damage occurs regardless of route of administration/absorption of arsenic. Arsenic seems to prefer tissues rich in oxidative enzymes such as intestine, liver and kidneys. The most sensitive cells seem to be the capillary epithelial cells of these organs, the effect being relaxation of capillaries and an increase in capillary permeability.

Clinical signs : Poisoning due to arsenic is usually acute with major effects on gastrointestinal tract and cardiovascular system. Because of its direct effect on capillaries. there is transudation of plasma, loss of blood and hypovolemic shock.

In acute cases, profuse watery diarrhoea (rice water) sometimes tinged with blood is characteristic alongwith severe colic, dehydration, weakness, depression, weak pulse and cardiovascular collapse. The onset is rapid and signs are usually seen within few hours (or upto 24 h). In peracute poisoning animals may simply be found dead.

In subacute cases, the animals may live for several days. Signs of poisoning include colic, anorexia, depression, staggering, weakness, diarrhoea with blood and/or mucosal shreds in faeces, polyuria and then anuria, dehydration, thirst, partial paralysis of hind limbs, trembling, stupor, cold extremities, subnormal temperature and convulsions (occasionally).

Chronic cases are rare and are characterized by wasting, poor condition, thirst, brick-red mucous membranes, normal temperature and a weak and irregular pulse.

Post-mortem lesions: In peracute toxicosis, no significant lesions may be seen. Inflammation and reddening of GI mucosa (local or diffuse) may occur, followed by oedema, rupture of blood vessels and necrosis of epithelial and subepithelial tissue. Necrosis may progress to perforation of gastric or intestinal wall. GI contents are often fluid, foul smelling and blood tinged. They may contain shreds of epithelial tissue.

There is diffuse inflammation of liver, kidneys and other visceral organs. Liver may have fatty degeneration and necrosis and kidneys have tubular damage. Lungs may be oedematous and congested. In case of cutaneous exposure the skin may exhibit necrosis and be dry or leathery.

Diagnosis :
(i). History
(ii). Clinical signs
(iii). Post-mortem lesions
(iv). Chemical examination of arsenic in tissues (liver or kidney) or stomach contents provides confirmation. However, it is difficult to state a diagnostic level of arsenic in animal tissues. Liver and kidneys of normal animals rarely contain > 1 ppm arsenic (wet weight basis); toxicity being associated with a concentration of > 3 ppm. The determination of arsenic in stomach contents is of value usually within first 24-28 h after ingestion. The concentration of arsenic in urine can be high for several days (upto 14 days) after ingestion.

History of sudden onset of severe colic, bloody/watery diarrhoea containing mucosal shreds alongwith post-mortem findings of gastroenteritis and degenerative changes in liver and kidney may be suspected as a case of arsenic poisoning. No other metal or metalloid causes such a speedy onset of gastrointestinal (GI) damage.

Differential diagnosis: Large number of poisons cause diarrhoea and colic, but only materials such as caustics, irritant plants, urea, chlorate, pesticides and enteric diseases are likely to cause GI signs similar in severity to those seen in typical cases of arsenic poisoning.

Lead poisoning (if severe) can mimic effects of arsenic poisoning but the spectrum of clinical signs is different from those of arsenic. For example, nervous and behavioural signs are minimal/absent in arsenic poisoning but very prominent in lead and pesticides poisoning.

Treatment : In animals with recent exposure and no clinical signs, emesis should be induced (in capable species) followed by activated charcoal alongwith a cathartic. This should be followed by oral administration of GI protectants (small animals, 1-2 h after charcoal), such as kaolin-pectin and fluid therapy as needed. In animals already showing clinical signs excessive fluid therapy, blood transfusion (if needed) and administration of BAL (dimercaprol / 2,3 dimercaptopropanol) @ 4-5 mg/kg, deep intramuscular, three times a day for 2-3 days or until recovery, is recommended.

$$
\begin{array}{ccc}
CH_2{-}SH & & \\
| & & \\
CH{-}SH & + & \text{(Arsenic-lipoic acid complex)} \\
| & & \\
CH_2{-}OH & &
\end{array}
\longrightarrow
\begin{array}{c}
CH_2{-}S \\
| \quad\quad \rangle As{-}R \\
CH_2{-}S \\
| \\
CH_2OH
\end{array}
+ \ \alpha\ \text{lipoic acid}
$$

Dimercaprol   Arsenic-lipoic acid complex   Arsenic mercaptide   α lipoic acid

In large animals, thiooctic acid (lipoic acid) alone (50 mg/kg; as a 20% solution; IM, three times a day) or in combination with BAL (3 mg/kg, deep IM; every 4 h for first 2 days, four times a day on third day, and twice daily for next 10 days or until recovery), is recommended. BAL should be administered as a 5% solution in 10% solution of benzyl benzoate in arachis oil. The dose of BAL is crucial and an overdosage may result in signs of toxicity such as deep respiration, salivation, vomition, lameness, stiffness, tremors, convulsions, coma and death.

In large animals, sodium thiosulphate may also be used orally or intravenously. The treatment in horse and cattle consist of giving 8-10 g in the form of 10-20 % solution intravenously and 20-30 g orally in about 300 ml of water. The dose in sheep and goats could be 1/4 and in small animals, 1/40 - 1/10 of the dose in large animals.

The water-soluble analogs of BAL, 2, 3 dimercaptopropane -1-sulphonate (DMPS) at 100 mg/kg/day for 10-12 days and dimercaptosuccinic acid (DMSA) at 30 mg/kg/day for 5 days are considered more effective than BAL and could be given orally.

D-penicillamine, an effective chelator having wide margin of safety could be used at 10-50 mg/kg, orally, 3-4 times a day for 3-4 days. Supportive therapy should be given greater weightage, particularly when cardiovascular collapse is imminent. Kidney and liver functions should also be monitored.

Organic arsenicals :

These include aliphatic organic arsenicals (cacodylic acid and acetarsonic acid, monosodium methanearsonate (MSMA) and disodium methanearsonate (DSMA), aromatic arsenicals (thiacetarsamide and arsphenamide) and phenylarsonic / benzene arsonic compounds (arsinilic acid, roxarsone and nitarsone). Phenylarsonic compounds are less toxic than inorganic or other type of organic compounds. In phenylarsonic compounds, arsenic is in pentavalent form.

Sources of poisoning : Toxicosis results from an excess of arsenic containing additives in pig or poultry diets. The severity and rapidity of onset are dose-dependent. Signs of toxicity may be delayed for weeks after incorporation of 2-3 times the recommended (100 ppm) levels or may occur within days when fed with more than 10 times the recommended levels. Persistence of MSMA or DSMA in soil and their tendency to accumulate in plants is a potential for arsenic poisoning, especially in grazing animals. In calves, large dosage of arsinilic acid may cause typical signs of inorganic poisoning. Phenylarsonic compounds are involved in toxicosis in swine and poultry when these are used as feed additives to improve production.

Toxicokinetics: The benzene arsonic compounds are poorly absorbed from GI tract and are thus excreted mainly in faeces. After absorption, the compounds are distributed throughout the body and are excreted in urine without being metabolized to a great extent. The elimination of parenterally injected compounds is nearly complete in 24-48 hours, but several days are required for its elimination after absorption from the gut.

Toxicodynamics : The demylenating disease or polyneuritis that results from overdosage of these compounds (benzene arsonic acid in particular) is very similar to that of vitamin B-complex deficiency. However, it has not been conclusively proved and the mechanism remains unclear.

Clinical signs : In pigs, the earliest signs of toxicity may be reduction in weight gain, followed by incoordination, posterior paralysis and eventually quadriplegia. The affected animals remain alert and retain good appetite. Blindness is characteristic of arsinilic acid intoxication and not of other arsenicals. In ruminants, phenylarsonic toxicosis is similar to inorganic arsenic poisoning.

Post-mortem lesions : Usually there are no specific lesions present in phenylarsonic poisoning. Demyelination and gliosis of peripheral nerves, optic tract and optic nerves are usually seen on histopathology.

Diagnosis :
(i).   History.
(ii).   Clinical signs- neurotoxic symptoms.
(iii). Post-mortem lesions particularly peripheral nerves.
(iv).   Analysis of liver and kidney cortex of affected animals and feed for arsenic may confirm the poisoning.

Differential diagnosis: The poisoning due to organic arsenicals should be differentiated from salt poisoning (particularly in chicken and pigs), insecticide poisoning and pseudorabies. In cattle, it should be differentiated from lead poisoning, insecticide poisoning and infectious bovine viral diarrhoea. Polyneuritis in swine or poultry due to vitamin deficiency or infectious disease may also be excluded before confirming diagnosis.

Treatment and preventive measures : There is no specific treatment or antidote. The neurotoxic effects are usually reversible if offending feed is withdrawn within 2-3 days of the onset of ataxia. Once paralysis occurs, the nerve damage is irreversible but the affected animals retain their appetite and weight gain is good if competition for food is eliminated. Recovery is generally doubtful when exposure is for longer duration and onset of intoxication is slow.

2.2 LEAD

In veterinary medicine, lead is one of the most common causes of metallic poisoning in dogs and cattle. Poisoning in other species is limited because of reduced accessibility to lead, more selective eating habits or lowered susceptibility. Intake of lead causes acute and chronic toxicosis.

Factors affecting lead toxicity :
(i).   **Age:** Young animals are considerably more sensitive than old ones.
(ii).   **Species and individual variation :** Goats, swine and chicken are more resistant. Individual variation to susceptibility and retention of lead are also known.
(iii).   **General health and reproductive state :** Poorly nourished and debilitated and parasitized animals are more susceptible. Pregnant ewes are reported to be more susceptible than non-pregnant ewes.
(iv).   **Route of entry :** Only 1-2 % of the ingested lead may be absorbed. The organic forms of lead (tetramethyl lead and tetraethyl lead) can also penetrate the intact skin. Large amounts of lead ingested in a short period of 1-2 days may be fatal but smaller amounts ingested over a prolonged period may not be detrimental.
(v).   **Form of lead :** Soluble salts are more readily absorbed.
(vi).   **Condition of the gastrointestinal tract :** Presence of ingesta may delay or reduce the absorption of lead. From the damaged intestinal mucosa, absorption of lead is comparatively better.
(vii).   **Past-exposure :** Previous chronic low level exposure may present a fair degree of tolerance to some of the biological effects of lead. Previous acute exposure may reduce the subsequent amount of lead required to produce lethal effects.

Sources of poisoning : Lead poisoning can result when curious animals ingest lead -based paints. Lead in paints is available as lead tetraoxide, carbonate or sulphate. In cattle, many cases of lead poisoning are associated with seedling and harvesting activities.

When used engine oil and battery disposal is handled improperly, it becomes a major source of lead poisoning. Vegetation grown in lead smelter areas where plants accumulate lead and contamination of vegetation on highways by exhaust fumes (petrol contains tetraethyl lead as contaminant) are other important sources of lead poisoning. Animals particularly sheep are reported to acquire a taste for lead and frequently graze in high contamination areas. Feeding on crops sprayed with lead insecticides (lead arsenate) may also result in lead poisoning in animals.

Toxicokinetics : Lead salts are sparingly soluble. The absorption of lead from GIT is very limited (1-2 %) and therefore, about 98% of lead is eliminated in the faeces. After absorption, a large proportion of lead in blood (85-90% in sheep and 65-70% in cattle) is carried to erythrocytes membranes. The remainder is bound to serum albumin and only a small proportion (< 1%) is actually free (unbound). This form is in dynamic equilibrium with lead bound to erythrocytes and serum albumin. Distribution of lead to various soft tissues takes place from unbound form. Free form of lead may also be excreted in milk in dangerous levels and may pose great threat to public health.

As the portal circulation carries lead through liver, much of it is excreted in bile from where it is reabsorbed (*enterohepatic biliary cycle*). In the process liver tends to contain large amounts of lead. Kidneys (kidney cortex in particular) present concentrations even higher than that of liver. Lead readily crossess the blood - brain barrier and the placental barrier.

From soft tissues, there is redistribution of lead to bones. The bone is considered to be a 'sink' for lead and it may contain 90-98% of the total body burden of lead. Thus, bone sink is an important detoxification mechanism under conditions of chronic exposure to small amounts of lead. Bones, however, can not store indefinite amounts of lead and when bones get saturated, signs of toxicosis may suddenly appear because of rising blood and soft tissue concentrations of lead. However, if exposure to lead ceases before the onset of signs of poisoning, lead is generally lost from bones. Withdrawal times in food producing animals may be months in such situations.

Toxicodynamics : The acute oral lethal dose of lead (in the form of lead salts) in various species has been considered as : calves, 50-400 mg/kg; cattle, 600-800 mg/kg; fowl, 160-600 mg/kg; horses, 500-700 g; sheep, 30-40 g and swine and dogs, 10-20 g (total dose). Chronic toxicosis can arise when lead is ingested over a period of days, weeks or months. A daily intake of 6-7 mg/kg of lead constitutes cumulative fatal dosage for cattle and 2.4 mg/kg for horses. A dose of 10 mg/kg/day of lead carbonate can kill a dog in 5 weeks and 3 mg/kg in 5 months.

Lead produces central nervous system effects and is also irritating, immunosuppressive, genotoxic, teratogenic, nephrotoxic and toxic to the haematopoietic system. Some of these effects can be explained as under :

After crossing the blood-brain barrier, lead exerts its effects at cellular level and produces ultrastructural changes. It is reported that these changes occur alongwith decrease in local concentrations of essential trace metals like copper, iron and zinc. It is further postulated that lead interferes with these metals as prosthetic groups of enzymes in the mitochondria, thus affecting vital metabolic pathways such as cellular respiration, oxidative phosphorylation and ATP synthetase complex formation. This could explain the loss of structural integrity of nerve cells, leading to visible lesions and functional disorders (behavioural and locomotor effects).

Anaemia is a constant finding in most of the affected animals. It could be due to destruction of erythrocytes (due to increased fragility), depressant action on bone marrow (deposition of lead) and inhibition of haeme synthesis. Lead may affect haeme synthesis at a number of steps as shown in Figure 2.1 :

(i).   Lead depresses amino levulinic acid (ALA,) dehydratase enzyme (a copper containing enzyme) resulting in increased serum level of *delta* aminolevulinic acid (δ-ALA) and its excretion in urine. This is considered as the major step involving lead.

(ii). Lead inhibits the conversion of coproporphyrinogen III to protoporphyrin IX. This effect is only 1/25 to 1/50 of that of ALA metabolism.

(iii). Lead appears to inhibit haeme synthetase, a thiol containing enzyme which is required to incorporate iron in the haeme molecule. It also prevents the entry of iron from cytosol to mitochondria.

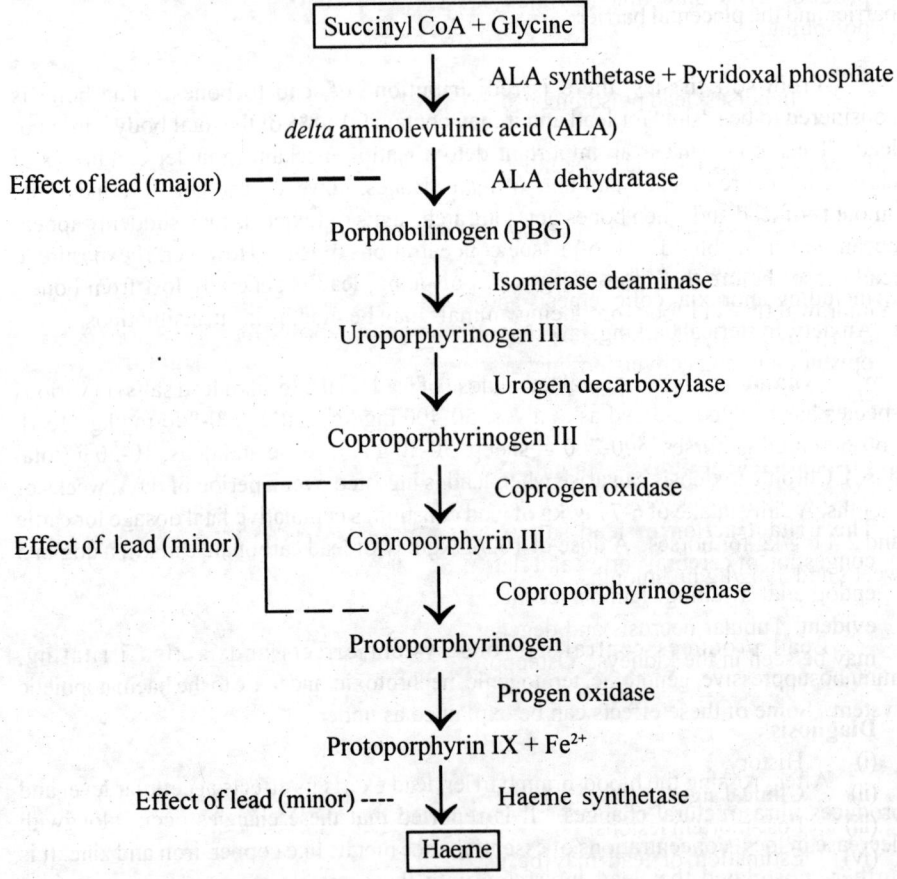

Fig.2.1 : Flow diagram showing effect of lead on different steps in haeme synthesis.

Other mechanisms, proposed to explain the biological effects of lead include the effect of lead on sulfhydryl group containing enzymes and tissues rich in mitochondria. It is argued that changes in cellular copper, iron and zinc concentrations, cell ultrastructure, myelination and cell and organ functions might be the result of sulfhydryl group inhibition. Elevation in serum transaminases (AST and ALT) activities and decrease in serum alkaline phosphatase and lactic dehydrogenase (LDH) activity indicate cell damage in the intoxicated animals, has also been described.

Clinical signs : The prominent clinical signs are associated with the gastrointestinal and nervous systems. In cattle, signs appear within 24 h of exposure and include ataxia, blindness, salivation, spastic twiching of eyelids, jaw champing, bruxism, muscle tremors and convulsions.

Subacute lead poisoning, usually seen in sheep and older cattle is characterized by anorexia, rumen stasis, colic, dullness and transient constipation, frequently followed by diarrhoea, blindness, head pressing, bruxism, hyperaesthesia and incoordination. Chronic lead poisoning which is occasionally seen in cattle, may produce syndrome that has many features in common with acute or subacute poisoning.

In horses, lead poisoning usually produces a chronic syndrome characterized by weight loss, depression, weakness, colic, diarrhoea, laryngeal or pharyngeal paralysis (roaring) and dysphagia frequently resulting in aspiration pneumonia.

In avian species, the prominent sings of lead toxicosis include anorexia, ataxia, loss of condition, wing and leg weakness, anaemia, gastrointestinal abnormalities including anorexia, colic, emesis and diarrhoea or constipation may be seen in dogs. Anxiety, hysterical barking, jaw champing, salivation, blindness, ataxia, muscle spasms, opisthotonus and convulsions may also be seen in intoxicated animals. Some dogs may show CNS depression rather than excitation.

Post-mortem lesions : Animals dying from acute lead poisoning may show few observable gross lesions. Oil or flakes of paint or battery may be seen in the GI tract. The irritant action of lead salts causes gastroenteritis. In the nervous system, congestion of cerebral cortex and flattening of cortical gyri are present. Histologically, endothelial swelling, laminar cortical necrosis and oedema of white matter may be evident. Tubular necrosis and degeneration of intranuclear acid fast inclusion bodies may be seen in the kidneys. Osteoporosis has been described in the affected lambs.

Diagnosis :
(i)   History.
(ii)   Clinical signs.
(iii)   Post-mortem lesions.
(iv)   Estimation of lead may be useful to evaluate excessive accumulation and to reflect the level or duration of exposure, severity and prognosis of toxicosis.
(v)   Concentrations of lead in blood at 0.35 ppm against normal level of 0.05 -0.25 ppm (lead being ubiquitous in nature), liver or kidney cortex at 10 ppm are of diagnostic value in lead poisoning in most species.
(vi)   Haematologic abnormalities which are indicative of lead poisoning include anaemia, basophilic stippling of RBCs, anisocytosis, poikilocytosis, polychromasia and hypochromia.

(vii)   Blood or urinary ALA and protoporphyrin levels are sensitive indicators of lead exposure. In poisoned animals, ALA levels above 500 µg/100 ml may be observed against the background (normal) level of 140 µg/100 ml.

Differential diagnosis: Lead poisoning may be confused with other diseases that cause nervous or GI abnormalities. In cattle such diseases/conditions may include polioencephalomalacia, nervous coccidiosis, tetanus, hypovitaminosis A, hypomagnesemic tetany, nervous acetonaemia, urea and insecticide poisoning, arsenic and mercury poisoning, brain abscess or neoplasia, rabies, listeriosis and *Haemophilus* infections. In dogs, rabies, distemper and hepatitis may appear to be similar to lead poisoning.

Treatment : As a first step, magnesium sulphate (400 mg/kg, *per os*) or rumenotomy (gastric lavage in case of dogs) may be useful to remove the unabsorbed lead from GI tract. Barbiturates or tranquilizers may be useful to control convulsions. However, if tissue damage is extensive particularly of CNS, treatment may not be rewarding. Specific antidotal treatment that may be useful includes :

(i)   Administration of calcium disodium edetate (CaEDTA) ; in livestock, it is given intravenously or subcutaneously (110 mg/kg/day) divided in two treatments daily for three days. The treatment should be repeated two days later. In dogs, a similar dose divided into four treatments a day is administered subcutaneously in 5% dextrose for 2-5 days. If clinical signs persist, additional 5 day treatment, after a gap of one week is recommended.

      Calcium EDTA chelates lead to render it (lead) inactive/nontoxic and the complex (lead EDTA) so formed is rapidly excreted (50 times the rate of excretion of lead). Only the lead stored in bones is excreted and not the one in soft tissues.

Calcium disodium EDTA   Lead disodium EDTA complex

(ii)   Thiamine (2-4 mg/kg/day, SC) alleviates clinical manifestations and reduces tissue deposition of lead. Combined CaEDTA and thiamine treatment may produce most beneficial response.

(iii) D-penicillamine can be administered orally to dogs (110 mg/kg/day) for two weeks. However, side effects such as emesis and anorexia are associated with this treatment. Also, it is not recommended for livestock.

(iv) Calcium phosphogluconate by intravenous route is also recommended in lead toxicosis to check intestinal spasms (colic). The recommended dosage is 250-500 ml for cattle and horses and 50-100 ml for sheep and goat.

2.3 MERCURY

Mercury exists in a variety of organic and inorganic forms. Toxic effects of the two forms are largely dissimilar. Organic mercurials are less toxic than inorganic compounds. The replacement of common mercurial compounds viz. antiseptics, diuretics and fungicides by other agents has decreased the likelihood of mercurial toxicosis; however, the possibility of exposure to environmental sources exists. Both forms of mercury, arising as affluents from industrial processes are converted into soluble methyl mercury in the lakes and rivers. This is then carried down to sea, where it is taken in by living organisms, notably fishes, the concentration factor being as high as 300 folds. This entails serious risk to creatures, including human beings, living largely on fish.

Sources of poisoning : Sources of inorganic mercurial poisoning include elemental mercury and salts such as mercuric chloride (corrosive sublimate), mercurous chloride (calomel), yellow mercuric oxide, red mercuric iodide, and mercuric nitrate. Dogs and cats are particularly susceptible to the toxic effects of mercury from mercurial ointments, both due to direct absorption and licking of parts to which the ointment has been applied. Poisoning in cattle has also been recorded after the use of ointments containing mercuric iodide, mercuric nitrate etc. Mercury vapours appear to be extremely toxic to sheep and cattle and are absorbed through respiratory tract.

Main sources of organic mercurial toxicity include fungicides (ethyl mercuric chloride and hydroxide) and methyl mercuric dicyandiamide and methoxy ethyl mercuric silicate used as seed dressing agents in agriculture. Antiseptics (mercurochrome, thiomersal, phenylmercuric acetate and nitrate) and diuretics (mersalyl) have also been implicated in toxicity. Secondary poisoning may arise from the ingestion of flesh of animals which have been fed on mercurial fungicides.

Toxicokinetics: Soluble mercuric salts are rapidly absorbed from the gut while insoluble salts (such as calomel) are absorbed only to a slight extent, and are relatively nontoxic unless given in large doses or when the toxicant remains in the gut for a prolonged period during which it gets converted into mercuric salt.

Once absorbed, mercury gets distributed throughout the body and is stored mainly in the liver and kidneys. Absorbed mercury is eliminated very slowly, chiefly in the urine, but to some extent in the faeces, saliva, sweat and milk. Considerable amounts are retained in the tissues indefinitely. Organic mercurials have the tendency to readily bioaccumulate in the brain; aryl mercurials (e.g. phenyl mercury fungicide), however, are less prone to bioaccumulation. The excretion is slow, the half life being 70 days. Two important outbreaks in human have been on record from Japan following consumption of shell fish caught from the sea rich in methyl mercury. These outbreaks have been reported as Minimata disease from Minimata bay area (1953 - 1960) and Niigata disease from Niigata bay area (1964-65).

Toxicodynamics : The toxic doses of mercuric chloride by oral route in different species are : horse and cattle, 8 g; sheep, 4 g; dog, 0.2-0.4 g. The compound is reported to cause high mortality in chicks when given at 250 ppm in drinking water. With mercurous chloride, the toxic dose is almost double to that of mercuric chloride.

The symptoms of acute inorganic mercury poisoning are largely those of corrosive poisoning where in shock or collapse is imminent. The mechanism of action of inorganic mercurials is similar to that of inorganic arsenicals wherein metal combines with sulfhydryl or disulphide groups thereby causing interference with the functioning of glutathione. Interference with protein synthesis has also been indicated.

Methyl mercury may interact with DNA and RNA and binds with-SH groups resulting in changes in the secondary structure of DNA and RNA.

Clinical signs: Clinical signs of inorganic mercurials poisoning include vomiting, diarrhoea and colic. Polydipsia, albuminuria and anuria (severe cases) is common. In cattle, death may occur rapidly after the ingestion of mercury with signs of colic and subnormal temperature as the only symptom. Inhalation of mercury vapours causes dyspnoea and coughing, nasal discharge, fever, loss of appetite and condition and sometimes bleeding of mucosa.

A single large dose may also lead to delayed poisoning, death being delayed for several days, and being finally due to severe damage to gut or kidneys. Cattle may develop paralysis under such circumstances. In rare cases of such poisoning, the CNS effects resemble to those of organic mercury poisoning. In man, cumulative poisoning due to continued absorption of small amounts over a long period leads to profuse salivation, swelling of gums, loosening of teeth and necrosis of jaw bones. These symptoms constitute what is known as mercurial ptyalism.

Animals poisoned with organic mercury compounds exhibit symptoms which are quite different from those of inorganic compounds. Symptom of CNS stimulation and locomotor abnormalities are seen after a lengthy latent period of weeks. Signs of toxicity may include blindness, excitation, abnormal behaviour and chewing movements, incoordination and convulsions. Cats show hind leg rigidity, hypermetria, cerebellar ataxia and tremors. Neurological signs may be irreversible once they develop.

Post-mortem lesions : The lesions of mercury toxicosis include gastroenteritis, stomatitis, gingivitis and acute parenchymatous nephritis. There is oedema of lungs, hydrothorax, hydropericardium, haemorrhages in epicardium and endocardium following inhalation of mercury vapours. Microscopic changes include necrosis of convoluted tubules of kidneys.

In animals affected due to organic mercurials, there are no characteristic lesions in liver and kidneys. Specific changes are seen in the brain which include degeneration of neurons (sensory in particular) and perivascular cuffing in the cerebrocortical grey matter, cerebellar atrophy of granular layer and damage to Purkinje cells.

Diagnosis :

Inorganic mercurials poisoning: In addition to history, clinical signs and post-mortem lesions, confirmation of poisoning may rest upon the detection of mercury in the stomach contents and of abnormal amounts in the kidneys (cortex) and liver. This will differentiate from other corrosive poisons producing similar symptoms and lesions.

Organic mercurials poisoning: Besides neurologic signs, laboratory diagnosis is required to confirm diagnosis. It must demonstrate normal concentrations of mercury in tissues (especially blood, kidneys and brain) and feed (< 1 ppm) and concentrations associated with poisoning. Differential diagnosis includes conditions with tremors and ataxia, and prominent signs, such as in other metals and insecticides and cerebellar lesions due to trauma or feline parvovirus.

Treatment :

In the treatment of acute mercury poisoning, speed is very important. As a first aid measure, raw white of egg or milk may be followed by gastric lavage with saturated sodium bicarbonate solution or sodium formaldehyde sulfoxylate (5%). When poisoning is due to topical application of mercury preparations, the skin should be thoroughly cleaned with soap and water. Fluid / electrolyte therapy may also be instituted.

Chelation therapy with dimercaprol (3 mg/kg, deep intramuscular, every 4 h for first 2 days, every 6 h on the third day and every 12 h for the next 10 days or until recovery is complete) has been used beneficially. The water soluble less toxic analog of BAL like DMSA is the chelator of choice for organic mercury poisoning. N-acetyl- DL-penicillamine (NAP, 15-50 mg/kg, *per os*) may be used only after gut is free from ingested mercury and renal function has been established.

2.4 SELENIUM

Selenium functions both as a toxin and an essential element. As a micronutrient (0.1-0.3 ppm), it is added to the diet to prevent several deficiency disease states of cellular degeneration and cell membrane damage such as white muscle disease in cattle and sheep, hepatosis dietetica in swine and exudative diathesis in chicken. Selenium deficiency is known to exert negative effects on milk yield and milk somatic cell count and immune response in animals. Dietary allowance of selenium above the requirements during pregnancy has been reported to improve selenium status of the newborn calves and enhances calf vitality and immune response; thus suggesting that selenium is one of the very important micronutrients in the feed/diet of animals.

Selenium is present in inorganic and organic forms and all animal species are susceptible to selenium toxicosis. Poisoning is, however, more common in forage eating animals when dietary selenium level exceeds 5 ppm. Severity of disease depends upon the oxidation state of selenium (selenide, selenite or selenate), quantity ingested, duration of exposure and individual and species susceptibility. Pigs from selenium deficient farms are reported to be more susceptible than the normal animals from the similar farm. Toxicity ranking of various forms of selenium is : natural organo-selenium (as it occurs in plants and grains) twice > selenate (Se^{6+}) = selenite (Se^{4+}) > selenide (Se^{2-}) > synthetic organoselenium compounds. Elemental selenium is almost non-toxic.

Selenium poisoning in animals occurs in acute or chronic form. Acute selenium poisoning under field conditions is less frequent because animals usually avoid the offending plants because of their offensive odour; however, when pasture is limited during periods of feed scarcity, selenium accumulator plants may be the only food available to the starving animals.

A number of plants have been reported to accumulate selenium from the soil. The obligate accumulators, also called selenium indicator plants require selenium for their growth. Such plants may contain as much as 15000 ppm of selenium and examples of such plants are *Astragalus, Stanleya, Oonopsis* and *Xylorrhiza*. The facultative accumulators do not require selenium but they may take up the element from the soil under favourable conditions. Selenium content in these plants may reach around 1500 ppm and may include species of *Aster, Atriplex, Machaeranthera, Sideranthus* etc. Another category of importance is of passive accumulators which may contain 20-60 ppm of selenium. Passive accumulation occurs in crop plants such as corn, wheat, oats, barley and grass simply because the selenium content is high in soil. The real danger to animals arises from accumulators by directly eating them or when they decompose in soil, the changed form is absorbed by cereal and other crops. Levels in excess of 5 ppm are dangerous and may cause toxicity.

Acute intoxication can occur in any species as a result of consumption of highly seleniferous forage or grains. Chronic selenium poisoning is of two types- *alkali disease* and *blind staggers*. *Alkali disease* (name first attributed to the consumption of alkali waters) results from prolonged ingestion (weeks) of forage or grain containing 5-40 ppm of selenium. *Blind staggers* develop after consuming moderate amounts (> 30 ppm of selenium) of seleniferous forage or grains over a period of days or weeks. It has been described in cattle and sheep but not in horses and swine.

Sources of Poisoning : Plants containing high selenium concentrations are the most important source of acute selenium poisoning in cattle, sheep and horses. Poisoning has been reported to occur in swine and poultry consuming grains raised on seleniferous soils or due to error in feed formulation. Parenteral selenium products are quite toxic, especially to young animals. Medicated shampoos (containing selenium disulfide) used for the treatment of some types of dermatitis and against dandruff have been implicated in selenium toxicosis in dogs and cats. Selenium as antioxidant in lubricating oils, funigicides and insect repellents is also involved in toxicity.

Toxicokinetics : Selenium is readily absorbed from the gut and distributed throughout the body, particularly in liver, kidneys and spleen. Chronic exposure results in large concentrations in hair and hoof of affected animals. Selenium can cross the placental barrier and can also pass into the eggs. It does not penetrate the intact skin but may penetrate abraded or diseased skin. In blood, the erythrocytes to plasma ratio of selenium is 3 : 1.

Organo-selenium, after absorption in the form of selanomethionine, may be oxidized from selenide state to selenite or reduced from inorganic selenate form to selenite. It is the selenite form which (selenium in selenite state may substitute sulphur in the synthesis of aminoacids and proteins) is responsible for toxicosis. Variable amounts (25-70 %) of dietary selenium may be eliminated within 2 days of ingestion of a large dose.

Toxicodynamics : In general, a single oral dose of selenium in the range of 1-5 mg/kg is lethal to most animals. Deaths have been reported in baby pigs at a dose of 1 mg/kg. Chronic toxicosis has been reported with dietary organic selenium concentrations of 7-15 ppm in different species.

Gastrointestinal signs and lesions in acute selenium toxicity are, in part, due to the irritant nature of selenium in large concentrations. However, the biochemical effects could be explained as under :

(i)   Replacement of sulphur of aminoacids such as cysteine and methionine resulting in the synthesis of abnormal proteins and enzymes. This may explain hoof and hair defects of chronic selenosis.

(ii)   Inhibition of SH - containing enzymes, such as succinic and other dehydrogenases, may result in decreased ATP synthesis. As a result of this, oxygen utilisation is decreased in liver, kidneys and brain. Inhibition of cell oxidation may also occur due to interference with haeme containing selenoprotein found in muscle tissue of selenium - treated animals.

(iii)   A dramatic reduction in the tissue glutathione (GSH) concentration occurring in the affected animals could be brought about by direct complexion of two molecules of GSH with every selenite ion. There is, perhaps, a competition between selenium and sulphur in the regeneration or synthesis of GSH, stimulation of liver and erythrocyte GSH-peroxidase activity (a selenium containing enzyme) or a combination of the two.

Clinical signs :

Acute selenium poisoning : In acute poisoning death usually follows within few hours after the consumption of highly seleniferous feeds/forage. Poisoning in ruminants is characterized by abnormal posture, unsteady gait, peculiar "rooted-to-one spot" stance with head and ears lowered, diarrhoea, polyuria, fever, mydriasis, abdominal pain, increased pulse and respiratory rates, blood tinged frothy nasal discharge, prostration and death. Sheep, instead, may become depressed and die suddenly without showing much signs of toxicity.

Chronic selenium poisoning : It is of two trypes

Blind staggers: This condition in cattle is manifested in three stages :

(i)   Affected animals intend to wander and may walk into objects. Usually, the body temperature is normal. The vision becomes impaired and the animal looses its appetite.

(ii)   The wandering increases, the front legs become weak and the vision becomes further impaired.

(iii) The throat and tongue become paralysed, body temperature becomes subnormal and the animal dies from respiratory failure.

In sheep, these stages are less clearly differentiated.

Alkali disease : Characteristic signs of alkali disease include cracking of hooves, lameness, stiffness of joints, dullness and lack of vitality, emaciation and loss of hair. In horses, loss of long hair from the mane and tail usually is the first clinical sign and is followed by cracking of the hoof at the coronary band. New growth of the hoof pushes the dead tissue downward and causes sloughing. In cattle, deformed hooves, 15-18 cm long and turned upward, may be seen. Pigs show breaks in the hoof similar to those seen in cattle. In sows, conception rate decreases and mortality of piglets at birth increases. Eggs with > 2.5 ppm of selenium have low hatchability and embryos are usually deformed, without beaks and with ropy feathers. Biochemical changes in blood in selenium toxicosis include decreased fibrinogen levels and prothrombin activity, increased serum alkaline phosphatase, ALT, AST and succinic dehydrogenase activity.

Post-mortem lesions : Animals which die of acute toxicosis show pulmonary congestion and oedema and degenerative changes in liver and kidneys. In blind staggers, necrosis and cirrhosis of liver, enlargement with localised haemorrhagic areas on the spleen, congestion of renal medulla, epicardial petechiae, hyperaemia and ulceration of abdomen and small intestine and erosion of the articular surface (particularly of tibia) are seen at necropsy. Ascites is almost a common finding. In alkali disease, lesions almost resemble those of blind staggers, the more pronounced being atrophy of heart and atrophy and cirrhosis of liver.

Diagnosis : Diagnosis is based on history, clinical signs, necropsy findings and laboratory confirmation of presence of selenium levels in animal's diet (feed, forage, grains) and blood or tissues (liver, kidney). Selenium levels beyond 5 ppm in diet are indicative of selenium toxicity. Severe signs of toxicity with 10-25 ppm of selenium confirm selenosis. In acute toxicosis, blood selenium concentration may reach upto 25 ppm, and in chronic toxicity, it may be between 1-4 ppm. Kidneys or liver may contain 4-25 ppm in both, acute and chronic poisoning. Content of selenium in the hair or urine may also be an indication of toxicosis.

Differential diagnosis: The diseases /conditions that may be confused with acute toxicosis or blind staggers include pneumonia, anthrax, infectious hepatitis, enterotoxaemia and pasteurellosis. Conditions that mimic aspects of chronic selenosis (alkali disease) include ergotism, molybdenosis, fluorosis and laminitis. The odour of rotten garlic or rotten horseradish in a fresh carcass is suggestive of acute toxicosis but absence of such an odour may not completely rule out this condition because the volatile selenides may escape quickly. However, a test for selenium in tissues may confirm the diagnosis.

Treatment and preventive measures. There is no specific treatment or antidote for selenium toxicosis except for eliminating the source and exposure. Symptomatic and supportive care of affected animals should be started as early as possible. A high protein diet, linseed meal, sulphur, arsenic, copper and cadmium have been reported to reduce selenium toxicity in laboratory species, but use of all these under field conditions needs confirmation. Arsenic has been recommended in drinking water (5 ppm of arsenic as sodium arsenite) or arsenic salt (containing 25 ppm of arsenic) or as arsinilic acid at 0.02% to reduce the incidence of selenium toxicity.

2.5 COPPER

Copper is an essential component of the animal system and plays an important physiological role in haematopoiesis, myelin formation, phospholipids formation, connective tissue metabolism and enzyme systems.

Copper toxicity has been encountered in most parts of the world. Sheep are affected most often, although other species are also susceptible. Acute poisoning is usually seen after intake of excessive copper salts being employed in agriculture and veterinary practice.

Many factors that alter copper metabolism influence chronic copper poisoning by enhancing the absorption or retention of copper. Low levels of molybdenum or sulphate also cause copper toxicosis. Primary chronic poisoning is seen most commonly in sheep when excessive amounts of copper are ingested over a prolonged period. Phytogenous and hepatogenous factors influence secondary chronic copper poisoning. Phytogenous chronic poisoning is seen after ingestion of plants that produce mineral imbalance and result in excessive copper retention. The plants that are hepatotoxic contain normal amounts of copper and low levels of molybedenum. Their ingestion for several months may cause hepatogenous chronic copper poisoning. These plants contain hepatotoxic alkaloids, which result in retention of excessive copper in the liver.

Sources of poisoning : Toxicity may occur due to copper sulphate, copper subacetate, copper chloride, oxychloride and carbonate, and copper edetate. Cases of acute poisoning occur when animals have mistakenly been given too large a therapeutic dose and, occasionally, when they get access to food contaminated with copper salts. Losses in lambs occur after treatment for foot-rot, either from contamination of ewes udder or from drinking of foot-rot bath water. Copper edetate used in the treatment of swayback disease in lambs has been reported to cause fatalities in undersized animals even at recommended dose levels (50 mg/kg).

Chronic copper poisoning occurs due to intake of successive non-toxic doses of copper. Grazing of animals in orchard sprayed with copper salts (copper sulphate), salt-licks and copper-supplemented diets result in chronic toxicity. Ponds treated with copper sulphate for algicidal effect, chicken manure (being a good source of non-protein nitrogen) mixed with silage or other feed and contaminated forage in the vicinity of mines or smelters could be other sources that could result in chronic toxicity.

Damage to liver by grazing plants like *Heliotropium europeum* or *Senecio* sp. can lead to abnormal accumulation of copper and cause haemolytic crisis in chronic copper poisoning. In certain areas, soil may be normal in copper contents but special climatic conditions may favour either the growth of non-graminous plants like *Trifolium subterraneum* which accumulate copper. Development of a high copper : molybdenum ratio within plants is considered to be involved in chronic copper poisoning since it is well established that a low molybdenum intake enhances the storage of copper in liver.

Toxicokinetics : After oral intake, copper is absorbed from the intestine and then enters a carrier state in the blood. In the blood, it is present in the erythrocytes as well as serum. Liver removes most of the copper from blood, but other soft tissues also store some copper. The storage of copper in the liver is variable in different species. The liver excretes copper in the bile but it gets reabsorbed from the intestine.

Toxicodynamics : The toxic dose of copper sulphate is in the order of 25-50 mg/kg in lambs and 130 mg/kg in sheep. In adult cattle, the toxic dose is about 200 mg/kg.

Chronic poisoning in sheep may occur with daily intake of 3.5 mg/kg when grazing pastures that contain 15-20 ppm (dry matter basis) of copper and low level of molybdenum. Swine given feed additive containing > 250 ppm of copper are dangerous. Low dietary intake of molybdenum and sulphur (sulphate) may enhance these levels.

Based on clinical observations and reasoning, sequence of events leading to chronic copper toxicity can be postulated. The term 'chronic' however, is misleading because the ingestion of copper may be chronic but the onset of toxicosis is acute due to the sudden release of copper from the liver.

It is known that liver (and other tissues) can not store copper to an indefinite level. Also, it is the large concentrations of copper saturating the cells that inhibit essential metabolic enzymes. The inhibition of vital enzymes may lead to liver dysfunction in days and weeks, and this may initiate liver necrosis ultimately leading to its (liver) inability to excrete copper. As a consequence, there is increase in serum glutamic oxaloacetic acid transaminase (SGOT), lactic dehydrogenase (LDH), plasma arginase and plasma bilirubin. Ultrastructural changes in mitochondria, endoplasmic reticulum, golgi apparatus and sinusoidal border are also seen.

The necrosed liver starts releasing large amounts of copper into blood stream. The copper enters the erythrocytes and is excreted in urine 24 hrs before the onset of haemolytic crisis. The erythrocytes can not release copper easily; therefore, the concentration of copper becomes large in these cells.

In the erythrocytes, normal ferrous (Fe^{+2}) state of haemoglobin (Hb) is converted to methaemoglobin (met Hb) by cupric ions (Cu^{+2}). As much as 35% of Hb may be converted to met Hb in sheep. Though significant, it is not an extraordinary concentration of met Hb. Simultaneously, glutathion (GSH) concentration in the erythrocytes drops suddenly. This is due to the monovalent state of copper (Cu^+) interacting with sulfhydryl (SH) group of GSH system. Monovalent copper is produced following the reduction of divalent copper (Cu^{+2}) during the conversion of Hb to met Hb. The reduction in concentration of GSH makes erythrocytes very fragile (susceptible to lysis by osmosis or chemical forces). The oxidative factors may also cause the formation of Heinz bodies (from met Hb) which get attached to the cell membrane of erythrocytes to render it less deformable. All this leads to sudden and massive lysis of erythrocytes called the haemolytic crisis. This crisis is absent in nonruminants, perhaps because they store less copper than ruminants.

Just before the onset of haemolytic crisis, other associated events that occur include a marked increase in packed cell volume (PCV) due to decrease in blood volume, an increase in erythrocytes size and possibly an increase in the number of erythrocytes. Also, there is an increase in copper and iron concentration in the kidneys. Cytoplasmic inclusions appear in the renal tubules.

As a consequence of haemolytic crisis, kidneys fail due to clogging of renal tubules with haemoglobin and necrosis of tubules and glomeruli brought about by excessive amounts of copper in the kidneys. Also, there is elevation of plasma creatinine phosphokinase (CPK) concentration indicating damage to skeletal muscles. This damage is thought to be due to anorexia and some other factors released during haemolysis.

Sheep died of copper poisoning may show lesions in their brains and these have been attributed to increase in blood urea and ammonia concentration, or alterations of glial transport mechanism by late-occurring inhibition of ATPase. The animals that survive haemolytic crisis may die of uremia.

Clinical signs : Symptoms of acute copper poisoning are nausea, vomition (in capable species), salivation, purgation, violent abdominal pain, dehydration, tachycardia, shock and collapse, ending in death. The faeces of affected animals contain mucus and are of deep green colour due to the presence of a copper-chlorophyll compound.

Chronic copper poisoning takes places in three stages. In the first stage, which may last for 2 to 3 months, there are no apparent clinical signs except a decrease in ruminal fermentation and ruminal stasis. In the second stage which may last for 14-25 days, there is impairment of liver function which is characterized by symptoms of anorexia, depression, weakness, thirst and diarrhoea. In the third (haemolytic crisis) stage lasting for 2-5 days, there is generalised icterus, haemoglobinemia, haemoglobinuria and recumbency. Severe hepatic insufficiency is responsible for deaths. Animals that survive acute episode may die of subsequent renal failure. Herd mortality is often < 5% though more than 75% of the affected animals die. Losses may continue for upto two months after dietary problem has been rectified.

Post-mortem lesions :

Acute copper poisoning:
(i)   Severe gastroenteritis with erosions and ulcerations in the abomasum of ruminants.
(ii) A characteristic feature is that blood is found to have coagulated at the time of death.
(iii) Icterus develops in animals that survive beyond 24 h.

Chronic copper poisoning:
(i)   Generalised icterus is the main lesion.
(ii) The liver is enlarged, yellow in colour and friable.
(iii) Gall bladder is distended with thick greenish- brown bile.
(iv) Swollen gunmetal coloured kidneys (showing haemorrhagic mottling when capsule is removed).
(v)   Port wine-coloured urine
(vi) Enlarged spleen with dark, brown-black (blackberry jam) parenchyma.
(vii) Histologically, there is centrilobular hepatic and renal tubular necrosis and the brain may manifest spongy degeneration and astrocyte damage.

Diagnosis : Copper is a normal constituent of tissues and blood. The diagnosis of chronic copper poisoning would, therefore, depend upon finding markedly elevated concentrations of copper in the tissues of animals having history, clinical signs and lesions of copper toxicosis.

Blood and liver copper concentrations are increased during the period of haemolytic crisis. Blood levels often rise to 5-20 µg/ml compared to the normal level of almost 1 µg/ml. Liver concentrations of > 150 ppm (wet weight) are significant in sheep.

Evidence of blue-green ingesta, deep green-coloured faeces and increased faecal (8000 - 10000 ppm) and kidney (> 15 ppm, wet weight) levels of copper are considered significant in acute copper toxicosis.

Differential diagnosis: Other clinical conditions that mimic copper toxicosis include haemolytic diseases, hepatitis and poisoning by other haemolytic agents. The levels of copper in tissues must be determined to rule out other causes of haemolytic disease.

Treatment and preventive measures : Often, treatment is not successful. Gastrointestinal sedatives and symptomatic treatment of shock may be useful in acute toxicity. Penicillamine or calcium versenate may be useful, if administered in early stages of toxicosis. Experimentally, intravenous administration of ammonium tetrathiomolybdate has proved effective for the treatment and prevention of copper poisoning. This could be due to interaction between copper and tetrathiomolybdates resulting in the formation of insoluble-complex in the rumen and poor absorption. The complex may decrease the availability of copper and inhibit copper-dependant enzymes. Daily administration of ammonium molybdate (50-500 mg) and sodium thiosulphate (300-1000 mg) reduces losses in affected animals. Such a treatment can reduce liver copper concentration to two third in six weeks, and to half in further six weeks.

Dietary supplementation with zinc acetate may be useful to reduce the absorption of copper.. Reducing access to plants that cause phytogenous or hepatogenous copper poisoning is desirable. Primary chronic or phytogenous copper poisoning can be prevented by top-dressing of pastures with 30 gm/acre (70 g/hectare) molybdenised super phosphate or by molybdenum supplementation or restriction of copper intake.

2.6 MOLYBDENUM

Molybdenum is an essential micronutrient. It is an essential component of metalloenzymes (xanthine oxidase and aldehyde oxidase) that are necessary for proper health of all animals. In ruminants, dietary intake of excessive molybdenum causes, in part, a secondary hypocuprosis. Non- ruminants are less susceptible, the resistance being 10 times that of cattle and sheep. The susceptibility of ruminants to molybdenum toxicity depends on a number of factors described below:

(i) Copper and inorganic sulphate of the diet and their intake by animals: Tolerance to molybdenum toxicity decreases as the content and intake of copper decreases. High dietary sulphate with low copper exacerbates the condition, while low dietary sulphate causes high blood molybdenum levels due to decreased excretion.

(ii) Chemical form of molybdenum: Water soluble molybdenum in growing herbage is most toxic.

(iii) Species of animal: Cattle are less tolerant than sheep.

(iv) Age: Young animals are more susceptible than adults.

(v) Season of year: Plants start concentrating molybdenum during beginning of spring (April) and becomes maximum in September-October.

(vi) Composition of pasture : Legumes take up more of elements than other species.

In feeds of cattle, copper : molybdenum ratio of 6 : 1 is considered ideal; 3 : 1 is borderline, and less than 2 : 1 is toxic. Dietary molybdenum concentrations of > 10 ppm can cause toxicity regardless of copper intake; as little as 1 ppm may be hazardous if copper content is < 5 ppm (dry weight basis).

Sources of poisoning : Grazing of cattle on pastures with high molybdenum content leads to toxic syndrome of molybdenosis. Top dressing of pastures with sodium molybdate has been responsible for poisoning in livestock. Industrial molybdenosis due to contamination of pasture with molybdenum emanating from industrial undertakings engaged in the manufacture of ferro-molybdenum alloys or aluminium alloys has been described.

Toxicokinetics : Molybdenum compounds are both rapidly absorbed and excreted. The storage is transient and throughout the body tissues but is greatest in kidneys and bones. The level is low in new born, increases upto 20 years and then starts declining. The retention or excretion of molybdenum is considerably affected by the amount of copper and inorganic sulphate in the diet. Increasing inorganic sulphate intake promotes the excretion of large quantities of molybdenum, particularly in urine, and the level of molybdenum in the blood falls.

Toxicodynamics : The toxic actions of molybdenum are complex and may involve various mechanisms:

(i)   Higher molybdenum and lower copper levels in the liver may inhibit various enzyme systems necessary for skeletal development. There may be competition between molybdenum and phosphorus for deposition in the bone. Molybdenum can even cause the displacement of phosphorus. All this may explain the occurrence of skeletal lesions.

(ii)   Persistent scouring with passage of liquid faeces full of gas bubbles (teart) could be due to complex formation between molybdenum and catechols, which, like other phenols, are bacteriostatic and control the activity of bacteria in the gut. But when there is complex formation, the bacterial activity increases markedly and then results in scouring diarrhoea.

(iii)   Anaemia (hypochromic) is another characteristic of molybdenum poisoning. It is attributed to copper deficiency occurring as a result of reduction in activity of sulphide oxidase in the liver.

Clinical signs : Most of the clinical signs attributed to molybdenum toxicity arise from impaired copper metabolism and resemble to those of copper deficiency. Clinical signs of toxicity in cattle appear within 1-2 weeks of grazing on affected pastures and toxicity is characterized by persistent and severe scouring with liquid faeces full of gas bubbles (peat scours or tearts). Depigmentation resulting from fading of the hair coat is most noticeable in black animals and especially around eyes, which gives a spectacled appearance. Other signs of toxicity include unthriftiness, anaemia (hypochromic), emaciation, joint pain (lameness), osteoporosis and decreased fertility. The affected animals show abnormal pacing gait (called *pacing disease*). Sheep develop pica. Sheep and young animals in particular show stiffness of the back and legs with reluctance to rise; the condition is called *"enzootic ataxia"* in Australia while the syndrome is named as 'Swayback disease' in U.K. Ewes become anaemic with stringy wool whereas lambs are severely incoordinated, ataxic and blind.

Effects on reproduction, particularly in heifers, include delayed puberty and reduced conception rates. Recent studies suggest that levels of molybdenum may exert these effects on certain metabolic processes, particularly reproduction that is independent of alterations in copper metabolism.

Diagnosis: A provisional diagnosis can be made if diarrhoea stops within a few days of oral dosing with copper. Diagnosis can be confirmed by demonstrating abnormal concentrations of molybdenum and copper in blood or liver and by a high dietary intake of molybdenum relative to copper in the diet/forage. Molybdenum level beyond 0.1 ppm in blood (normal level, 0.05 ppm) and >5 ppm in liver (normal level, 3-4 ppm) is indicative of molybdenosis.

Treatment and preventive measures : Scouring can be controlled by daily administration of copper sulphate 1 g for calves and 2 g for cattle. Copper glycinate injectable (subcutaneous) at a dose of 60 mg for calves and 120 mg for cattle has been used successfully as an adjunct to therapy.

In areas where, molybdenum content of forage is < 5 ppm, the use of 1% copper sulphate has provided effective control of molybdenosis. With higher levels of molybdenum, 2% copper sulphate has been successful; upto 5% copper sulphate has been used in regions presenting very high levels of molybdenum. In areas where, for various reasons, cattle do not consume mineral supplementation, the required copper may be supplied as weekly drench, or as a top-dressing to the pasture.

CHAPTER 3

TOXICITY OF NON-METALS

Satish K. Garg

3.1 PHOSPHORUS

Phosphorus alongwith calcium is an integral constituent of the animal body and is universally distributed in soil and plants. About 80% of the total body phosphorus is deposited in bones and teeth in combination with calcium while rest is primarily in organic combinations. One of the most fundamental functions of phosphorus is the transfer of biological energy, particularly through ATP. Deficiency of phosphorus in diet resulting into rickets and osteomalacia is a common syndrome, however, hyperphosphatemia is comparatively of less common occurrence. Feeding of excess of wheat bran which is very rich in phosphorus and deficient in calcium causes bran disease in horses. Similarly, meat high in phosphorus and low in   calcium causes hyperphosphatemia in dogs and cats. Severe nephritis also results into secondary hyperphosphatemia as it interferes with the excretion of phosphorus. Blood phosphorus levels are elevated 2-4 times which in turn results in hypocalcemia and secondary parathyroid hyperplasia. Cattle, sheep, dogs and game birds are affected but most cases of poisoning occur in swine.

Sources of poisoning : Though phosphorus poisoning is of rare occurrence in farm animals because of lack of exposure, but accidental ingestion of fertilizer whilst still in sacks or immediately after spreading (from clumps of fertilizer on cultivated land), fire works, baits containing lumps of white phosphorus for rats, pets or ants kept on the pastures or ingestion of rats poisoned with rodent baits or grazing of animals on the battle fields where certain explosives have been used. Sometimes, in finely divided form, phosphorus is mixed with fats and oils to promote their absorption; over consumption of such fats and oils also results into phosphorus poisoning. Red phosphorus is inert and nontoxic but white or yellow phosphorus is toxic. Toxicity potential of phosphates is very low. These need to be ingested in considerable quantities before toxicity signs appear.

Mechanism of toxicity: Exact mechanism of phosphorus - induced toxicity is not known. However because of its local caustic action, phosphorus causes severe irritation of the gastrointestinal mucosa and induces gastroenteritis and diarrhoea. Whatever phosphorus is absorbed into the blood stream, it circulates in the blood first as element and ultimately oxidised to phosphate which causes hepatic degeneration.

Toxic doses vary greatly depending on in which state phosphorus is ingested. White phosphorus in finely divided form causes toxicity in horses and oxen (0.5- 2 g), pigs (0.05-0.3 g), dogs (0.05 - 0.10 g) and fowl (0.02 g).

Clinical signs : In per-acute toxicity, animals die after showing intense abdominal pain, violent convulsions, severe CNS depression and coma while in acute poisoning salivation, nausea, vomition, severe diarrhoea with mild abdominal pain, fever, polydipsia and polyuria are the main signs. Most of the poisoned animals also exhibit symptoms of jaundice, haematuria, oliguria, followed by delirium, convulsions, coma and death. Signs of gastroenteritis appear within 1-2 hrs. and course of illness is 3-5 days depending on the severity of toxicity. Pigs vomit violently and the vomitus is luminous in dark and gives a characteristic garlic odour. However, salivation in horses and paralytic like weakness in fowls without any other definitive signs are observed.

Chronic toxicity in animals is rare, however, as a professional hazard, inhalation of fumes containing phosphorus in industrial plants, human beings are commonly affected.

Post mortem lesions:
(i)   Garlic like odour of the gastrointestinal tract contents,
(ii)   Liver is enlarged, pale and yellowish in colour.
(iii)   Spleen is small and atrophied.
(iv)   Congestion and haemorrhagic inflammation of gastrointestinal tract.
(v)   In some of the cases, hydrothorax and oedema of other parts of the body.
(vi)   Fatty degenerative changes with centrilobular necrosis in liver.
(vii)   Fatty changes are also observed in kidneys and heart.
(viii) Inflammation of the mucus membrane of stomach and intestines.
(ix)   Extravasation of blood into subcutaneous tissues and muscles.
(x)   In birds, visible fumes of phosphorus can be appreciated on opening the gizzard.

Diagnosis :
(i)   History.
(ii)   Clinical symptoms particularly acute gastroenteritis.
(iii)   Garlic odour of the vomitus and intestinal contents.
(iv)   Post-mortem lesions.
(v)   Estimation of phosphorus in the blood, vomitus, intestinal contents and faeces.

Note: No preservative should be added to the specimens to be sent for laboratory investigation.

Differential diagnosis :
(i)   Inorganic poisonings (arsenic, lead, mercury) causing gastroenteritis and diarrhoea.
(ii)   Organophosphate compounds.

Treatment : No specific antidote is available for phosphorus poisoning. However, treatment is aimed at removing the poison from the body by giving activated charcoal, emetics or purgatives. Oily purgatives should not be given as these favour absorption of phosphorus, therefore, saline purgatives should be used. In addition, give symptomatic treatment, like- gastrointestinal demulcents and astringents to take care of gastroenteritis and excess of fluid therapy to replenish the lost body fluids and electrolytes. In conditions of shock, give cardiac stimulants and intravenous glucose infusion.

3.2 NITRATES AND NITRITES

Nitrates and nitrites are closely linked as cause of poisoning in animals, however, it is difficult to differentiate between these poisonings as nitrates can be converted into nitrites and both the forms are toxic. Thus, nitrates and nitrites are discussed together as the mechanism(s) involved in poisoning is the same.

Nitrates are relatively non-toxic as ruminants convert nitrate of the plants into nitrite in the rumen and then to ammonia by the action of ruminal and intestinal microbes. Ammonia so produced is utilized as a source of nitrogen by the rumen microflora for protein synthesis. However, if the rate of reduction of nitrate to nitrite exceeds to that of nitrite to ammonia, then excessive nitrite accumulates in the rumen and gets absorbed into the blood stream and produces toxicity as shown below:

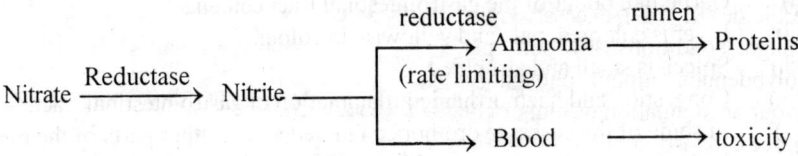

Reduction of nitrates depends on the supply of fermentable energy as carbohydrates supply hydrogen ions for reduction of nitrates and production of nitrite in the rumen. Poor feeding probably reduces the biotransforming capacity of ruminal microbes and thus increases the production of nitrite ions. However, the rate of reduction of nitrates in the digestive tract of horses is comparatively limited.

Ruminants are most commonly affected, though human beings, horses, pigs and dogs are also poisoned. Pigs, however, are most susceptible to nitrite poisoning. Non ruminants are susceptible to nitrite but not nitrate poisoning due to absence of rumen microflora.

Toxicity of nitrates/nitrites not only depends on the dose but also the time e.g. single higher dose is more toxic compared to the same or even higher dose administered in divided doses over a period of time say 24 hours. Toxic doses are difficult to quantify precisely as toxicity varies depending on variation in susceptibility, production of nitrite from nitrate and food deprivation. Food deprivation increases the overall toxicity while concentrate rich or cereal based feeds reduce it. Oral LD_{50} values for nitrates and nitrites are as given below Table 3.1.

Source of poisoning : Main source of poisoning in animals, particularly ruminants is excessive ingestion of certain nitrates containing plants. More than 80 specific nitrate containing plants are known to cause nitrite poisoning, a few of the important ones are given in Table 3.2. Roots and stems usually contain more nitrate than leaves. Nitrates are present in high concentrations in soil, ground water from deep wells, plants, animal excreta and silage etc. Plants / crops grown on nitrate rich soil and / or water accumulate higher concentrations of nitrates.

Table 3.1: Oral median lethal dose (LD_{50}) values (mg/kg) of nitrates and nitrites in different species of animals.

Species	Nitrate	Nitrite
Rat	584-656	60.0
Cattle	409-750	88-110
Sheep	409-547	40-50
Pig	19,000	88-60
Horse	61,000-152,000	--
Dog	3,000	---

Use of chemical fertilizers like ammonium and potassium nitrates, weedicide/ herbicides (2, 4 D), excessive rain, insufficient light, excessive and persistent frost and drought conditions, low soil pH, low soil temperature and aeration, deficiency of molybdenum, sulphur, phosphorus, iron etc. in soil and viral infection of plants etc. favour accumulation of higher concentration of nitrates in plants. In addition to plant sources, ingestion of nitrate fertilizers (ammonium or potassium nitrate), consumption of nitrate rich water from shallow surface well and ponds, animal wastes, sewage, preserved food left at room temperature for long time, silo juices, industrial effluents and lime stone deposits etc. may also result in nitrate/nitrite toxicity. Animals receiving roughages alone are more susceptible to toxicity compared to those on concentrates feed rich in carbohydrates. Young animals are more susceptible to poisoning compared to adults. Hypovitaminosis A, pre-existing anaemia, methaemoglobinemia and fasting aggravate the toxicity of nitrates and nitrites. Monensin facilitates the conversion of nitrate to nitrite in the rumen and may result in poisoning in cattle or sheep on high nitrate fodder.

Table 3.2: Some of the nitrate containing plants

Amaranthus retroflexus (red root or pigweed)	*Avena sativa* (common oats)
Astragalus hamosus	*Beta vulgaris* (sugar beat)
Brassica napobrassica (turnip)	*Brassica rapa* (turnip)
Chenopodium album (lamb's quarter)	*Cucurbita maxima*
Datura sp.	*Eupatorium* sp.
Franseria discolor (white ragweed)	*Ipomoea* sp.
Linum usitatissimum (linseed)	*Melilotus officinalis*
Pennisetum clandestinum	*Solanum* sp.
Tribulus sp.	*Zea mays* (maize)

Mechanism of toxicity : Nitrates *per se* are relatively nontoxic and are irritant or caustic on the alimentary mucosa, kidneys and urinary tract. However, nitrites produced/ liberated in the gastrointestinal tract are toxic. Nitrite is 6-10 times more toxic than nitrates.

Nitrate is reduced to nitrite in the rumen. Excessive nitrite ions are absorbed into the blood circulation. Acute poisoning is as a result of two actions of nitrite ions :

(a) Interaction of nitrite ions with haemoglobin : One mole of absorbed nitrite reacts oxidatively with two moles of haemoglobin (Hb), and in the process, there is loss of an electron and ferrous (Fe^{2+}) form of iron in Hb is converted to ferric (Fe^{3+}) form resulting in methaemoglobin formation. Methaemoglobin has severely reduced oxygen carrying capacity. The process of met- Hb production and its conversion back to Fe^{2+}- Hb is endogenously regulated by two reducing enzyme systems in the blood, namely -NAD-dependent Diaphorase I and NADP-dependent diaphorase II as shown below:

$$Fe^{2+}Hb \rightleftharpoons Fe^{3+}Hb(met\ Hb)$$

Diaphorase I and II

Thus implying that nitrates are converted to nitrites in the gastrointestinal tract and nitrites to nitrates in the blood. But if nitrates are ingested in excess and reductases (Diaphorase I and II) fail to reduce met-Hb to Hb, nitrites produce toxicity due to methaemoglobinemia. Clinical signs of toxicity appear on conversion of 20 per cent of the Hb to met-Hb and severity of toxicity progresses as more and more of Hb is converted to met-Hb in the blood, as a result tissues are starved of oxygen. Death is almost certain with 80 per cent methaemoglobinemia due to tissue hypoxia.

(b) Relaxation of the vascular smooth muscles : It involves the formation of nitric oxide (NO) from nitrite. Nitric oxide results in vasodilatation, systemic arterial hypotension and decreased cardiac output, though it is not the main action. Nitrite ions also have a direct effect on the nervous system, the exact mechanism of which is not known.

Clinical signs : Onset of acute nitrite poisoning symptoms is abrupt and characterized by dyspnoea with violent respiratory efforts or gasping. Rapid respiration is the predominant sign. Salivation, voiding of colourless urine, vomition, diarrhoea, colic, rapid and weak pulse, accelerated heart rate, progressive development of cyanosis, muscular weakness, ataxia, recumbency, terminal anoxic convulsions and death within 2-24 hours depending on the content of nitrite consumed. Bluish discolouration of the mucous membranes and unpigmented areas of the body and chocolate brown colour of blood are quite characteristic of nitrite poisoning, however, there is no change in the coagulation time. In pregnant animals it may induce abortion as nitrite ions cross the placental barrier and convert foetal-Hb to Met-Hb.

Chronic nitrite poisoning is although rare, but sometimes observed in poorly nourished animals drinking high nitrate content water contaminated with chemicals and organic fertilizers, domestic waste waters, industrial effluents etc. or regular and frequent ingestion of plants or fodder rich in nitrates (e.g. rape seed, cabbage, brassica, rye grass) or of hay contaminated with wild plants (e.g. *Amaranthus*, *Chenopodium* or fat hen). Characteristic clinical features of chronic nitrite poisoning are abortion, poor development of mammary tissue, infertility in adults and lower birth weight in newly born lambs or calves if dam is exposed to nitrite during pregnancy. Ewes show hypothyroidism and hypovitaminosis A. Other common signs are ataxia, tremors, shaking, immunosuppression, increase in the incidence of mastitis, metritis and diarrhoea in calves and piglets etc.

Post-mortem lesions:
(i)   Dark chocolate brown or coffee brown colour of blood and it clots poorly.
(ii)   Brown stained tissues.
(iii)   Congestion and inflammation of intra-abdominal organs.
(iv)   Petechial and large haemorrhages on the serous surfaces.
(v)   Dilatation of the blood vessels.
(vi)   Generalized cyanosis.
(vii)   Blood stained pericardial fluid.

Diagnosis :
(i)   History.
(ii)   Clinical signs.
(iii)   Chocolate brown colour of blood and tissues.
(iv)   Post mortem findings if there is mortality.
(v)   Analysis of the stomach/ruminal and intestinal contents for nitrate /nitrite detection.
(vi)   Detection of nitrite in the urine.
(vi)   Estimation of met-Hb and serum nitrite levels (serum positive diphenylamine spot test; 20 µg/ml nitrate). In dead animals, positive diphenylamine spot test with ocular fluid indicating > 50 µg/ml nitrate for confirmation of diagnosis.
(vii)   Response to methylene blue treatment.

Note : Specimens for laboratory examination should be collected within 1-2 hours of the death of animals and include blood for met-Hb estimation, ingesta and suspected plants or water; chloroform or formalin be added to these specimen samples to prevent reduction of nitrates by bacterial fermentation. However, if there is some time lapse after the animals have died, aqueous humour of eye and cerebrospinal fluid should be collected for chemical assay.

Differential diagnosis :
(i)   Hydrocyanic acid poisoning
(ii)   Chlorate poisoning
(iii)   Carbon monoxide poisoning,
(iv)   Hydrogen sulfide poisoning
(v)   Acute pulmonary oedema
(vi)   Hydrosulfide emphysema
(vii)   Anaphylactic reaction
(viii) Hypotensive agents

The important differential points amongst some of the above mentioned poisonings are summarized in Table 3.3.

Treatment : Remove the source of poison. Objective of treatment in nitrite poisoning is to reduce met-Hb to Hb by using a suitable oxidizing agent. Methylene blue (1%w/v) solution in isotonic salt solution is the specific antidote used and be administered by slow IV injection @ 8.8 mg/kg to sheep and cattle and 4.4 mg/kg to other species and be repeated after 15-30 min if clinical response is not satisfactory. The half life of methylene blue in tissues is about 2 hours. Further repetition of treatment is recommended at 6-8 hours interval, if necessary.

Methylene blue (oxidizing agent) is reduced to leucomethylene blue by NADPH$_2$-dependent system which reduces met-Hb to Hb as shown below.

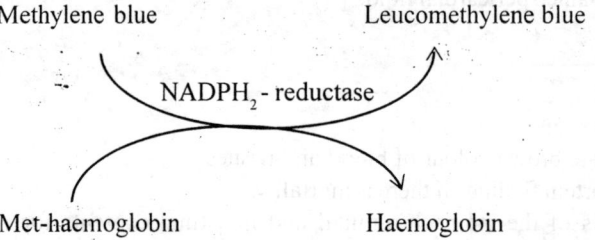

Leucomethylene blue also activates Diaphorase I and II systems. However, dose of methylene blue is very important and critical. If methylene blue is in high concentrations, reducing system NADPH$_2$ - reductase may be saturated and as a result whole of methylene blue may not be reduced to leucomethylene blue. Thus, methylene blue may oxidize Hb to met-Hb and aggravate the toxicity. Ascorbic acid (5-20 mg/kg), another reducing agent, administered intravenously may also be useful.

Further approach be to sooth the gastrointestinal tract lining and reduce the rate of absorption of nitrite by administering mineral oil like liquid paraffin or saline purgatives, to hasten the elimination from gastrointestinal tract. Large doses of antibiotics in excess of water be administered orally to inhibit microbial conversion of nitrate to nitrite by gastrointestinal flora.

Supportive and symptomatic treatment includes oxygen therapy, blood transfusion, shock therapy etc. to take care of hypotension. Dietary supplements like vitamin A, D and E are also recommended.

Preventive measures :
(i) Stop access of the animals to offending fields or fodder.
(ii) Provide adequate carbohydrates in feed.
(iii) Supplement the diet with chlortetracycline (30 mg/kg of feed) or sodium tungstate or monensin for two weeks to suppress gastrointestinal flora to reduce conversion of nitrate to nitrite in the rumen.

Table 3.3 : Differential diagnosis of some of the common toxicants which may be confused with nitrate/nitrite poisoning.

Differential features	Nitrate/nitrite	Sodium chlorate	Cyanide	Silo gases (Nitrogen dioxide; NO_2) Nitric oxide (NO)	Carbon dioxide	Carbon monoxide
Blood colour	Dark brown to chocolate	Tarry or brown black	Bright red	Slight brown	Dark	Cherry red
Mechanism	Methaemoglobin	Methaemoglobin	Blocks cytochrome oxidase system	Slight methaemoglobin	Displaces oxygen	Carboxyhaemoglobin
Treatment	Methylene blue	Methylene blue	Sodium nitrite and sodium thiosulfate	Methylene blue	Oxygen, fresh air	Fresh air, oxygen + 5% carbon dioxide, respiratory stimulants

3.3 CHLORATES

Sodium chlorate, a strong oxidizing agent, is used as a weedicide and defoliant in the form of a spray or dry salt. Though chances of poisoning with chlorates are rare, yet because of its salty appearance and taste, salt hungry animals may consume lot of sodium chlorate or it may be used in animal feed in place of common salt by misidentification. Cattle are more often poisoned than sheep. The lethal doses of sodium chlorate for cattle, sheep and poultry are 1, 2 and 5 g/kg body weight, respectively. However, lethal dose for dogs is 1.5 - 3.5 g/kg. Death usually occurs within 6-48 hours. It is readily absorbed from the gut and is excreted largely unchanged.

Mechanism of toxicity : Chlorate is a strong oxidizing agent and produces methaemoglobinemia by the similar mechanism as described in nitrite poisoning, though compared to nitrite poisoning, the process of met-Hb formation is slow and continues even after the death of the animal so long as chlorate is present in the blood stream as it is not inactivated. It also causes severe haemolysis.

In addition to methaemoglobinemia, chlorates also result in the formation of low concentrations of sulfhaemoglobin which is a partially oxidized and denatured mixture of the pigments due to nonspecific oxidation and there is no mechanism in the erythrocytes for the reversal of sulfhaemoglobinemia. But it never exists in life-threatening concentrations. Chlorate is directly irritant to gastrointestinal tract and causes purging and abdominal pain.

Clinical signs : Except for the delayed onset, the general signs of chlorate toxicity are similar to that of nitrite poisoning. The gastrointestinal symptoms like vomition, diarrhoea, colic etc. appear earlier and the respiratory distress is observed later in chlorate poisoning while respiratory signs are first to be observed in nitrite poisoning. Further, haematuria, haemoglobinuria and effusing out of dark tarry blood, which clots readily, from natural orifices of the body is quite characteristic of chlorate poisoning. In acute cases, death may occur even without any apparent signs of toxicity.

Post Mortem lesions :
(i)   Generalized cyanosis.
(ii)   Blood is of dark chocolate colour.
(iii)   Brownish discoloration of the organs and tissues.
(iv)   Exudation of blood from natural orifices.
(v)   Gastroenteritis.

Diagnosis :
(i)   History.
(ii)   Clinical symptoms.
(iii)   Dark chocolate coloured blood.
(iv)   Musculature is dark or almost black and cut surfaces become some what lighter on exposure to air.
(v)   Liver is almost black and heart is flabby and dark.
(vi)   Oozing out of dark tarry blood from the natural orifices.

(vii) Post mortem findings, if death.

(viii) Analysis of blood, urine, tissues and ruminal contents for chlorate detection.

(ix) Estimation of met-Hb in blood during clinical signs period, on mortality and after death. The concentration of met-Hb increases with the passage of time as the process of met-Hb formation continues even after the death of animal.

(x) Response to treatment.

Differential diagnosis :

Since in chlorate poisoning respiratory distress is one of the main symptoms followed by death in acute cases, it should be differentiated from other poisonings summarized below with similar symptoms, some of the important differential points have already been summarized in Table 3.3.

(i) Nitrite poisoning - No oozing out of blood from the natural orifices.

(ii) Cyanide poisoning : Blood is bright red in colour.

(iii) Hydrogen sulphide : Blood is dark in colour.

(iv) Carbonmonoxide : Blood is bright red in colour.

(v) Poisoning due to other haemolytic agents like copper, dimethyl sulfoxide, gossypol, snake-venoms and certain drugs.

(vi) Warfarin poisoning.

Treatment : Objective of treatment in chlorate poisoning is similar to that in nitrite poisoning i.e. to reduce the chlorate-induced met-Hb formation from Hb. methylene blue is the drug of choice. Details of the mode of action, dosages and precautions in the use of methylene blue have been explained in nitrite poisoning. However, compared to nitrite poisoning, methylene blue need to be administered more frequently and repeatedly for a longer period because production of met-Hb continues in the body so long as chlorate remains in the body. Next approach should be to remove the poison from the gut by gastric lavage or / and saline purgatives.

3.4 FLUORINE

Chemically, fluorine is so reactive that it is found in combination with many minerals/elements like aluminium, iron, calcium, silicates, phosphates etc. and thus is an important constituent of many of the rocks and ores. Fluorides are emitted from industries involved in the manufacture of aluminium, steel, phosphate fertilizers, brick kilns, potteries, ferro-enamel, fused tricalcium phosphate etc. The fumes or effluents coming out of the industries may settle on fields, pastures or water reservoirs. Similarly, gases or smoke dust coming out of volcanic eruptions also result in contamination of fields, pastures or water, thus, fluorine is present in varying concentrations in soil, water, atmosphere and plants.

Both humans and animals exposed to very low concentrations of fluorine for prolonged periods result in accumulation of fluorides in body particularly bones and teeth without exhibiting any clinical signs in the beginning. The signs, however, become apparent after long periods when too much damage has been produced in the target organs. Thus, acute fluorine poisoning is not commonly observed except in accidental ingestion/exposure to high quantity of fluorine. But chromic fluorine poisoning, also termed as fluorosis is a very serious and important syndrome both in animals and human beings.

Acute Fluorine Poisoning :

Acute fluorine poisoning generally occurs due to accidental ingestion of large quantities of fluorine containing salts e.g. sodium fluoride, sodium fluoroacetate, sodium fluorosilicate or excessively contaminated water, feeds and fodder. Sodium fluoride is used as a vermifuge for the control of round worms in pigs and lice in poultry. Sodium fluoroacetate is used as a rodenticide. Accidental ingestion of the baits containing fluoroacetate or the poisoned animals may also be responsible for acute fluorine poisoning. Sometimes even excessive licking of phosphate rocks as mineral supplements also results in fluorine poisoning in animals.

Pigs are most commonly affected. About 4-5 per cent sodium fluoride in feed is lethal for pigs. Lethal dose of fluorosilicate is 100 g for equines and 200 g for bovines.

Mechanism of toxicity : Exact mechanism of action is not known, however, fluorides, being strongly irritant produce gastroenteritis. Fluoride-ions increase permeability of blood vessels and cause coagulation defects, haemorrhages, congestion and oedema of organs, particularly brain. Fluoride-ions being very reactive inhibit a number of enzymes, namely-preglycolytic, phosphatases and cholinesterase and, thus inhibit glycolysis and increase sensitivity of the body system to acetylcholine.

Being too much irritant, fluoride ions produce too much gastric irritation and vomitions. However, probability of toxicity is more in those species where vomition reflex is absent.

Clinical signs : Vomition, anorexia, salivation, ruminal stasis, abdominal pain, gastroenteritis, diarrhoea, urination, weakness, constant chewing, dyspnoea, excitability, muscular tremors, pupillary dilatation, tetany, clonic convulsions, sudden collapse, coma and death due to respiratory and cardiac collapse.

Postmortem lesions :

(i)   Haemorrhagic gastroenteritis.
(ii)   Congestion of viscera, particularly liver and kidneys.
(iii) Bone and dental lesions are absent.

Diagnosis : It is very difficult, however, history and circumstantial evidences may be helpful.

Differential diagnosis :
(i)   Inorganic poisoning e.g. arsenic
(ii)   Cholinesterase inhibitors
(iii)   Warfarin poisoning

Treatment :   No specific antidote is available. Though prognosis of poisoning is poor, yet symptomatic treatment may be given.
(i)   Gastrointestinal sedatives.
(ii)   Intravenous infusion of calcium salts.
(iii)   Intravenous infusion of glucose solution.

Chronic fluoride poisoning (Fluorosis) :

Fluorosis is chronic fluorine intoxication and often observed after prolonged ingestion of small but toxic amounts of fluorine in the diet/feed or fodder contaminated by industrial pollution from aluminium smelting factories, steel works, cement factories, brick kilns, coal burning electric power stations, glass manufacturing and phosphate processing units etc. Consumption of contaminated water or use of fluoride containing minerals or phosphate rocks as feed supplements for animals may also be responsible for fluorosis. Though it is a non-fatal syndrome, but productivity of the affected animals goes down.

Animals normally ingest low levels of fluorine throughout the life as fluoride is a normal constituent of forages, especially legumes. The fluorine gets deposited in bones and teeth without any apparent signs of toxicity. There is very long latent period probably due to gradual saturation of bones and teeth with fluorine as these tissues act as sink for fluorine. Out of a total body load of fluorine, 95-96% is deposited in bones and teeth. Once these structures get saturated, fluorine ions start exerting general toxic effects. Unabsorbed fluoride is eliminated through faeces.

Ingestion of fluoride in fodder (ppm of dry weight) at levels less than 30 ppm in young cattle and milking cows, 40 ppm in beef cattle, 50 ppm in sheep, 70 ppm in pigs, 90 ppm in horses and 150 ppm in chickens do not produce clinical signs of poisoning.

Sources of poisoning : Feeds, fodder, water, mineral supplements rich in fluorine and top dressing of pastures with phosphate lime stone are the common sources of poisoning. These conditions have been reported from the areas adjoining industrial units emitting fluorine containing gases or dusts or the areas in the vicinity of volcanos. Animals ingesting vegetation/crops or grazing on the pastures where industrial effluents or the smoke and gases coming out of industrial units or volcanos settle are also poisoned.

Factors affecting toxicity : Fluorosis depends on solubility of the compounds (e.g. sodium fluoride is more toxic than calcium fluoride), amount ingested, duration of exposure, rate of excretion, age of the animal, nutritional status, general health conditions of the animals, individual susceptibility and resistance, stress factors, species variation etc. Though almost all species of animals including human beings are susceptible to fluorosis, species susceptibility is as follows : calves, dairy cows, beef cattle, sheep, horses, pig and poultry. Thus, poultry seems to be most resistant to fluorosis. Compared to cattle, susceptibility of sheep is less because of comparatively lower biological availability of fluoride ions in sheep.

Mechanism of toxicity : The exact mechanism of fluoride-induced chronic toxicity is not known. Fluoride ions probably interfere with excretion of calcium or osteoclast activity. Fluoride ions replace hydroxyl radicals in the apatite crystals thus resulting in abnormal osteoid which is further responsible for poor bony matrix and irregular mineralization. Because of damage to blast cells, mineralization of pre-enamel, predentine, precementum and preskeletal matrices may be delayed i.e. fluoride ions by affecting osteoblasts may cause abnormal mineralization of bones. In addition, it also inhibits enzymes involved in bone and teeth formation.

Deposition of fluoride in bones occur throughout the life but in teeth only during the formative stages. Thus, it affects only the developing teeth. Therefore, dental lesions may be absent even in very severe fluorosis cases, if the animals are adult.

In bones, deposition of fluoride is highest on the periosteal surfaces of long bones where exostoses develop. It also produces degenerative changes in bone marrow, kidneys, liver, adrenal glands, heart muscles and CNS.

Dental and bone lesions are indicators of fluorosis, however, teeth lesions are the earliest and most severe in young and growing animals. Generally, two forms of fluorosis- dental fluorosis and osteofluorosis have been appreciated and characterized by mottling and abrasion of teeth and intermittent lameness, respectively.

Dental fluorosis :

Clinical signs : Dental lesions are painful. The earliest and mildest sign is mottling, consists of light yellow, green, brown or black spots or bands arranged horizontally across the teeth. Occasionally, vertical bands are also seen where pigment is deposited along the enamel fissures. Mottling generally occurs on incisors, however, molars and premolars are worst affected which are difficult to examine. Such teeth give dull, opaque white and chalky appearance.

If animals continue to consume fluoride containing feeds or water, mottled areas become pits, teeth become brittle and break unevenly and sometimes are reduced to the level of gums or even there is shedding of teeth in worst affected cases. In young animals, eruption of permanent incisors is delayed, there are oblique eruptions or hypoplasia of teeth with wide gaps between the teeth.

Due to uneven surface or shedding of teeth, mastication becomes difficult and even impossible in worst affected cases, thus there is poor growth in young and growing animals and acetonaemia in adults. Affected animals like to drink cold water to avoid pain while ingesting fodder. Other general signs are anorexia, dry and rough hair coat, ruffled fur or feathers, emaciation and reduced milk production.

Osteofluorosis :

Clinical signs : Sudden onset of lameness, unthriftiness, stiffness, painful gait and posture are the first characteristic signs of osteofluorosis often observed by the farmer. Lameness is observed in both the fore limbs and hind limbs. It is moving and diagonal, observed first in one leg and then in other leg. Most marked effects are observed in loins, hip joints and hind limbs. Bones are palpably and visibly enlarged and thickened. Such lesions are observed on the medial surface of the proximal third of metatarsal bones followed by thickening and enlargement of mandible, deformation of the jaw, sternum, metacarpal, ribs and spine. Pressing of affected limb bones give indication of pain. Thickening of the periosteal layer results in lateral exostoses of long bones of the legs. Articular surfaces are not involved, there is spurring and bridging of joints leading to rigidity of the spine. Other bone lesions are hyperostosis, osteoporosis, enlargement, chalky white appearance and roughening with intermittent limping. Bones become more prone to fractures.

As the condition worsens, there is cachexia and death. Other supplementary general signs of fluorosis are loss of weight, intermittent diarrhoea, polydipsia, polyuria, poorly concentrated urine, aplastic anaemia due to diminished bone marrow cavity, reduction in milk and wool production, anestrous etc. Though, fluoride ions are known to cross the placental barrier, but no apparent toxic effects on the foetus are known.

Post mortem lesions :

(i)   Normal ivory colour of bones is changed to chalky white, surfaces of bones are roughened, diameter enlarged and lateral exostoses of the long bones.

(ii)   Mottling, cavities on teeth, which are unevenly broken.

(iii)   Bone marrow cavity is diminished and shows gelatinous degeneration and aplastic anaemia.

(iv)   Microscopically, bony trabeculae are thickened and have a dense appearance with sharp heavy outlines. Degenerative changes are observed in kidneys, liver, adrenal glands, heart muscles and central nervous system. In kidneys, degeneration and disintegration of the tubular epithelium, slight glomerular changes, thickened arterioles and fibrosis are also observed.

(v)   There is atrophy of spongiosa, defective and irregular calcification of newly formed osseous tissue, hypoplasia of enamel and dentine.

(vi)   Severe gastroenteritis is absent.

Diagnosis :

(i)   History.

(ii)   Clinical signs.

(iii)   Post mortem findings.

(iv)   X-ray examination reveals sclerosis, perosis, hyperostosis or a combination of these. There is increased density of abnormal porosity, periosteal feathering and thickening.

(v)   Microscopically, thick cortex due to periosteal hyperostosis, uneven mineralization, zones of immature bone, excess of osteoid tissue and atrophy of spongiosa.

(vi)   Fluoride assay of feed, water, blood, urine, bone, teeth and faeces.

Differential diagnosis :

(i)   Vitamin D deficiency.

(ii)   Deficiency of phosphorus and calcium

(iii)   Parathyroid disease.

Treatment : Prognosis is very poor as removal of fluorides from teeth and bones is difficult. No specific antidote is available. There is no remedy except for the symptomatic or supportive treatment.

(i)   Remove the animal(s) from the contaminated pastures.

(ii)   Change the feed, mineral supplement and drinking water.

(iii)   Correct the mineral deficiency in the food/rations, particularly calcium and phosphorus. Add vitamin A and D to the feed.

(iv)   Symptomatic treatment using steroids, analgesics, antibiotics, fluid therapy, vitamin C etc.

(v)   Intravenously give glucose and calcium fluid.

(vi)   Supplement the feed of affected animals with aluminium sulphate (30 g/day) or calcium carbonate. These help in reducing the absorption as form complex with fluorides in the intestines which gets excreted through faeces.

3.5 COMMON SALT (SODIUM CHLORIDE)

Common salt is an essential nutrient and added to the feed and ration of animals but it is the quantity consumed which makes it toxic. Paracelsus stated "what is there that is not a poison" ? All things are poisons and nothing is without poison. The dose solely determines that a thing is not a poison. Thus, excessive ingestion of sodium chloride causes toxicity and the condition is also termed as water deprivation syndrome. All species of animals including human beings and poultry are poisoned but poultry and pigs are most susceptible to salt poisoning. In poultry too, young chicks are most susceptible due to indiscriminate feeding behaviour, poor sense of taste, low plasma proteins in chicks and decreased glomerular filtration area in the kidneys of chicks compared to mammals.

Sources of poisoning : Feeds containing high quantities of common salt, accidental over ingestion of common salt or excessive licking of salt licks kept on the premises particularly when the animals have been on restricted salt supply or after a period of salt deprivation for quite some time then the animals may develop craving for salt or salt hunger, excessive consumption of salty meat or meat flavoured brines by carnivores, swill feeding to pigs if it contains residues from bakeries, brine from butcher's shop, salt whey from cheese factories or salted fish waste, oil fields as salt water is an effluent from oil production, change from low salt ration to high salt ration, low vitamin E and sulphur containing aminoacids. Sometimes overdosing of animals with sodium sulphate or some other sodium salts also results in salt poisoning. However, one of the important determinants of salt poisoning is the salt hunger and availability or restriction of water supply as over-consumption of salt can be tolerated by the animals if sufficient quantity of water is made available to the animals immediately after ingestion of salt. The surplus sodium ions (Na^+) are excreted primarily by the kidneys and little by skin and alimentary tract. It is difficult to say something about the threshold toxic dose of sodium chloride as consumption of sufficient quantity of water facilitates excretion of sodium ions, yet 2.2 g/kg of salt by mouth is the toxic dose for ruminants, horses and pigs, 6 g/kg for sheep and 4 g/kg for dogs. Toxic dose of salt as a % of the diet is 1.5 - 2.0 for cattle and sheep, 0.25 for pigs and 0.25-2.0 for poultry when water supply is restricted, however, when drinking water is freely available, toxic dose for pigs is 10.0 percent of diet while 5-10% for poultry.

Mechanism of toxicity : Exact mechanism is not known. But sodium ions and water balance are mainly disturbed. Excess of sodium ions in gastrointestinal tract cause mild irritation and secretion of water into the lumen of intestine and thus diarrhoea further resulting into dehydration.

Absorption of Na^+ results in hypertonicity of blood and hypernatremia, shrinkage of kidney tubules, deposition of sodium crystals in the tubules, anuria, uremia etc. following transient polyuria due to initial excretion of Na^+ in the urine. Decreased ability of sodium pump to remove Na^+ from the cells sets up an osmotic gradient. Excess of Na^+ in blood stream (extracellular hyperosmolarity) results in shrinkage of capillary vascular endothelial cells in the brain and meninges which in turn stimulate the capillary permeability and escape of water from blood to interstitial spaces (intracellular dehydration) and development of brain or cerebral oedema. Cerebral oedema in pigs is also associated with the accumulation of eosinophils in the brain tissue. As a result of cerebral oedema, there is increased cerebrospinal fluid pressure and reduced blood flow to the brain and thus hypoxia and the clinical signs and lesions are observed.

Clinical signs : General signs of salt poisoning are anorexia, excessive thirst, salivation, initially diarrhoea followed by constipation, polyuria followed by anuria, nasal discharge and weak pulse. Body temperature is normal but ear and skin are cold. There is rigidity of muscles, hyper-irritability, blindness, stumbling, walking backwards or in circles, pedalling of limbs, recurrent convulsive seizures, recumbency, coma and death in a few hours to few days.

Vomiting in dogs, profuse watery diarrhoea with colic in ruminants, diarrhoea with colic, mucus in faeces, knuckling of fetlocks, dehydration and prostration in lactating animals, pruritis, ataxia, dog sitting posture, blindness, convulsions, comatose and paddling in pigs while intense thirst, respiratory distress, fluid discharge from beak, weakness, wet faeces and limb paralysis in poultry are more marked.

Post mortem lesions:
(i)   Congestion and inflammation of gastrointestinal tract.
(ii) Faeces are fluidy and dark or dry.
(iii) Hydropericardium.
(iv) Severe acute inflammation of gastric and intestinal lining.
(v)   Oedema of tissues and body cavities.

(vi)   Renal congestion.
(vii)   Oedema of the cerebral cortex (polioencephalomalacia in cattle, eosinophiliç meningoencephalitis in pigs). Microscopically, large number of eosinophils in the distended perivascular space and meninges are almost pathognomonic in pigs but not in poultry. Other changes are degeneration of neurons and slight general gliosis, vacuolization, disruption of the area between cortex and white matter.
(viii)   Congestion of liver in chicks, hyperaemia of the organs and deposits of uric acid in kidneys, ureters and droppings in mature birds.
(ix)   Congestion of liver and kidneys and lungs are collapsed and full of blood in dogs.

Diagnosis :

(i)   History of salt ingestion.
(ii)   Clinical signs of poisoning including excessive thirst.
(iii)   Post-mortem lesions.
(iv)   Species involved.
(v)   Circumstantial evidences of relatively restricted water or salt supply.
(vi)   Laboratory investigations indicating plasma sodium levels of exceeding 150 mEq/L in live animals and CSF sodium level of > 145 mEq/L and brain sodium > 1800 ppm in dead animals.

Differential diagnosis :

(i)   Poisoning due to chlorinated hydrocarbons, but there is no thirst and hyperthermia.
(ii)   Poisoning due to organophosphate compounds - but there is hypothermia.
(iii)   Poisoning due to drugs or plants causing central symptoms of stimulation.
(iv)   Poisoning due to lead or other metals where gastric symptoms are more severe.
(v)   Injury to CNS.
(vi)   Pseudorabies
(vii)   Encephalitis.

Treatment : No specific antidote is available. Do not give strong natriuretic drug.

(i)   Remove the toxic feed/ or water.
(ii)   Salt free fresh water be made available, however, initially access of the animal to water should be restricted as large intake of water will kill the animal by aggravating cerebral oedema. Thus give small quantities of water but more frequently i.e. return to water slowly.
(iii)   Isotonic or hypotonic salt solution intraperitoneally daily for 2-3 days.
(iv)   Gastrointestinal tract sedatives.
(v)   Sedatives to counter the CNS stimulation.

3.6 UREA AND AMMONIA

Urea is used as a fertilizer on crop and pasture fields and sometimes as a substitute for common salt in melting of snow and ice in snow clad residential areas. However, its use as a cheap source of non-proteinous nitrogen (NPN) in ruminants ration started during the world war II. Urea is much better tolerated when mixed with sufficient amounts of other feeds. About 2-3 per cent of total ration (dry weight basis) can safely be urea but larger amounts are likely to produce poisoning. Ruminal microflora possess urease activity which hydrolyse urea into ammonia and water. Ammonia is assimilated for amino acids and microbial protein synthesis. Therefore, dietary requirement of proteins is decreased if NPN source is added to the ration of ruminants. For proper utilization of ammonia (NH_3), within certain limits, readily and sufficiently available soluble carbohydrates in the form of starch and glucose are essential. However, production of ammonia beyond handling capacity of liver results in urea poisoning.

Ruminants rapidly adapt to urea feeding. Animals should be accustomed to urea feeding by gradually increasing its level in the ration/feed as it has been found that on adaptation, cattle can assimilate even 400 g urea per day while without proper adaptation even as little as 50 g/day is toxic. Tolerance and adaptation to urea feeding is lost rapidly. If the animals receive no urea for three days, they become susceptible to urea poisoning. Tolerance is also reduced by starvation and low protein diet. Other factors, affecting urea toxicity are availability of soluble carbohydrate sources in the feed/diet, adaptation of the animals, abrupt change in the diet, lack of water, ruminal pH etc. Food/ rations based on concentrates reduce severity of intoxication, however, simultaneous feeding of soyabean meal potentiates urea toxicity due to liberation of excess of ammonia as soyabean meal contains urease enzyme.

Toxicity of urea depends on the rate of ammonia formation and absorption. All mammalian species are susceptible to ammonia poisoning, however, ruminants are most sensitive. Compared to ruminants, horses (equines) appear to be tolerant to relatively large doses of urea where its hydrolysis takes place in the caecum. However, pigs are quite resistant to even very large doses of urea. This is probably due to higher capacity of ruminants than non-ruminants to handle absorbed ammonia as the former have a greater hepatic urea synthesizing ability or a high glutamine synthetase activity in the spleen, liver and brain.

Oral LD_{50} value of urea in cattle and sheep is 1.0-1.5 g/kg and horse is 4.0 g/kg while toxic dose in cattle and sheep is 0.3 - 0.5 g/kg.

Sources of poisoning : Accidental ingestion of solid or liquid form of urea due to improper storage or spilage, feeding of large quantities of NPN urea-mollases feeds to unaccustomed animals, improperly mixed feed etc.

Mechanism of toxicity: Hydrolysis of urea occurs in the rumen. The rate of ammonia production depends primarily on the amount of ration ingested, the amount of urease present in the ruminal contents or the diet and pH of the ruminal contents. Toxicity of urea is due to ammonia absorbed from the stomach /rumen which depends on ruminal pH.

$$\text{Urea} + \text{water} \longrightarrow > CO_2 + NH_3$$

At rumen pH of < 6.2, majority of the released ammonia is in the form of ammonium ions (NH_4^+) which are highly water soluble and absorbed poorly i.e. low pH favours production of NH_4^+. When pH of rumen is 9, ratio of ammonia : NH_4^+ is 1, much of the ammonia is absorbed. If absorption of ammonia into blood is upto certain limits, body detoxifies ammonia as absorbed ammonia is normally incorporated into the urea cycle and excreted as urea in urine as shown in Fig. 3.1. Conversion of ammonia (NH_3) to urea occurs in the liver. Ruminants have a higher capacity to handle absorbed NH_3 than the non-ruminant species due to greater hepatic urea - synthesizing ability.

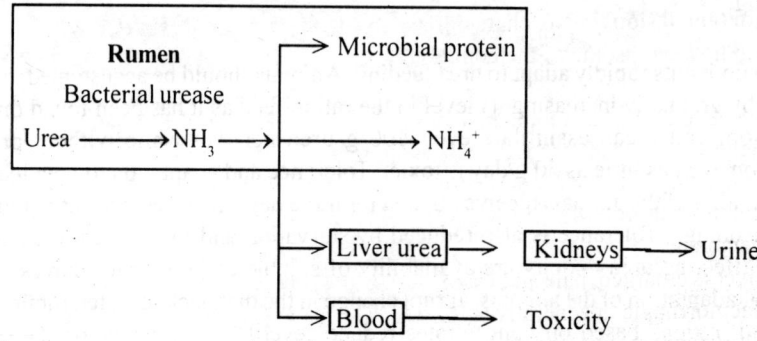

Fig. 3.1: Production and handling of urea in the body and its toxicity

When concentration of NH_3 in rumen exceeds 80 mg/dl, NH_3 appears in the peripheral blood and high NH_3 concentrations are built up in blood stream and thus NH_3 accumulates in tissue cells. When blood NH_3 nitrogen reaches 0.80 - 1.30 mg/dl, clinical signs of poisoning become apparent. Ammonia inhibits the citric acid cycle but the exact mechanism by which ammonia blocks citric acid cycle is not known. But saturation of glutamine-synthesizing system causes a backing up in the citric acid cycle and decrease in its intermediates. As a result, there is decrease in energy production and cellular respiration and thus cells begin to malfunction. Central nervous system is probably the first to be affected as it has a large requirement for energy. Since urea synthesizing cycle is dependent on citric acid cycle, ammonia handling capacity of body is further reduced. Impairment of TCA cycle results in cellular energy and respiration deficits and ultrastructural degenerative changes. There is an increase in anaerobic glycolysis, blood glucose and blood lactate. Liver dysfunction increases susceptibility to poisoning.

Clinical signs : Clinical signs and the intensity of signs vary from animal to animal and sometimes even the animals are found dead. Onset of the signs may be fast or delayed depending on the production of ammonia in the rumen and its absorption into the blood stream. But signs are most acute in nature and death generally occurs within half to four hours of ingestion.

Characteristic signs of poisoning are weakness, initial restlessness, salivation, prominent frothing at the mouth and nose, grinding of teeth, abdominal pain, bradycardia, marked jugular pulse, dyspnoea, bloat, forced rapid breathing, pulse and respiration progressively become weak and slow. Later severe groaning, shivering, twitching of eye lids, lips, tail due to effect of ammonia on nerve centres, ataxia, terminal tonic convulsions and death after violent struggling and bellowing followed by death generally when the blood ammonia concentration are > 5 mg/dl. However, there is no abnormal posturing or jumping on the unseen objects. Animals appear rigid rather than depressed between the convulsions. Colic is an outstanding sign, however, there is no hypermotility of intestines. In some of the cases, pulmonary oedema is also produced due to neurogenic-adrenergic factors which increase pulmonary capillary permeability.

Post -mortem lesions : No characteristic post-mortem lesions are present in urea poisoning, however, some of the commonly observed lesions are as follows:

(i)   Strong odour of ammonia in the rumen.
(ii)   Generalized haemorrhages, congestion and vascular injuries.
(iii)   Hydrothorax and hydropericardium.
(iv)   Haemorrhagic enteritis with oedema and ulceration of intestinal mucous membrane.
(v)   Liver is enlarged, pale and friable.
(vi)   Haemorrhagic degenerative changes in brain.
(vii)   Fatty degenerations of liver and kidneys.
(viii) Pulmonary oedema and acute catarrhal bronchitis, peribronchial and intra-alveolar haemorrhages.
(ix)   Encephalomalacia in pigs similar to that in salt poisoning except the eosinophilic aggregation.

Diagnosis :

(i)   History of access to urea.
(ii)   Clinical signs.
(iii)   Post mortem lesions.
(iv)   Laboratory investigations indicating high ruminal fluid and blood ammonia concentrations, ruminal pH of more than 7.5 and increase in blood urea nitrogen.
(v)   Feed analysis for urea, ammonium salts etc.
(vi)   Stomach / ruminal contents for urea or ammonical fertilization.

Differential diagnosis :

(i)   Arsenic poisoning.
(ii)   Strong caustics poisoning.
(iii)   Lead poisoning.
(iv)   Organochlorine pesticides.

(v)   Organophosphates toxicity.
(vi)   Nitrate and cyanide poisoning.
(vii) Encephaletic disease, enterotoxaemia, brain engorgement.

Treatment : There is no specific antidote for urea poisoning and generally the treatment is ineffective. First of all, remove the source of exposure. Treatment is aimed at reducing the concentration of ammonia in the blood either by reducing the production of ammonia and hastening the conversion of ammonia to urea. Weak acids, generally vinegar or 5% acetic acid in sufficient quantity of cold water is administered. The dose of acetic acid for sheep is 0.5 - 1.0 litre and cattle is 4.0 litres. It not only dilutes the ruminal contents but also reduces the production of ammonia by lowering pH of the rumen and slowing the rate of hydrolysis of urea by reducing urease activity and also promotes diuresis. Excessive gas accumulated in the rumen be removed by trocar and canula. However, the really effective treatment is emptying of the rumen by stomach tube or ruminotomy. Other line of treatment is generally symptomatic.

CHAPTER 4

TOXICITY OF PLANTS

4.1 INTRODUCTION TO PLANT TOXICOLOGY AND TOXIC PRINCIPLES OF PLANTS

M.A. Ayub Shah and V.P. Vadlamudi

Man and animals mostly depend on vegetable kingdom for their food. Plants by their metabolic activities besides being the source of feeds and fodder also elaborate other substances viz. alkaloids, glycosides, toxalbumins, essential oils, resins, bitter principles etc. which are important from medicinal and toxicological point of view. Many plants are categorized as poisonous plants. In India, there are more than 700 poisonous plant species belonging to more than 90 different families.

A 'toxic plant' may be defined as "one which detrimentally affects the health of man or animal when eaten in such amount as would be taken normally or under special circumstances like restriction of choice of diet or extreme hunger". A plant is called a toxic plant when through contact or ingestion it hinders or destroys normal processes leading to distressing symptoms, pathology or mortality. The toxic (active) principles present in the plants are called as phytotoxins.

Animals, particularly the grazing livestock are indiscriminate eaters. When hungry they ingest food as well as nonfood plants, particularly during the scarcity periods. Absence of specific grazing areas or pasture fields in our country confounds the problems of plant toxicity due to wide distribution of nonfood or toxic plants in the waste lands, the so called grazing areas for our livestock.

Some plants are well known for poisoning in man and animals and causing serious harmful effects or even death. Whereas, many plants, though do not cause fatal poisoning, may produce far reaching effects on the health and production of animals, ultimately resulting in great economic losses to farmers. Besides causing adverse effects in animals, the phytotoxins present in plants may enter the human food chain through animal products such as eggs, meat and milk. Hence, the veterinary students and field veterinarians should be well versed with the poisonous plants of the region, their deleterious effects and the measures to treat their poisoning and prevention of plant poisoning in livestock.

Toxic principles of plants and their chemistry: Literally, a plant is a storehouse of hundreds of chemicals of diverse biological activities. Most of the chemicals present in the plants are harmless and are necessary for the survival of both the plant and animal kingdoms. The simple sugar molecules (e.g. glucose) generated through the interaction of sunlight with cells of green plants during photosynthesis fuel the fire for all the food chains in nature. The basic framework of protoplasm consists of amino acids- the building blocks of proteins. All the proteins are synthesized from the same basic set of 20 amino acids which are essential for sustaining life. The pathway of metabolism that are essential to life are known as primary metabolism and compounds (such as glucose, amino acids etc.) which are directly involved in these pathways are referred to as primary metabolites. Incidentally, a variety of compounds other than the primary metabolites are also produced by plants through secondary pathways, and even some primary metabolites (e.g. sugar) may occur secondarily (e.g. as a part of glycosides). However, the functions of these secondary metabolites (e.g. alkaloids, glycosides etc.) are not clearly understood and they do not appear to be essentially related to the sustenance of life.

Plant toxins may be referred to as secondary plant metabolites which are toxic compounds. Most of these secondary toxic metabolites do not have any apparent function in the plant except for defence mechanisms or survival adaptations.

On the basis of their chemical nature, the toxic plant principles may be categorised as follows:
A)   Alkaloids
B)   Terpenes
C)   Glycosides
D)   Proteinaceous compounds
E)   Organic acids
F)   Resins and resinoids.

A) Alkaloids : Alkaloids are basic nitrogenous substances containing cyclic nitrogen which are insoluble in water but combine with acid to form water soluble salts. Most of the alkaloids occur in combination with plant acid(s) and widely found in the plant kingdom. Not all, but many of the plant alkaloids produce significant pathophysiological effects on human and animal health. Some of the important toxic alkaloids present in the plant species have been summarized in Table. 4.1.

Table 4.1: Toxic plant alkaloids.

Type of alkaloid	Chemistry	Examples
1. Tropane or atropine like	Atropine	Atropa belladona Datura (Jimsonweed) Hyoscyamus (Henbane) Mandragora (Mandrake) Erythroxylum (Coca tree)
2. Pyrrolizidine alkaloids	Retronecine	Senecio (Ragwort) Crotolaria sp. Heliotropium sp. Trichodesma sp.
3. Pyridine/piperidine alkaloids	Coniine	Conium (Hemlock) Lobelia (Indian tobacco)

(Continued)

4. Pyrrolidine-pyridine alkaloids

Nicotine

Nicotiana sp.(Tobacco)
Equisetum sp.(Horsetail)

5. Purine alkaloids

Caffeine

Coffea sp.(Coffee)
Theobroma sp. (Cocoa)
Camellia sp. (Tea)

6. Quinoline alkaloids

Quinine

Cinchona sp. (Quinine tree)
Echninops sp. (Globe thistle)

7. Isoquinoline alkaloids

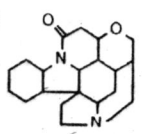

Morphine

Papaver somniferum
(Opium)
Sanguinaria sp.
Dicentra (Blood root)

8. Indole or indolizidine alkaloids

Strychnine

Claviceps (Ergot)
Psilocybe (Magic mushroom)
Astragalus sp. (Locoweed)
Gelsemium sp.
Strychnos (Strychnine)

9. Quinolizidine alkaloids

Anagyrine

Lupinus sp.
Laburnum sp.(Golden chain)
Baptisia sp. (False indigo)
Cytisus sp. (Scotch broom)

10. Steroidal glycoalkaloids

Solanidine

Lycopersicum sp. (Tomato)
Solanum sp. (Nightshades)

11. Steroidal alkaloids

Jervine

Veratrum sp.
Zigadenus sp.

12. Diterpenoid alkaloids

Aconitine

Delphinium sp. (Larkspur)
Aconitum sp. (Monkshood)

13. Phenylamine alkaloids

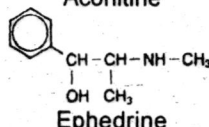

Ephedrine

Lophophora sp. (peyote)
Ephedra sp.

(Continued)

B) Terpenes : Terpenes are substances biosynthesized by plants and contain the branched 5-carbon skeleton of isoprene. On the basis of number of isoprene units present in the structure of the molecule, the terpenes are categorized into various groups like (1) monoterpenes (C10 compounds), (2) sesquiterpenes (C15 compounds), (3) diterpenes (C20 compounds) and triterpenes (C30 compounds). The important toxic terpenes are summarised in Table 4.2.

Table 4.2: Toxic terpenes present in plants

Type of terpenes	Examples
1. Monoterpenes	Cantharidine (an animal monoterpene from fly ash) Picrotoxin from *Anamirta cocculus* (Fish berries)
2. Sesquiterpenes	Coramyrtin from *Coriaria myrtifolia.* Geigerin from *Geigeria* Helenalin from *Helenium microcephalum*
3. Diterpenes	Andromedotoxin Mezerein Aconitine from *Aconitum* sp.
4. Triterpenes	Cicurbitacins (toxic principles of bitter gourd) Lantadenes from *Lantana* sp.

C) Glycosides: Glycosides are ether-like combinations of sugars with other organic structures (non-sugar aglycone or genin). Glycosides are relatively inactive when two parts of the molecules are connected; but when separated from the sugar moiety, the genin or aglycone becomes more active i.e. toxic. During digestion in animals, separation of aglycone takes place due to hydrolysis of the O_2 bond between the sugar and aglycone. Major toxic glycosides which are naturally occurring in plants are summarized in Table 4.3.

D) Proteinaceous compounds: Generally, proteins are harmless and often beneficial agents. Plant proteins in general and reserve proteins of seeds in particular are an important source of food. After ingestion, proteins get hydrolysed through various enzymatic reactions in the gastrointestinal tract and the amino acids are absorbed into the system for protein biosynthesis in the body. However, there are a number of proteins, peptides or amines which are of toxicological importance. Some of the important toxic plant proteinaceous substances are enumerated in Table 4.4.

Table 4.3: Naturally occurring glycosides in plants.

Type of glycoside	Chemistry	Examples
1. Cyanogenetic glycosides	Amygdalin	HCN present in *Hydrangea*, *Linum* (linseed), *Prunus* (Wild cherry) *Sorghum vulgare* (Jowar) *S. sudanese* (Sudan grass) Gossypol (Cotton seed) Amygdalin (Almond seed)
2. Steroidal glycosides i) Cardionilides/cardiac glycosides	Digitoxin	Digoxin from *Digitalis* sp. Oubain from *Strophanthus* Convallarin from *Convallaria* *Asclepias* (Milk weed) *Nerium oleander*
ii) Saponigenic glycosides		*Argostemme* (Corn cockle) *Aleurites* (Tung nut) *Phytolacca* (Poke weed) *Hedera* (English ivy) *Saponaria* (Soap wort)
3. Coumarin glycosides	Esculin	*Ausculus glabra* Moldy *Meliolotus* sp. (White sweet clover) *Ipomoea* sp. (Sweet potato)
4. Anthraquinone glycosides		*Cassia fistula* (Senna) *Aloe* sp.
5. Mustard oil glycosides	Sinigrin	Thiocyanates or isothiocyanates of mustard oil (*Brassica nigra*).

Table 4.4: Proteinaceous toxic principles of plants.

Type	Examples
1. Proteins/toxalbumins	Abrin from *Abrus precatorius* Ricin from *Ricinus communis*
2. Polypeptides	Amatoxins, phallotoxins and phalloidin from *Amanita* sp.

(Continued)

3. Amines	Aminotryptaline from seeds of *Sativus odoratus* *Phoradendron* sp. (Berries of mistle toe) Mimosine from *Mimosa pudica* and *Leucaena leucocephala* (Subabul) Indoscopine (*Indigofera spicata*) Canavanine from *Canavalia ensiformis* (Jack beans)

E) Organic acids: Some of the acids accumulated by the plants, particularly in their fruits (e.g. malic acid, tartaric acid, citric acid or ascorbic acid) are non-toxic, but other plant acids possess significant toxic properties solely due to their acidity e.g. plant oxalic acid and its soluble sodium, potassium or ammonium salts (Fig. 4.1). Details of plant oxalates are discussed elsewhere.

```
COOH            COOK            COONa
|               |               |
COOH            COOH            COONa

Oxalic acid     Potassium oxalate   Sodium oxalate
```

Fig. 4.1: Oxalic acids and soluble oxalates.

F) Resins and resinoids: Toxic plants resins are phenolic compounds. One of the most important naturally occurring phenolic resin in plants is tetrahydrocannabinol (THC, Fig. 4.2 a) and related compounds from *Cannabis sativa* (marijuana or hemp). The other phenolic resinoid molecules of the plants are urushiol from poison ivy, poison oak and *Rhus* sp. and hypericin from *Hypericum perforatum* (Fig 4.2b).

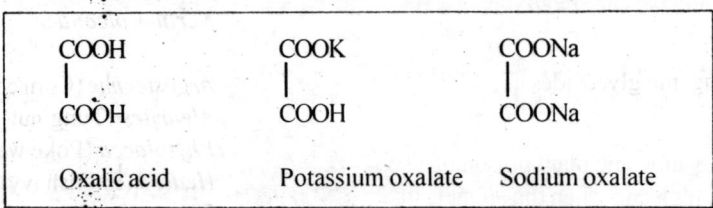

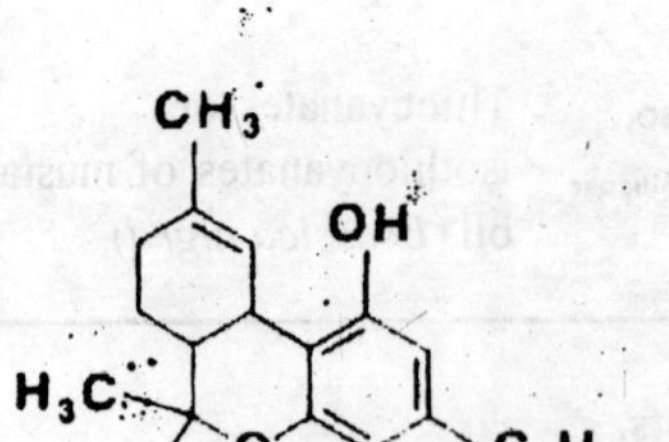

(a) Tetrahydrocannabinol 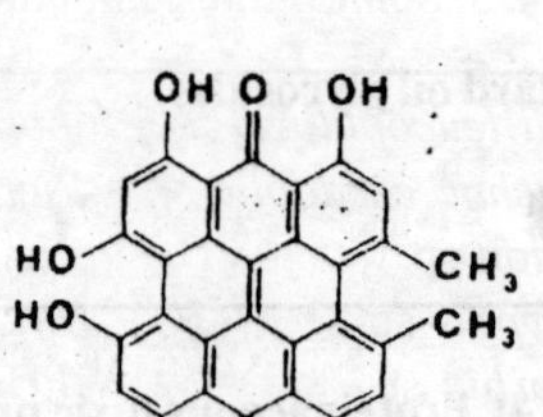   (b) Hypericin

Fig. 4.2 (a): Tetrahydrocannabinol from *Cannabis sativa* and (b) hypericin from *Hypericum perforatum*.

4.2 CYANOGENETIC PLANTS

V.P. Vadlamudi

The plants which contain hydrocyanic acid (HCN) or cyanogenic glycosides or cyanogenetic glycosides are called as cyanogenetic plants or cyanogenic plants. HCN owes its toxicity to cyanide. In plants, cyanide is present in two forms viz. a free form as HCN and a bound form as cyanogenic glycoside. Free HCN is released from the glycoside by physical disruption (mastication, trampling), stress (drought, frost) or the action of an enzyme (*beta* glycosidase and hydroxynitrile lyase), which may be present in the same plant or other plants or in bacteria in the gastrointestinal tract of man and animals. In ruminants, the hydrolyzing enzymes present in ruminal microflora liberate free HCN from the glycoside. Poisoning may also occur due to accidental ingestion of potassium cyanide or malicious poisoning of man and animals.

Cyanogenetic plant poisoning is one of the most common plant poisonings among the grazing livestock. Cattle and buffaloes are the most susceptible species. Sheep and goats are relatively less susceptible compared to cattle and buffaloes due to differences in the enzyme systems in the forestomachs of these species. Monogastric animals (pig and horse) are relatively resistant to cyanogenic plant poisoning due to destruction of the glycoside by the gastric acidity.

Cyanogenic plant poisoning usually occurs in animals when HCN content in the plants is high and when the animals ingest large amounts of the plant in relatively short period of time. Starved or underfed animals when let loose for grazing ingest large amounts of the toxic plants within a short period and succumb to toxicity. There have been instances of poisoning in starved sheep fed on linseed meal.

Sources: There are more than 120 plants containing sufficient quantities of cyanogenetic glycosides and responsible for poisoning in animals, however, some of the most commonly occuring cyanogenetic plants are listed in Table 4.5 :

Table 4. 5: Some of the commonly available cyanogenetic plants

Acacia leucophloea	*Lotus* sp.	*Sorghum vulgare*
Andrachne cordifolia	*Nerium oleander*	*Sorghum halepense*
Cynodon sp.	*Neyraudia medagascariensis*	*Sorghum sudanensis*
Eucalyptus sp.	*Phaseolus lunatus*	*Kalanchoe integra*
Euphorbia sp.	*Prunus* sp.	*Zea mays*
Linum usitatissimum	Sugarcane leaves/tops	*Triglochin maritima*

In India, the most common source of cyanogenic plant poisoning in livestock is feeding of green *Sorghum* fodder, particularly immature plants or young shoots and also accidental ingestion of the pods or leaves of *Acacia leucophloea* (Fig. 4.3) by sheep and goats. The HCN or the glycoside content of the plant(s) varies based on several factors.

Fig. 4.3 : *Acacia leucophloea*   **(see also Colour Plate I)**

Factors affecting cyanide poisoning:

i) Stage of growth of plants: Young or immature plants or the rapidly growing plants after a period of drought are highly toxic as these have higher levels of HCN. Similarly, wilted or frost bitten plants are also likely to be more poisonous than normal and mature plants.

ii) Climatic conditions: During drought years, cyanogenetic glycosides contents are highest in plants.

iii) Soil composition: High soil content of nitrogen and low phosphorus is favourable for high HCN content in plants.

iv) Use of fertilizers and weedicides: Nitrate fertilizers and use of weedicides (2,4-D) enhance the HCN content in plants.

v) Processing of the material: Drying of material or making of silage reduce much of the cyanide content in plant material.

The cyanogenic glycoside content in plants also varies between different parts of the same plant. For example immature pods of *A. leucophloea* contain dangerous levels than the mature pods or leaves.

More than 20 glycosides have been identified from different plants, but the important cyanogenic glycosides are linamarin (linseed meal), lotaustralin or lotusin (*Lotus* sp.), dhurrin (*Sorghum* sp.) and amygdalin (bitter almonds).

The plant material containing more than 20 mg of HCN/100 gram is considered toxic to animals. The oral minimum lethal dose of cyanide or potassium cyanide in sheep and cattle (even in most of other species) is about 2 mg/kg body weight.

HCN is rapidly absorbed from gastrointestinal tract following ingestion and from lungs through inhalation. HCN is rapidly detoxified in liver by conversion to thiocyanate through the action of a specific rhodanese enzyme system. The thiocyanate is excreted in urine. Some HCN is also eliminated through the lungs, giving a characteristic bitter almonds smell to the exhaled air.

Mechanism of toxicity: Acute cyanogenetic plant poisoning causes cytotoxic anoxia, which is the actual cause of death. Toxic effects of HCN are due to its affinity towards metalloporphyrin containing enzymes, more specifically cytochrome oxidase. The cyanide of HCN reacts with ferric ion (Fe^{3+}) of cytochrome oxidase resulting in the formation of CN-cytochrome oxidase complex. The CN^- inactivates the most essential enzyme system of the respiratory chain, impairing oxidation reactions at cellular level (Fig.4.4).

A cyanide concentration of only 33 μM can completely block electron transfer through mitochondrial electron chain, thus preventing oxygen utilization. Oxygen is retained in blood due to failure of its exchange and utilization by the tissues. Thus, CN^- causes a situation similar to lack of oxygen (anoxia) ultimately causing death of cells (cytotoxicity) i.e. death is due to cytotoxic anoxia not histotoxic hypoxia as thought earlier.

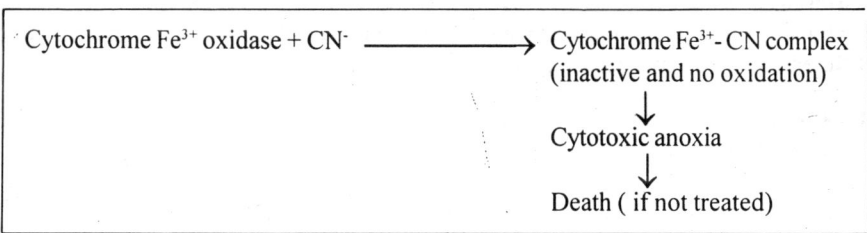

Cytochrome Fe^{3+} oxidase + CN^- ⟶ Cytochrome Fe^{3+}- CN complex
(inactive and no oxidation)
↓
Cytotoxic anoxia
↓
Death (if not treated)

Fig. 4.4: Biochemical processes of cyanide poisoning.

Clinical signs: Under field conditions most of the recognised cases of cyanogenic plant poisoning are of acute nature. The affected animals die suddenly (within 10 to 25 min) following ingestion of large quantities of the plant material without premonitory signs. Acute ingestion of HCN or KCN results in death of animals within a few minutes.

The onset of symptoms depends upon the quantity of plant material ingested and also the rate of liberation of HCN from the glycoside. If animal consumes large amounts of the plant it dies within a few minutes without showing much clinical symptoms except dyspnoea, anxiety, restlessness, stumbling gait, tremors, recumbency, terminal clonic convulsions and death in per-acute cases.

In acute cases, clinical manifestations such as excitement, staggering, muscle tremors, dyspnoea, hyperaesthesia, lacrimation, hypersalivation, bright red mucosa, dilatation of pupil, nystagmus and terminal convulsions are seen. The mucosae appear congested and cyanotic in terminal stages.

Chronic toxicity is believed to be sometimes observed in horses, cattle and sheep. Clinical signs are urinary incontinence, loss of hair and arthrogryposis, but not proved.

Post-mortem lesions:
(i)   Congestion and/or haemorrhages in abomasum, small intestines, trachea, lungs and heart.
(ii) Blood remains unclotted and bright red in colour.
(iii) When the stomach/rumen is opened a smell of 'bitter almonds' is felt.
(iv) Laboratory analysis of stomach or ruminal contents, liver and muscle reveals presence of HCN.
(v)   If postmortem is delayed for 1 or 2 days, muscle tissue must be sent for laboratory analysis.

Diagnosis :
(i)   History of sudden death following grazing in the vicinity or near the suspected plants.
(ii)   Acute anoxic syndrome.
(iii) Bright red coloured blood and mucous membranes.
(iv)   Post-mortem lesions.
(v)   Detection of HCN in the suspected plant material or ruminal/stomach contents by picrate paper test.

Differential Diagnosis: Toxicity due to cyanogenetic glycosides may be differentiated from many other poisoning conditions having almost similar clinical signs. Important differential points have been summarized in Table 3.3.

Treatment : The treatment of cyanogenic plant poisoning is specific and it must be initiated immediately after the diagnosis. For the treatment of cyanide poisoning, sodium nitrite and sodium thiosulfate, either former followed by the latter or both may be administered together. Rationality of treatment is as follows:

Sodium nitrite ($NaNO_2$) when given intravenously, the nitrite ions cause vasodilatation and counteract the CN^--induced vasospasm, thus improving the perfusion of vital tissues with blood. Nitrite ions also combine with haemoglobin (Hb) and convert it to methaemoglobin. Methaemoglobin (Fe^{3+}) competes with cyanide complexed cytochrome oxidase for HCN and binds to HCN to form cyanmethaemoglobin thereby providing an alternative for preventing combination of HCN with cytochrome oxidase. Only the ionic form of cyanide (CN^-) combines with methaemoglobin. Though the combination (cyanmethaemoglobin) is good and non-toxic but unstable and can give rise to cyanide toxicity again as cyanmethaemoglobin is not able to carry oxygen and thus, sodium thiosulfate should be given immediately which will detoxify cynamethaemoglobin and free cyanide present in the blood to form thiocyanate. Sodium thiosulfate given orally or intraruminally will fix HCN present in the rumen. As free cyanide in the blood decreases, additional cyanide dissociates from cyanmethaemoglobin and subsequently eliminated as shown in Fig. 4.5. Artificial respiration with oxygen should also be given alongwith sodium nitrite and sodium thiosulphate as it increases their efficacy.

$NaNO_2 + Hb(Fe^{2+})$ ⟶ Methaemoglobin (Hb^{3+})

Cytochrome Fe^{3+}oxidase - CN^-complex + Methaemoglobin (Fe^{3+}) ⟶ CyanmetHb + Cytochrome oxidase

NaS_2O_3 + CyanmetHb ⟶ SCN (thiocyanate*) + Hb + Na_2SO_4

Fig. 4.5: Biochemical mechanisms of reversal of the cyanide poisoning

*Thiocyanate is nontoxic, stable and easily excreted in urine.

4.3 PLANTS CONTAINING PYRROLIZIDINE ALKALOIDS

M.A. Ayub Shah

The pyrrolizidine alkaloids are toxic to animals and occasionally to human beings. This group of plant toxins possess hepatotoxic, pneumotoxic and nephrotoxic properties. However, liver is the most commonly affected organ followed by lungs (pneumotoxicity occasionally) and kidneys (nephrotoxicity rarely).

Sources of poisoning: The major sources of pyrrolizidine alkaloids poisoning is the ingestion of plants belonging to *Senecio* sp. (*Compositae* -most of the plants of this genera contain pyrrolizidine alkaloids), *Crotolaria* sp. (*Leguminosae*) and *Heliotropium* sp. (*Boraginacae*).

The genus *Senecio* (ragwort or groundsel) is of considerable toxicological importance. More than 1200 species are existing all over the world of which about 25 are demonstrated to be potentially toxic. Some of the plants of this genus with considerable toxicity potential include *Senecio burchelli, S. jacoboa, S. vulgaris, S. aquaticus, S. squalidus* etc. *S. vulgaris* has been found in Indian gardens and about 65 plants belonging to this genus are occurring in the mountainous tracts of India. Ingestion of 12-15 % of body weight for 1 or 2 days is sufficient to cause acute death and 4-8 % of body weight consumed over a period of 30-150 days may induce chronic toxicity.

More than 70 species of *Crotolaria* are found in India. *Heliotropium* sp. which are toxic to animals are *H. europeum, H. eichwaldi* or *Lasiocarpum* (Bithua or Atwin in Hindi), and *H. indicum* (Indian Heliotrope).

Toxic principles: Pyrrolizidine alkaloids are alkaline esters which yields an alkanolamine (a pyrrolizidine base- '*necine*') and a carboxylic acid (a '*necic acid*') on hydrolysis.

Senecio contains retrosine (a mixture of retrosine and isatinecic acid). *S. jacoboa* contains jacobine (retronecine and jacobinecic acid), senecionine, fluvine, jacinine and seneciphylline. *Crotolaria* sp. [*Crotolaria burkeana, C. globifera, C. dura* (wild lucerne), *C. spectabilis, C. acicularis, C. mucronata, C. sagittalis* and *C. incana*)] contains monocrotaline. The toxic principles present in *Heliotropium* are cynglossine (in roots and seeds *of H. europeum*) and heliotrine and lasiocarpine (in *H. lasiocarpum*). *Heliotropium indicum* also contains an alkaloid.

Mechanism of toxicity and pathogenesis: Both fresh and dried (hay) *Senecio* plants are toxic. The fresh plant is unpalatable and most animals avoid to graze. Most of the poisonings with *Senecio* result from ingestion of hay contaminated with ragwort/ groundsel.

The pyrrolizidine alkaloids-induced toxicity has often been termed as 'seneciosis' or 'pyrrolizidine alkaloidosis'. Seneciosis is worldwide in occurrence since time immemorial. The condition has been reported in almost all species of domestic animals including equines, cattle, yak, sheep, goats, pigs, chickens, quails and doves. However, sheep are comparatively resistant to pyrrolizidine alkaloidosis as the alkaloids are detoxified by bacterial enzymes in the rumen.

The exact mechanism by which toxic alkaloids induce toxicity is not precisely understood. Pyrrolizidine alkaloids produce their effects on the tissues by causing necrosis, inhibiting mitosis and causing megalocytosis or acting directly on the blood vessels and causing oedema and vascular disease. Worst affected organ is the liver which microscopically is characterised by haemorrhages, necrosis and cirrhosis; parenchymatous megalocytosis and veno-occlusion associated with cytoplasmic invaginations, inclusion globules, portal fibrosis, bile duct proliferation and/or tumours.

Oedema, fibrosis, epithelialization and/or emphysema occur in the lungs of horses, cattle, sheep and pigs following ingestion of pyrrolizidine alkaloid-containing plants.

The metabolites of pyrrolizidine alkaloids- pyrrolic acid or pyrroles are chemically reactive alkylating agents and are probably responsible for the cytotoxic effects and can cause alterations in liver, kidneys and lungs similar to that of pyrrolizidine alkaloids.

Clinical signs: Acute poisoning is rare. Clinical signs are dullness, rapid pulse and respiration, weakness, colic, jaundice and death in a few days to a few weeks. Violent nervous excitement may be noted in horses.

Senecio poisoning is generally chronic in nature and develops after several weeks or even months and is characterised by loss of condition, anorexia and constipation followed by jaundice and paleness of visible mucous membranes. Persistent straining may be observed in cattle.

Post mortem lesions: Post-mortem changes are enlarged and cirrhotic liver, ascites, lung oedema, petechiae in small intestines, heart and serous membranes. The kidneys may be congested and sometimes fatty degenerative changes may be noted. Gastroenteritis and oedema of the abomasal wall may also be observed.

Diagnosis:
(i)   History of feeding/exposure of the animal(s) to suspected pyrrolizidine alkaloid containing plants.
(ii)   Detection of pyrrolizidine alkaloids in feeds.
(iii)   Liver biopsy.
(iv)   Clinical pathology- liver function tests.
(v)   Post-mortem lesions.

Treatment: There is no specific treatment for pyrrolizidine alkaloids poisoning.
(i)   *Senecio* poisoned horses may be treated with 10-15 g crystalline methionine in one litre of 10% glucose saline injected intravenously.
(ii)   Methanogenesis inhibitors like chloroform or chloral hydrate retard pyrrolizidine alkaloid metabolism in the rumen of sheep and these agents may be employed as a protective measure for sheep grazing on plants containing these alkaloids.
(iii)   Liver protectants.

4.4 *ABRUS PRECATORIUS*

H.S.Panwar and Satish K. Garg

Abrus precatorius (Rathi, Jequirity bean, prayer bean plant; *Leguminosae*) found throughout the tropics is a trailing, twining perennial vine with yellow or red flowers and pod like fruits containing oval, shiny red and black seeds. Seeds are used as beads in jewellery and necklaces. Seeds of *Abrus precatorius* possess a powerul phytotoxin known as abrin.

The seeds are used for malicious killing of animals and rarely for homicide. Cattle are generally poisoned for revenge by leather workers to obtain the hides cheaply. The seeds are crushed to make powder and paste and made into needles of sharp pointed spikes are inserted into the wounds or under the skin of animals.

Abrin - the toxic principle of *Abrus precatorius* and a toxalbumen is similar to viper snake venom. Abrin-a, one of the four isoabrins from the plant is consisted of two chains of amino acids; chain A of 250 amino acids and chain B of 267 amino acids. Other active principles are abrine (N-methyl tryptophan) - an amino acid, haemagglutinin, abralin-a glycoside and a lipolytic enzyme. Median lethal dose value of injected abrin in mice is less than 0.1 µg/kg and thus abrin is considered as one of most toxic substances known.

Orally, the whole seeds have no toxic effect on cattle but these are toxic to fowls which die within a few days. The powdered seed is toxic by mouth to all species of animals. Broken seeds are highly toxic - one seed (approximately 0.5 cm in diameter) can kill a small child or a medium sized dog. Sixty gm of powdered seeds may kill a horse. However, compared to horses, cattle and goat are much more resistant to oral administration of seeds. Parenterally, aqueous extract of decorticated seeds (0.6 mg/kg) causes fatal poisoning in cattle.

Mechanism of toxicity : Abrin, a phytoprotein is neither degraded nor altered by gastric juices. It is a potent cytotoxin. It acts as a proteolytic enzyme and has the highest inhibitory effect on protein synthesis by acting on ribosomes of the cells. It is also known to cause agglutination (clumping) of red blood cells. It binds to surface glycoproteins of the erythrocytes, thus forming bridges between the cells.

Abrin, probably, because of peculiar binding potential is selectively transported (suicide transport) by the neurons.

Clinical signs : An initial latent period of several hours is there before the clinical signs appear. Salivation, nasal discharge, nausea, vomition, profuse haemorrhagic diarrhoea, watery faeces, dehydration and occasionally ulcerative lesions in the mouth and oesophagus. Stiffness of muscles, incoordination, ataxia, muscular spasms, trembling, convulsions, paralysis, coma and death. There may be excessive painful swelling around the area of implant and also enlargement of the regional lymph nodes. Body temperature may be elevated by 3-4 degrees.

Post mortem lesions :

(i)   Acute gastroenteritis characterised by mucosal and serosal gastric haemorrhages and accumulation of fluid in the lumen of intestine.

(ii)   Petechial haemorrhages throughout the body.

(iii)   Hepatic and renal necrosis.

Diagnosis :

(i)   History and circumstantial evidences.

(ii)   Clinical signs.

(iii)   Post mortem examination

Treatment : No specific antidote is available. Give only the symptomatic and supportive treatment.

(i)   Immediate removal of the poison from the gastrointestinal tract by emesis or gastric lavage, followed by activated charcoal, demulcents and saline purgatives.

(ii)   Intravenously infuse sufficient quantities of fluids and electrolytes.

4.5 *LANTANA CAMARA*

Om P. Sharma, Anita Singh and Sarita Sharma

Lantana camara (common name: lantana, wild sage, bunch berry; family *Verbenaceae*) is one of the ten most toxic weeds in the world. It was brought to India in early part of the nineteenth century as an ornamental plant. Now, this plant has naturalized in various parts of the country. Lantana occupies a sizeable part of land area in pastures, forests, so called waste lands and orchards. *Lantana camara* plants, bearing red, pink, white, yellow or intermediate coloured flowers are present in different parts of India (Fig.4.6).

Fig. 4.6 : *Lantana camara*   (see also Colour Plate I)

Lantana is a very hardy shrub and grows very luxuriantly, upto an elevation of nearly 2000 m above mean sea level in tropical and subtropical parts of the world. It is capable of prospering in adverse soil and weather conditions. Lantana plant can propagate rapidly by means of stumps or cuttings but natural propagation appears to be from seeds disseminated by the birds through their droppings or the faeces of moving flocks of sheep and goats. Lantana plant has a number of adverse interactions with the biosphere. It causes hepatotoxicity and photosensitization in grazing livestock and growth inhibition of the neighbouring vegetation due to allelopathic action which is rapidly changing the ecological balance. The plant foliage is rough, gives off smell and thus causes discomfort, itching and giddiness in the people working in lantana infested pastures and orchards.

Lantana poisoning in animals :   Field reports of lantana poisoning have been received from India, Australia, Cuba, Mexico, Kenya, Brazil, Fiji and New Zealand. Lantana is a very pungent plant and cattle will eat it only if other feed is scarce. *Bos taurus* cattle are more susceptible to lantana poisoning than *Bos indicus* cattle. The incidence varies from sporadic cases to heavy outbreaks during fodder scarcity due to drought or floods. Incidence of lantana poisoning has also been reported during transportation of animals from lantana free regions to lantana infested regions. The recorded cases provide only a small cross section of the problem because affected cases from far off places rarely come to the notice of scientific workers and clinicians. The estimates of economic losses in terms of mortality and morbidity of livestock due to lantana poisoning are not available for our country. However, a general observation as regards the effects of plant toxicosis is that mortality figures alone do not give the correct estimates of economic losses. Indirect losses due to morbidity, effect on weight gains, abortions and birth defects also require prime attention. Poisonous weeds like lantana act like double edged weapon : (a) they have competitive advantage over the other vegetation which serves as forage and cause loss of pastures, essentially leaving no grasses or other vegetation; and (b) grazing in lantana infested pastures results in toxicosis.

Chemical nature of the toxins : Lantana hepatotoxins are pentacyclic triterpenoids called lantadenes. Some major lantadenes are lantadene A, B, C and D which have a common core structure of 22-hydroxy-oleanonic acid (Fig. 4.7). Minor lantadenes like reduced lantadene A and reduced lantadene B are also present in lantana leaves (Table 4.6).   Lantadene A has been found to be toxic to sheep and guinea pigs and is

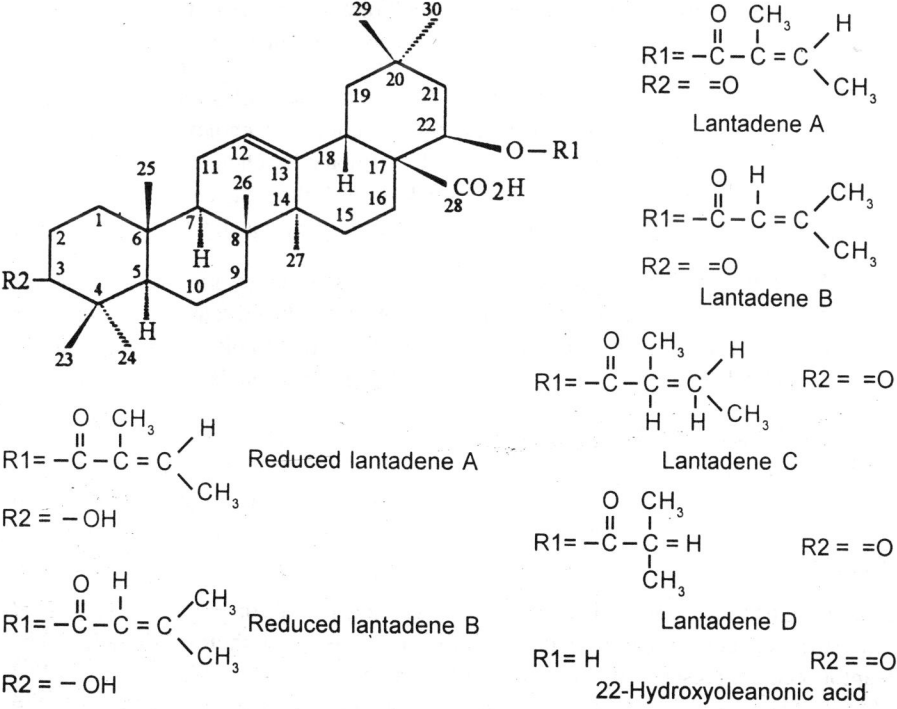

Fig. 4.7 : Chemical structure of lantadenes

considered the major hepatotoxin in the foliage of the toxic taxa of lantana plant. Oral administration of lantadene C induces severe toxicity in guinea pigs, however, effect of lantadene C has not been investigated in ruminants.

Table 4.6 : Quantification of lantadenes in lantana leaves

Lantadene	Amount (mg/100 g dry mass)
Lantadene A	721±18
Lantadene B	499±15
Lantadene C	642±38
Lantadene D	170±4
Reduced lantadene A	27±2
Reduced lantadene B	14±1

Values are mean ±S.E. (n=3). Data from Sharma *et al* (1997)

Mechanism of toxicity : Ingestion of lantana foliage causes hepatotoxicity and secondary photosensitization. Lantana manifests its toxicity in animals in three phases viz. gastrointestinal (GIT) phase, hepatic phase and the post-hepatic phase. Information on the release of toxins in different regions of the GIT is lacking. In lantana poisoned animals, the rumen contents become more toxic with the passage of time possibly due to the contents becoming more liquid, and thus enhancing absorption. Experiments on sheep showed that lantana toxins are absorbed from all parts of the gastrointestinal tract with the maximum absorption from the small intestine. Availability of bile in the alimentary tract is not important for the absorption of lantana toxins. The toxins are transported to the liver mainly in the portal blood. It is not known whether the toxins are absorbed in the native form or after biotransformation. Biotransformation of lantadene A and B has been observed in the caecum of guinea pig.

Interaction of the absorbed toxins with the biomolecules on/in hepatocytes followed by the cascade of biochemical reactions culminates in cholestasis constituting the hepatic phase of lantana poisoning. The primary biochemical event on the hepatocytes which trigger the toxic action is not known. Following ingestion of lantana, the sequence of biochemical events leading to lantana toxicity is depicted in Fig. 4.8.

Some evidence using guinea pig and female rats as laboratory animal models is available for the biotransformation of lantadene A and reduced lantadene A to polar metabolites. Ingestion of lantana toxins causes paralysis of gall bladder and closure of bile canaliculi which probably elicits decrease in bile flow during lanatana poisoning.

Cholestasis leads to regurgitation of bile which causes marked increase in the levels of bilirubin and phylloerythrin (the biodegradation product of chlorophyll) in blood. Both bilirubin and phylloerythrin bind proteins. They undergo photochemical reactions on exposure to light which causes photosensitization and associated skin lesions.

Clinical signs : Ingestion of lantana foliage by grazing animals or oral administration of lantana leaves powder causes hepatotoxicity, cholestasis and photosensitization. The heaptotoxins are present only in leaves of the lantana plant. The toxic dose is nearly 5 g (dry leaf powder)/kg body weight. The livestock used to grazing on lantana infested pastures rarely consume it as a matter of choice. However, very young animals which have not yet developed the instinct to avoid it, do consume lantana foliage and get poisoned. If the animals, kept on stall feeding for long periods are left for grazing in lantana infested pastures, they also browse upon lantana, usually, as a first choice. Similarly, animals not having the instinct to avoid lantana, consume its foliage from hedges or road sides during land transportation. The farmers in lantana infested localities are thoroughly aware of the symptoms of lantana poisoning and there are rarely cases of wrong reporting. There is ruminal stasis, animals go off-feed within a couple of hours of consumption of lantana foliage and become severely constipated. Within 24-48 hours these develop severe jaundice. The conjunctiva of eyes and mucous membranes of vagina and rectum become icteric. The eyelids become swollen and fissures appear on the muzzle, ear tips and nonhairy parts of the body. Terminal renal failure also occurs in affected animals. Exposure of light skinned animals to bright sunlight, photosensitization is also observed.

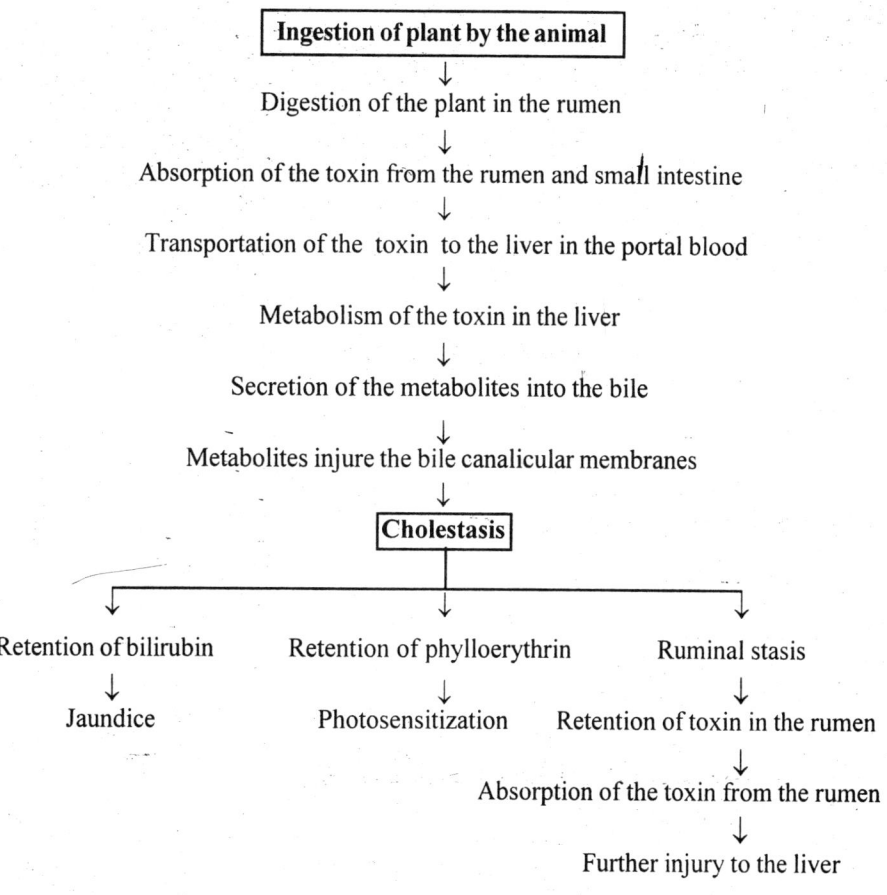

₢ Fig 4.8: Flow diagram showing sequence of proposed events in lantana toxicity

Laboratory animals models : Lantana poisoning in cattle, buffaloes, sheep and goats causes cholestasis, increase in the conjugated form of bilirubin and photosensitization. Oral administration of lantana leaves powder to guinea pigs of either sex elicits all the symptoms which typify toxicity in ruminants on experimental feeding of lantana leaf powder or in the field cases of lantana poisoning. Other parts of lantana plant (flowers, seeds or tender shoots) are nontoxic. Clinical signs of lantana toxicity are induced in rabbits , as well on oral administration of lantana leaf powder. Male rats are recalcitrant to the action of lantana toxins. On the other hand, female rats exhibit clinical signs of lantana toxicity on oral administration of reduced lantandene A. Guinea pig has been found to be the most suitable laboratory animal model for investigations on biochemical aspects of lantana toxicity.

Clinical pathology: Lantana intoxication of guinea pigs causes a decrease in liver dry weight, protein and DNA content while the amount of lipids in liver increases. Biochemical changes have been observed in liver mitochondria, microsomes, lysosomes, canalicular plasma membrane and cytosol of lantana poisoned guinea pigs. There is cholesterol enrichment of mitochondria, microsomes and plasma membrane. Activities of oxidative enzymes of mitochondria or cytosol are elevated. On the other hand, activities of microsomal enzymes associated with drug metabolism are decreased. Similarly, activity of cytosol glutathione- S-transferase is significantly decreased. The output of lipid peroxides by a number of tissues of lantana affected guinea pigs decreases and is attributed to enhanced activity of glutathione peroxidase and hyperbilirubinaemia. There is leakage of lysosomal enzymes from liver of lantana intoxicated guinea pigs. Significant alterations have been observed in different blood constituents of lantana intoxicated guinea pigs. There is an increase in plasma bilirubin, prominently of conjugated type, haemoglobin, urea and erythrocyte and leukocyte numbers.

Post mortem lesions : The liver is ochre coloured and greatly swollen. The gall bladder is greatly distended. The rumen contents are usually dry and undigested. There is impaction of faeces in the colon immediately near the spiral. Adrenals are enlarged and the thickened cortex turns yellow.

Diagnosis:

(i)   History.
(ii)   Clinical signs.
(iii) Clinical pathology.
(iv)   Post mortem lesions.

Differential diagnosis : One of the prime concerns of the clinician is the differential diagnosis of lantana toxicity from other hepatotoxicity and secondary photosensitization conditions. This is settled by correlating the clinical and biochemical picture with the flora present on the pasture where the animal grazed prior to intoxication. If the pasture where the animals were grazing had lantana bushes, the supporting evidence of clinical history and measurements of plasma bilirubin would support it to be a case of lantana poisoning. It has been observed that *Lantana camara* var . *aculeata* bearing red flowers (various shades of yellow to red) is the most toxic variety. Lantana plants bearing pink flowers have been found to be usually nontoxic.

Treatment : No specific antidote is available. Only symptomatic treatment need to be given :

(i)   Stop further exposure of the animals to noxious weed.

(ii)   Keep the animal in a well shaded areas away from direct sunlight, if photosensitization develops.

(iii)   Administer intravenously excessive amounts of glucose saline solution.

(iv)   Give hepatoprotective agents to tone up the liver. Systematic studies on the effect of antihepatotoxic or hepatoprotective agents in lantana poisoning have not been done.

(v)   Remove the toxic ruminal contents by rumenotomy.

(vi)   Give saline purgatives to facilitate removal of gastro-intestinal contents.

(vii)   Replace the ruminal contents with a suspension containing electrolytes, chaffed forage and rumen liquor from a healthy animal.

(viii) Administer a single dose of activated charcoal (5 g/kg) to bind the toxin in the rumen and prevent further absorption. Lantana poisoned cattle can be successfully treated by giving 2.5 kg of powdered activated charcoal in 20 litres of multiple electrolyte solution by stomach tube while in sheep, 0.5 kg charcoal in four litres of fluid is enough.

(ix)   Administer H_1 antihistaminics and antibiotics to take care of the photosensitization lesions and secondary bacterial infections.

Aversion therapy : Aversion therapy involves conditioning of the animals before they are left for grazing in the pasture infested with poisonous plants. This approach has been investigated for larkspur (*Delphinium barbeyi*) toxicity. Yearling heifers were intraruminally infused with lithium chloride while the animals were consuming tall larkspur. The animals got conditioned to associating the lithium chloride- induced stress with the ingestion of larkspur foliage.

Vaccination : Antibodies against lantadene A and B could be detected in sheep and cattle after injection of the conjugates of these compounds with suitable proteins. On administration of lantadene A or lantana to the vaccinated animals, there was decrease in the severity of toxicity but the protection was weak.

Biodegradation of lantadenes : Long term solutions to the problem of plant toxicosis are being sought by modification of rumen microflora for biodegradation of the plant toxins. The strategy involves isolating a microbe (anaerobe/aerobe), by the enrichment techniques, with the capacity for the degradation of the toxin, identification of the metabolic pathway and the genome and transferring it to the rumen microflora by the techniques of biotechnology. Success has been achieved by these techniques for the degradation of plant toxins like mimosine, fluoroacetate and indospicine. Lantadenes are nonpolar compounds and hence highly recalcitrant to microbial degradation. Our research group has obtained two organisms *Pseudomonas picketii* and *Alcaligenes faecalis* by enrichment from soil using lantadene A as the sole carbon source. Both these organisms utilized lantadene A, and probably it elicited mineralization when used as the sole carbon source. These organisms are rather slow degraders of lantadene A. Future research is needed for isolation of microbes with better capacity for degradation of toxic lantadenes, identification of the genome involved and transferring it to rumen microflora.

4.6 *IPOMOEA*

V. P. Vadlamudi

Several species of *Ipomoea* (*Convolvulaceae*) viz. *I. asarifolia*, *I. calobra*, *I. carnea*, *I. fistulosa*, *I. muelleri*, *I. purga*, *I. violacea (morning glory)*, *I. coccinea*, *I. multifida*, *I. quamoclit*, *I. tricolor* and *I. purpurea* are considered as poisonous plants. In India *I. carnea* is the most commonly encountered species. *I. carnea*, originally native to South America was introduced in our country as an ornamental plant (Fig. 4.9). The plant is a large straggling shrub widely grown in many parts as a hedge and for green manuring. Easy propagation and abundant growth make it easily accessible to the hungry grazing livestock. Livestock do not normally graze on this plant. However, in fodder scarcity areas, especially under drought conditions and lean periods, they do browse on it and succumb to toxicity. The plant is also recognised as toxic to livestock in Sudan, It is also believed to be hazardous to man.

Fig. 4.9 : *Ipomoea* sp.   (see also Colour Plate I)

There have been several reports from field veterinarians about *I. carnea* poisoning in goats and sheep in Maharashtra. Observation of lumbar paralysis in goats has also been attributed to ingestion of *I. carnea* leaves. Acute *I. carnea* poisoning in 12 adult goats was recorded in Maharashtra. Several acute or subacute cases of *I. carnea* poisoning in goats were recorded in Veterinary Polyclinic, Parbhani (Maharashtra). Extensive experimental toxicological investigations of *I. carnea* were also conducted among young ruminants, including goats, sheep and calves in Sudan.

Sources of poisoning: The major source of *I. carnea* poisoning is accidental ingestion of its leaves by the animals. The milky exudate from the plant is believed to cause blisters on hands in human beings, when they pluck its leaves. The smoke from the burning stem of *I. carnea* (when used as a fire wood or mosquito repellent) causes respiratory and eye ailments in man.

There is no information on toxicokinetics of *I. carnea*. The poisoning in animals results from ingestion of leaves of the plant. Hence, the phytotoxins present in leaves are well absorbed from the gastrointestinal tract in ruminants.

Mechanism of toxicity : The exact nature of the phytotoxins of *I. carnea* is not known. A number of lysergic acid alkaloids, some of which are hallucinogenic have been isolated from *Ipomoea* species. *I. purga* also contains a resinous cathartic. *I. carnea* leaves are known to contain toxic saponins. *Ipomoea* species also accumulate very high concentrations of nitrates from the soil. Ipomeanols (furanoterpenoid mycotoxins produced by *Fusarium* fungi, which grow on garden refuse) are also the cause of poisoning in sheep and goat.

Clinical Signs : Acute and subacute toxicity following ingestion of *I. carnea* leaves is reported in goats. Under field conditions subacute toxicity is more commonly encountered than the acute toxicity. The important toxic signs observed are salivation, diarrhoea, mydriasis, shivering, incoordination, staggering gait, prostration, ataxia, paralysis of limbs, lateral recumbency, hypotension and death. The haematological alterations noted in subacute *I. carnea* toxicity in goats were anaemia (normocytic and normochromic or hypochromic), leucopaenia and enhanced erythrocytic fragility. The affected animals also revealed elevated serum glutamate oxaloacetate transaminase and arginase activities and hypophosphataemia. Pathological changes recorded in experimentally-induced *I. carnea* toxicity in sheep and goats were degenerative changes in liver, oedema of lungs, congestion in kidneys and subendocardial haemorrhages.

Diagnosis : No pathognomonic signs indicative of *I. carnea* poisoning are there. However, diagnosis may be made based on:
(i)   History of grazing near *Ipomoea* hedges;
(ii)   Clinical signs.
(iii)   Presence of the plant material in the stomach.

Treatment : There is no specific line of treatment. General line of symptomatic and supportive treatment may be instituted as follows:
(i)   Remove the source of poisoning.
(ii)   Give saline purgatives to remove or go for the rumenotomy to remove the unabsorbed ingesta.
(iii)   Give activated charcoal to facilitate the removal of toxic principle(s).
(iv)   The affected animals should be administered hepatotonics, nerve tonics and haematinics.
(v)   Sodium acid phosphate can be administered to overcome hypophosphataemia.
(vi)   Success in treating the poisoned goats with intravenous 5 per cent glucose saline (250 ml) and vitamin C (50 mg) has also been reported. However, the therapeutic basis of this treatment is not explained.

4.7 *NERIUM OLEANDER*

H.S.Panwar and Satish K. Garg

Nerium oleander (oleander, kaner, rose laurel; family *Apocynacea*) is a small evergreen perennial shrub with green bark and dark green leathery leaves and sweet scented white, pink or other colours of flowers. It is an ornamental plant and grown in lawns and gardens in temperate climates. Another species of oleander are *Nerium indicum* (kaner; Fig. 4.10) and *Thevetia peruviana* (yellow oleander, pilla kaner; Fig. 4.11).

Fig. 4.10 : *Nerium indicum*   **(see also Colour Plate II)**

Sources of poisoning : Almost all species of animals are poisoned. Usually, horses, cattle and sheep do not eat it, but if cuttings of this plant get mixed with green fodder or hay, animals get poisoned. Ingestion of branches, stem or leaves during drought or periods of food scarcity or consumption of contaminated water by ruminants or horses or chewing of leaves may be responsible for toxicity in dogs. Ingestion of 30-60 gram leaves of oleander is fatal for horses, cattle and sheep. *Nerium indicum* and *Nerium oleander* appear to be equally toxic. All parts of the plants are toxic, the leaves and flowers may be less toxic than the seeds.

Lethal dose of fresh green leaves of *N. oleander* in cattle is 30-60 g, dried leaves in dogs and pigeon is 3 g and in cats is 2.5 g. Toxins are readily absorbed from gastro-intestinal tract and persist in the body due to enterohepatic circulation. Nerium has been known for its extreme toxicity since ancient times. Even the honey made from

Fig. 4.11 : *Thevetia peruviana* **(see also Colour Plate II)**

flowers of oleander, meat roasted on oleander sticks or milk from a cow that has consumed oleander foliage has been found to produce prostrating symptoms in human beings -salivation, vomiting, diarrhoea, tingling of skin, muscular weakness, visual difficulties, coma and convulsions etc.

Tubers of oleander contain two bitter principles - neriodorin and neriodorein which are potent cardiac poisons, a glycoside - rosaginine, an essential oil-neriene, tannic acid and wax. Leaves contain an alkaloid oleandrine, a glycoside-pseudocurarine and also neriene and neriantine. In general, oleander is known to contain heterosides related to digitalis and oleandroside, nerioside and nerianthoside.

Mechanism of toxicity: Cardiotoxins present in *Nerium oleander* are believed to interfere with the Na^+-K^+-dependent-ATPase system which results in a decrease in intracellular potassium. The normal electrical conductivity on the myocradium is reduced which results in conduction block, arrythmias and eventually complete loss of myocardial contractility or asystole. Sympathetic nerves are also paralysed.

Clinical signs : Often the animals are found dead. Two type of clinical signs are mainly observed :

(i) Gastrointestinal signs : Following acute exposure, these are the first to be observed within a few minutes to 1-2 hours and are characterized by anorexia, nausea, vomiting, abdominal pain, diarrhoea and tenesmus.

(ii) Cardiovascular signs : Slowing of heart, arrythmias, ventricular premature systoles, paroxysmal tachycardia, followed by complete heart block; as a result extremities are cold while the general body temperature is raised and pulse is rapid.

Both the rate and depth of respiration are increased, followed by tremors and tetanic stiffness, paralysis, convulsions and death. Duration of symptoms is generally less than 24 hours, and death is due to ventricular fibrillation.

Post mortem lesions:

(i)   Severe gastroenteritis.
(ii)   Petechial or ecchymotic haemorrhages on the gastric and intestinal mucosae, heart, gall bladder, meninges etc.
(iii)   Generalized congestion.
(iv)   Blood stained or clear fluid in serous cavities.
(v)   Microscopically, indications of hepatitis and nephrosis.
(vi)   Cortical tubules exhibit-coagulative necrosis and haemoglobin stained albumin casts.

Diagnosis :

(i)   History of feeding/ingestion oleander.
(ii)   Clinical signs.
(iii)   Post-mortem examination reveals presence of obnoxious material/plant in the stomach/rumen.
(iv)   Detection of cardiac glycosides in urine or tissues, but difficult.

Treatment : Generally, treatment is not possible as the animals are often found dead. No specific antidote is available.

(i)   Treatment should be aimed at reducing the level of toxicant in the gastrointestinal tract by administering emetics (apomorphine), osmotic or saline cathartics (magnesium sulphate, sodium sulphate), activated charcoal etc.
(ii)   Give gastric demulcents.
(iii)   Administer atropine in conjunction with propranolol.
(iv)   Administer sedatives and tranquilizers.
(v)   Provide artificial respiration.
(vi)   Give phenytoin or lidocaine to improve atrio-ventricular conduction.

Contraindications : Infusion of calcium containing fluids is contraindicated.

4.8 *DATURA* AND RELATED PLANTS

H.S. Panwar and Satish K. Garg

Certain plants of different genera belonging to the family *Solanaceae* have been found to contain the alkaloids having almost similar pharmacological / toxicological action i.e. these have anticholinergic effects. For convenience of the readers, such plants alongwith their alkaloids have been enlisted in table 4.7.

Table 4.7: Datura and related plants and their active principles.

Name of the plant	Common name(s)	Active principle(s)
Datura stramonium	Jimson weed, thorn apple, mad apple, stink weed	Hyoscyamine Atropine Hyoscine
Datura innoxia *Datura metel*		Scopolamine (hyoscine) Hyoscyamine Meteloidine Atropine
Hyoscyamus niger	Henbane	Hyoscyamine Scopolamine Atropine
Hyoscyamus muticus	Egyptian henbane	Hyoscyamine
Atropa belladona *Solanum tuberosum* *Solanum nigrum*	Deadly nightshade Potato Black nightshade	Atropine Tropane alkaloids (activity similar to that of atropine)

Datura stramonium is indigenous to India (Fig. 4.12) and grows abundantly throughout the temperate Himalaya from Kashmir to Sikkim. Deadly nightshade (*Atropa belladona*), henbane (*Hyoscyamus niger*) and thorn apple (*Datura stramonium*) have localized distribution and are unattractive to animals. Thus, ingestion of fresh plant and incidences of such poisoning are rare as the plants have a strong and unpleasant odour. However, poisoning of animals due to contamination of hay, ingestion of green berries or seeds of *Atropa bellodona*, *Datura* sp. or vines, green skin or sprouts of *Solanum* sp. are sometimes encountered. Most of the poisoning cases in animals may be due to overdosages of atropine like alkaloids used in therapeutics.

There is considerable variation in the toxicity of belladona or atropine depending on the route of administration and the species of animal. Herbivores are usually much more resistant than carnivores. Cats, dogs and birds are sensitive to its action, horses are less while cattle, sheep, goats, pigs and rabbits are said to be comparatively resistant. Rabbits are quite resistant to diet of belladona leaves since an esterase (atropinase) present in the liver rapidly hydrolyses and inactivates atropine, thus,

Fig. 4.12 : *Datura stramonium*   **(see also Colour Plate II)**

rabbits can tolerate larger doses of atropine. However, flesh of rabbits fed on such diets may prove toxic, if eaten by dogs, cats or man. Horses, cattle and goats are relatively resistant to belladona when it is administered orally, however, poisoning on ingestion occurs more often in swine. The resistance of ruminants may be correlated with a destructive maceration of atropine in the rumen. Both cattle and horses are equi-susceptible to subcutaneously injected atropine.

Sources of poisoning : All parts of the plants are toxic as these contain atropine like alkaloids. Though *Datura* sp. is unpalatable, but if eaten, has a considerable potential to cause toxicity. Seeds may also contaminate grains of the feed of animals. However, green leaves and berries of *Atropa belladona* or seeds or leaves of *Datura stramonium* may be responsible for poisoning in animals. Domesticated herbivores, birds and humans are sometimes poisoned. Drying of plant does not reduce its toxicity. Oral toxic or lethal doses are not known in all species of animals. Toxic dose of hyoscyamine is 2.7 mg/kg for pigs.

Mechanism of toxicity : Atropine and other related alkaloids are parasympatholytic agents and interact with muscarinic receptors of the effector cells and thus by occupying these receptors inhibit ACh-induced muscarinic action i.e. physiological responses to parasympathetic nerve impulses are thereby attenuated. In extremely high and nontherapeutic doses, atropine nonspecifically can also block the nicotinic receptors at the autonomic ganglia and the motor end plate of skeletal muscles.

Clinical signs : Signs of overdosage of atropine or poisoning with *Datura* sp. are similar in all species of animals. These include depression, dryness of mouth and mucous membranes, thirst, anorexia, dysphagia, ruminal atony, constipation, mydriasis, visual disturbances characterised by loss of accomodation, occasional blindness, cycloplegia, flushed skin, tachycardia, hyperpnoea, restlessness, delirium, hypothermia, insensible wandering, ataxia and muscle trembling may be seen; followed by convulsions, motor and sensory paralysis, respiratory depression, relaxation of sphincters and death from asphyxia within minutes to hours.

Toxicity in equines though rare is mostly characterized by the above symptoms and also narcosis and polyuria. Inhibition of salivary and related secretions leads to extreme thirst and consequently polyuria. The affected animals give an appearance as red as a beet, dry as a bone, mad as a wet hen and crazy as a newborn calf.

Post mortem lesions : Although there are no specific lesions, yet some of the commonly observed ones are :
(i)   Catarrhal inflammation of stomach and small intestine in pigs.
(ii)   Marked congestion and oedema of lungs and hydrothorax.
(iii)   Congestion of the meninges and dilation of the ventricles.
(iv)   Haemorrhages in the brain, stomach and upper intestine.
(v)   Presence of plant material in the stomach or forestomachs.

Diagnosis :

(i)   History.
(ii)   Clinical signs.
(iii)   Post-mortem lesions.
(iv)   Presence of plant parts in the stomach / rumen contents.
(v)   Response to treatment.
(vi)   Test for suspected materials containing atropine like alkaloids: A drop of urine from a suspected case of atropine poisoning causes mydriasis when placed in the eye of a cat. Also the tested pupil will not constrict when exposed to light, while untreated eye will constrict.

Treatment : No specific antidote is available, provide symptomatic care only.
(i)   Remove the animals from source of exposure / poison.
(ii)   If poisoning is due to ingestion of a plant material, give gastric lavage or induce vomition, if possible to remove the toxic substance from the gastro-intestinal tract.
(iii)   Parasympathomimetic agents like physostigmine, pilocarpine etc. are beneficial as pharmacological antagonists. But these may be harmful if used in large doses.
(iv)   Artificial respiration with a mixture of oxygen and carbondioxide may be quite helpful.
(v)   Animal should be kept moving in the early stages.
(vi)   Pentobarbitone, tranquilizers, sedatives etc. may be used to control CNS excitement and convulsions.
(vii)   Saline purgatives may be given to increase peristalsis.

Precaution : Do not give atropine antagonists.

4.9 STRYCHNOS NUXVOMICA

H.S.Panwar

Strychnine poisoning is an uncommon occurrence in farm animals and usually occurs as a result of accidental ingestion of seeds of the plant *Strychnos nuxvomica* (*Loganiaceae*) or powdered form of nuxvomica used as a bait to kill dogs, foxes or rats etc. In the past, strychnine was used in human and veterinary medicine as an analeptic, nervine tonic and ruminotoric, however, now almost deleted from therapeutics and is sometimes used as a rodenticide. It was introduced as rodenticide in 1690.

Sources of poisoning : Poisoning generally occurs in dogs and cats due to consumption of seeds of *Strychnos nuxvomica*, baits containing strychnine kept for foxes, rats, mice etc. or ingestion of birds or rats poisoned with strychnine or in other animals and human beings due to ingestion of strychnine treated seeds of peanuts, wheat etc.

Powdered seeds or beans constitute nux vomica. Seeds may contain 1.53 to 3.24 per cent of the total alkaloids of which 50% is strychnine. These also contain the glycoside loganin in addition to other alkaloids such as vomicine, *alpha*-colubrine, *beta*-colubrine, pseudostrychnine etc. Leaves contain brucine, strychnine and strychnicine while bark contains chiefly brucine and only traces of strychnine. The average amount of strychnine and brucine in nux vomica seeds is 1.23 and 1.55%, respectively. Toxic doses of nuxvomica are about one hundred times the corresponding oral lethal dose of strychnine. However, parenteral doses are 2-10 times less.

Strychnine - an indole alkaloid found in the seeds of *Strychnos nuxvomica* is extremely stable and persists in the food and environment and can be detected in specimens many years even after the death of animals. Strychnine is rapidly absorbed from the gastrointestinal tract and on injection. Significant amount of it is found in the liver and kidneys and only a fraction of it reaches the central nervous system. It is metabolized by hepatic mixed function oxidases and primarily excreted in urine. Excretion also takes place through saliva and excretion is generally complete within 10 hours. Most of the lethal dose is eliminated within 24 hours. This fact forms one of the important bases of therapy i.e. if the animal can be made to survive the first day or so, chances are there that no further treatment is needed.

Cattle are more susceptible to strychnine administered parenterally compared to oral dosing. Probably strychnine is destroyed in the rumen. Lethal dose of strychnine varies from species to species. Parenterally 30-60 mg of strychnine is fatal for cattle. Lethal doses of parenterally administered strychnine hydrochloride are 200-250, 300-400 and 15-50 mg for horses, cattle and pigs, respectively. Compared to mammals, birds are more resistant. The approximate acute oral lethal dose of strychnine in different species of animals are presented in Table 4.8.

Table 4.8: Lethal doses of strychnine in different species of animals and birds

Species	Lethal dose (mg/kg)
Cattle	0.50
Horses	0.50
Pigs	1.0
Dogs	0.75
Cats	2.00
Rats	3.00
Fowl	5.00

Mechanism of toxicity : Major site of action of strychnine is the recurrent inhibitory interneurons (Renshaw cells) of the reflex arc in the spinal cord and medulla. Strychnine is a competitive glycine receptors antagonist and thus interferes with glycine receptors mediated post synaptic inhibition at these sites.

Strychnine is thought to block glycine receptors by attaching to ligand binding sites or by interfering with the function of ion channels. Thus, the motor inhibitory effect of glycine in the reflex arc are prevented leading to uncontrolled excitation of spinal reflex, stimulation of extensor muscles, extensor rigidity and tonic seizures.

Clinical signs : Clinical signs appear within 10 minutes to two hours of ingestion. The initial signs of poisoning are nervousness, restlessness, muscle tremors, stiffness and convulsive motion of certain muscles, hyperirritability and intermittent tonic spasms in response to noise or other external stimuli like touch, bright light i.e. animals are extremely sensitive to external stimuli. As the condition progresses, the muscular twitching becomes more pronounced leading to spontaneous and nearly continuous tetanic seizures with marked rigidity. Skeletal muscles contract antagonistically i.e. limbs are extended and neck is curved upward and backward (opisthotonus) because the extensors are stronger than the flexor muscles. Convulsions are intermittent in the beginning with intervening periods of partial relaxation. During the period of relaxation, any slight external stimulus elicits excitation and general tetanic spasms. Body gives a typical saw horse stance or posture. Pupil are widely dilated. Body temperature may be elevated. Death results from asphyxia due to prolonged paralysis of respiratory muscles usually within a few hours but may be delayed for as long as 48 hours. Poisoned animals are conscious until near death. Vomiting is rare and dogs often retain stomach contents in fatal intoxications.

Post mortem lesions :
(i)   There are no specific post- mortem lesions, except petechiae resulting from anoxia due to arrest of respiration during the spasms.
(ii)   Venous blood is dark.
(iii)   Lungs and cerebral meninges are engorged.
(iv)   Stomach may be empty or may contain remnants of last meal.

Diagnosis :
(i)   History.
(ii)   Clinical signs.

(iii)   Laboratory investigation - myoglobinuria.
(iv)   Post mortem lesions.
(v)   Detection of strychnine in urine samples.
(vi)   Detection of strychnine in stomach contents, liver, kidneys or suspected baits.
(vii)   Onset of rigor mortis is rapid.

Note : Diagnosis can be confirmed by inoculating suspected material (e.g. urine or other) in the dorsal lymph space or the peritoneum of frogs. Appearance of typical tetanic convulsions on external stimuli like touch confirms strychnine poisoning.

Differential diagnosis :
(i)   Organophosphates poisoning.
(ii)   Chlorinated hydrocarbons poisoning.
(iii)   Carbamates poisoning.
(iv)   2, 4- D, nicotine, zinc phosphide, fluoride poisoning/toxicity.
(v)   Tetanus toxin, carbonmonoxide, metaldehyde poisoning etc.

Treatment : Strychnine poisoning is most likely to be encountered in dogs. There is no specific antidote available. Prognosis is good if the animal is treated rapidly and vital functions are maintained for 24-48 hours. General line of treatment is as follows :
(i)   Animals should be maintained undisturbed in a warm and quiet environment away from external stimuli.
(ii)   In the early stage of poisoning before onset of convulsions, attempt should be made to remove and make the unabsorbed strychnine nontoxic. For this purpose, stomach must be washed with :
(a)   2% aqueous tannic acid or dilute hydrochloric acid;
(b)   1 : 1000 potassium permanganate solution;
(c)   1 : 250 dilution of tincture of iodine; followed by final wash of stomach with plain water and intravenous drip of isotonic glucose or saline to accelerate urinary excretion of strychnine;
(d)   Gastric lavage or saline cathartics following administration of activated charcoal (2 g/kg). Gastric lavage without prior endotracheal intubation is contraindicated;
(e)   To control the convulsions, to calm the animal and to reduce the irritability, administer pentobarbitone @ 30 mg/kg I/V in case of small animals and 7 per cent chloral hydrate solution intravenously in large animals;
(f)   Glyceryl guaiacolate (110 mg/kg) or methocarbamol @150 mg/kg initially, followed by 90 mg/kg may be used as muscle relaxant. Mephenesin, xylazine and guaifenesin might also be effective. These reduce the strychnine-induced central respiration depressions. Diazepam, which is a sedative, tranquilizer and muscle relaxant may also be used as it mimics glycine in its activity to displace strychnine from the glycine receptors; or/and
(g)   Artificial respiration may be provided to prevent development of cyanosis.

Contraindication : Gastric lavage with sodium bicarbonate is contraindicated as it may enhance gastric absorption of strychnine. Ketamine and morphine are also contraindicated in strychnine poisoned animals.

4.10 RICINUS COMMUNIS

H. S. Panwar and Satish K. Garg

Castor bean (*Ricinus communis*; family *Euphorbiaceae*) is an ornamental plant (Fig.4.13) and is commonly found in jungles and also cultivated in India and other temperate regions. Seeds of this plant contain about 50% of the fixed oils and 26% proteins.

Fig. 4.13 : *Ricinus communis* **(see also Colour Plate III)**

Sources of poisoning : Poisoning may be due to deliberate mixing of castor beans or castor cake in the feed of animals as a protein supplement or accidental ingestion of large quantities of beans or cake by the animals. After extraction of oil from seeds under pressure, the residual cake contains two *lectins* - ricin-I and ricin -II, the toxic active principles of castor beans. Ricin II is more toxic. Ricin II is consists of two chains (chain A and chain B) of amino acids of the molecular weights 30,000 and 33000 daltons, respectively and joined by a disulfide bond.

Ricin, the toxic protein of castor beans, is one of the most powerful phytotoxins known. One mg of pure toxin can be isolated from 1 gm of seeds. Ricin resembles bacterial exotoxins as the animals can be hyperimmunized to it. As in case of bacterial toxins, heat (about 56 ° C) destroys the toxic fraction (toxophore) and leaves an immunizing fraction (haptophore). Boiling of the meal or whole seeds makes them nontoxic. Immunity can be developed in cattle and calves by feeding increasing amounts of castor beans. Ricin is absorbed from the digestive tract. It is toxic when administered by different routes, however, toxicity following parenteral route is less. Parenteral toxic doses are much higher (500 times) compared to oral routes.

All species of animals are susceptible to ricin toxicity. Horses are most susceptible; dogs, sheep, cattle, goats and pigs are comparatively resistant and ducks and hens are still more resistant as is evident from the oral lethal dose (g/kg) of seeds or cake of *Ricinus communis* (Table 4.9). There appears to be a wide individual variation in susceptibility to ricin poisoning as it has been found in horses that ricin may be lethal at a dose as low as 0.007 g/kg in certain animals while in others 0.3 g/kg may not be lethal.

Table 4.9: Oral lethal doses of *Ricinus communis* in animals and birds.

Species	Oral LD$_{50}$ (g/kg body weight)	
	Seeds	Oil cake
Horse	0.10	3.00
Rabbit	1.00	2.00
Sheep	1.25	2.50
Pig	1.30	5.00-6.00
Piglet	2.40	---
Cattle	2.00	3.00
Calf	0.50	----
Goat	5.50	----
Dog	0.6-5.00	3.50
Goose	0.40	----
Hen	14.00	40.00

Mechanism of toxicity : Exact mechanism of toxic action of ricin is not known. But ricin appears to induce hydrolytic fragmentation of ribosomes and inhibits protein synthesis. Chain A of ricin inactivates the 60S ribosomal subunit of cells and blocks protein synthesis and is endocytosed into the cell cytosol after chain B binds to a terminal galactose residue on the cell membrane. Ricin attacks the lining of digestive tract, blood and ultimately kidneys and induces hepatitis and nephrosis. Lectins in general are known to damage the gut epithelium, resulting into defective digestion and absorption and increased permeability of intestinal mucosa.

Clinical signs : There is a characteristic latent period between ingestion of castor beans or cake and the onset of clinical signs. In horses, the latent period may range from several hours to 2-3 days and in dogs, it may be 12-72 hours depending on the amount consumed and absorbed. During the latent period, dullness, loss of appetite etc. may be observed. The general signs of poisoning in different species are nausea, vomiting, signs of abdominal pain (grinding of teeth, humping of back) and gradually development of diarrhoea. It is followed by severe gastroenteritis, bloody diarrhoea, dehydration and icterus in fatal cases.

Main clinical signs are dullness, incoordination, profuse sweating, tetanic spasms of muscles, watery diarrhoea and abdominal cramps in horses; bloody diarrhoea and abortion in 6-8 weeks pregnant animals in cows; vomition, diarrhoea (not bloody) and cyanotic skin of ears, flanks and hams in pigs and vomiting, violent and protracted haemorrhagic diarrhoea with colic and cramps, oliguria, prostration, convulsions and death within 2-3 days in dogs. However, in pullets, dropping of wings, ruffled feathers, greyish coloured wattles and comb are the main signs of intoxication.

In experimental poisoning following parenteral administration of ricin, the animals begin to loose weight, followed by diarrhoea, convulsions and flaccid paralysis (alternating with convulsions) and finally death of the animals within an hour.

Post mortem lesions:
(i)   Contents of the gut are fluidy or semifluidy.
(ii)   Severe, intense protracted gastroenteritis.
(iii)   Haemorrhages of the mucosae of gastrointestinal tract.
(iv)   Mesenteric lymph nodes swollen.
(v)   Generalized congestion.
(vi)   Degeneration of the liver and kidneys, there are also foci of necrosis in the spleen, lymph glands, intestine and stomach.
(vii)   Pulmonary haemorrhages, oedema and emphysema and trachea and bronchi are filled with frothy oedematous fluid.
(viii) Microscopically, there is necrosis of the gastrointestinal epithelium in the affected areas.
(ix)   Liver shows hydropic and fatty degenerative changes.
(x)   In fowl, there is inflammation of the oesophagus, crop and true stomach.

Diagnosis :
(i)   History of feeding ricin containing feed stuffs.
(ii)   Clinical signs.
(iii)   Post-mortem lesions.
(iv)   Presence of ricin in animal body or feed stuffs by various biological methods.

Treatment :
(i)   Best treatment is the anti-ricin serum from a previously hyperimmunised animal, if available.
(ii)   Remove the toxic material from the gastrointestinal tract by gastric lavage, saline purgatives (sodium sulphate), vomition or adsorption on mucilage containing suspensions of activated charcoal.
(iii)   Symptomatic treatment : To restore the circulation, give blood transfusion or plasma expanders, adequate intravenous infusion of electrolytes containing fluids to prevent dehydration and also to promote urinary excretion of ricin. If oliguria, give mannitol and to prevent, haemoglobin precipitation in the tubules, give urinary alkalizer (sodium bicarbonate).

4.11 PLANTS CONTAINING OXALATES

M.A. Ayub Shah

Oxalates poisoning is of great concern in livestock and human beings all over the world. Oxalic acid is an organic dicarboxylic acid (Fig. 4.14) that readily forms insoluble salts with cations like calcium and magnesium. However, its sodium, potassium or ammonium salts are highly soluble and can induce toxicity in animals.

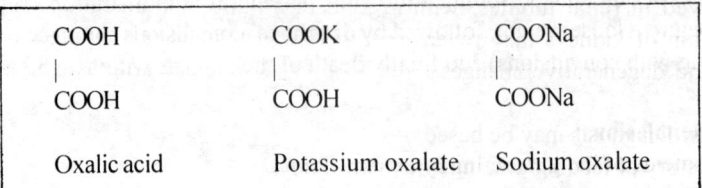

Fig. 4.14 : Oxalic acid and its soluble salts

Sources of poisoning: Oxalate poisoning in livestock primarily results from ingestion of large quantities of oxalate containing plants. Oxalate containing plants are palatable and frequently form bulk of ruminant's ration. For causing toxicity, the plants should contain more than 10% oxalic acid on dry-weight basis. The form in which oxalate is present in the plant may be of importance in terms of toxicity, that is whether it exists mainly as acid oxalate (e.g. in *Oxalis* sp., *Rumex* sp.) or oxalate ion (e.g. in certain *Chenopodiacae*). Some of the common oxalate containing plants are listed in Table 4.10.

Table 4.10 : Some of the important oxalate containing plants

Amaranthus retroflexus	*Rheum rhaponticum*
Atriplex sp.	*Rumex* sp
Beta vulgaris (beet)	*Sarcobatus vermicularis* (Grease wood)
Brassica hyssopifolia	*Spinacia aleracea* (Spinach)
Calandrinia sp·	*Salsola kali* sp.
Chenopodium album	*Setaria sphacelata*
Halogeton glomeratus	*Threlkeldia proceriflora*
Oxalis sp.	*Trianthema* sp.
Portulaca sp.	

Mechanism of toxicity : Normally, after ingestion of the plants, calcium present in the stomach may react with the oxalates to form insoluble salts which can not be absorbed and eliminated in the faeces. This process is more effective in ruminants than simple stomach animals and consequently, ruminants can consume large amounts of oxalate containing plants without any apparent signs of toxicity. However, when very large quantities are ingested that exceeds the capacity of digestive tract to convert the soluble oxalates into calcium oxalate, the soluble oxalates get absorbed through intestinal mucosa and are available to interact with blood calcium causing acute hypocalcemia. Further, calcium oxalate crystals may get accumulated in the kidney tubules causing severe renal damage. Calcium oxalate may also crystallize in the brain.

Clinical signs: Dullness, lowering of the head, loss of appetite and remaining isolated from the herd are initial evidences of oxalate poisoning seen after 4 hours of feeding oxalates-rich plants. These observations are followed by excessive salivation with frothing, progressive incoordination and coma with deep irregular respiration. Convulsive episodes may also be noted.

Post-mortem lesions: Lungs may be filled with dark-red or purplish coloured blood as most of the deaths in oxalate poisoning are associated with asphyxia. Petechiae haemorrhages and cyanosis may be seen at various locations. Crystalline masses may be observed in renal tubules, with a concentration in the cortices. Microscopic examination of kidneys may reveal presence of calcium oxalate crystals, ruptured tubules and degenerative changes.

Diagnosis: Diagnosis may be based on :
(i)   History of feeding/grazing on oxalates-rich plants.
(ii)   Clinical signs.
(iii)   Post-mortem lesions.

Treatment:
(i)   Discontinue feeding of oxalate containing plants.
(ii)   Administration of calcium ions to promote the elimination of oxalates as calcium oxalate. Dicalcium phosphate is very effective in aiding the elimination of oxalates (as calcium oxalate) from the gastrointestinal tract.

Prevention: As the oxalate ions react readily with Ca^{2+}, feeds containing high levels of calcium (calcium phosphate or dicalcium phosphate) may be fed to the animals so that it interacts with oxalic acid to form insoluble calcium oxalate and eliminated through faeces.

4.12 *PARTHENIUM*

V. P. Vadlamudi

Parthenium hysterophorus L. (*Compositae*), originally native to south and central America was accidentally introduced into India. Its seeds as contaminant of imported wheat from USA gained access into our country during the early sixties. The weed first appeared in Pune (Maharashtra) from where it spread like wild fire and later established in many regions of our country. Over the years it naturalized itself in an endemic nature, especially in southern states, posing a serious agricultural problem and health hazard in both man and animals.

Widespread invasion of *Parthenium* weed (Fig.4.15) in agricultural fields under contrasting ecological conditions has resulted in severe losses in crop yields. The plant, being a heavy feeder, exhausts the soil nutrients and make the farm fields barren. It is remarkably adaptable and grows abundantly in crops and fodder fields and waste lands throughout the year.

Fig. 4.15 : *Parthenium* sp.   (see also Colour Plate III)

Allergic contact dermatitis has been recorded in farm labourers engaged in its manual removal from crop fields. Its endemic invasion of pastures and waste lands has also been recognized as a health hazard in grazing livestock. *Parthenium hysterophorus* has been recognised as a problematic weed due to agricultural losses, hazards in farm workers and toxicity in livestock.

Hazards in farm labourers : Many species of the family *Compositae* have been reported to produce allergic dermatitis in human beings. Contact dermatitis attributed to *Parthenium* weed was first recognised during the year 1930 in USA. In India, the first incidence of *Parthenium*-induced contact dermatitis in farm labourers around Pune (Maharashtra) was reported in the year 1971. Adult males were found to be highly susceptible to dermatitis. A study among 300 persons occupationally exposed to *Parthenium* in Karnataka for periods ranging from 3 to 12 months revealed that 4 per cent of the persons developed contact dermatitis on exposed parts of the body, while 56 per cent were found sensitized to the weed without apparently exhibiting dermatitis. Similarly, another study among farm labourers also in Karnataka, involved in manual eradication of the weed reported itching and erythematous papular rashes on face, 'V' of the neck and arms. A typical case of severe contact dermatitis in an animal attendant engaged in collection and feeding of *Parthenium* weed to buffaloes (in an experimental study) was observed at the Veterinary College, Parbhani (Maharashtra).

Hazards in livestock : The weed also poses health hazard in animals, especially the grazing livestock. Cattle, buffaloes and goats graze on the weed where the fodder fields are heavily infested with the weed. There are no exact reports indicating the incidence of *Parthenium* toxicity in animals. Several experimental studies revealed development of dermatitis and liver pathology among livestock. The milk from the cows grazing on *Parthenium* weed was found to be bitter in taste.

Sources of poisoning : The major source of *Parthenium* toxicity in animals is inadvertent grazing on the weed, especially during fodder scarcity periods.

The toxic principles of *Compositae* are sesquiterpene lactones which cause allergic dermatitis. The major sesquiterpene lactone of *Parthenium* weed is parthenin. There are no reports on absorption, distribution, metabolism and excretion of parthenin in animals. The bitter taste of milk of cows may be due to secretion of parthenin into milk.

Mechanism of toxicity: The major sesquiterpene lactone of *Parthenium* weed i.e. parthenin is a photodynamic substance. Hence, as seen with other photodynamic agents, ingestion of *Parthenium* weed results in primary photosensitization, causing liver pathology and skin reactions.

Clinical signs: The information on symptomatology and pathology of *Parthenium* toxicity in animals is documented from experimental studies. There are no reports of acute toxicity due to ingestion of *Parthenium* weed. In general, toxicity in cattle and buffaloes develops following ingestion of the weed over a period of seven or more days. Initially the animals show diarrhoea followed by cutaneous lesions characterised by itching, erythematous eruptions on tip and base of ears, neck, sides of thorax, abdomen, knee, hock joint and brisket region. Depigmentation in patches may also occur in the above areas. Oedema around eyelids and facial muscles is also observed. The dermal syndrome is commonly seen in buffaloes. Retarded growth is noted in buffalo calves continuously fed on greens mixed with *Parthenium* weed. Such calves also showed elevated serum transaminase activity suggesting liver damage.

The pathological changes noted in buffalo calves due to continuous feeding on *Parthenium* weed are degenerative changes in liver, kidneys, lungs and intestines, such as necrotic foci in liver, cloudy swelling of tubular epithelium of the kidney, congestion and oedema in lungs and catarrhal inflammation in intestines.

Diagnosis :   There is no specific method of confirmation of *Parthenium* toxicity. Yet, *Parthenium* toxicity may be diagnosed based on the following :
(i)   History of grazing in fodder fields heavily infested with the weed.
(ii)   Appearance of dermatitis lesions.

Treatment : The line of treatment should be aimed at treating the cutaneous lesions by giving antipruritics and antiseptics and to improve liver function by giving hepatotonics. The animals should immediately be shifted to normal fodder.

4.13 OAK

Satish K. Garg

Oak (*Quercus* species; Fig. 4.16) belongs to the family *Fagaceae*. Oak poisoning primarily occurs in cattle, however, sheep and horses are also poisoned. Goats are capable of surviving much greater intake of *tanin* than cattle or sheep because of greater concentration of tannase enzymes in their ruminal mucosae. However, pigs are resistant to pyrogallic and tannic acids. Cases of oak poisoning are observed in hilly regions during the spring months due to ingestion of buds and autumn due to ingestion of acorns (fruits). During spring season, grasses and other forages are not available in plenty and oaks are among the first plants to show signs of new growth. The young, immature and tender leaves and buds seem to attract the animals. Poisoning, though in much milder forms, has also been documented in mid and late summer due to consumption of leaves under extreme drought conditions when other fodder are not available.

Fig. 4.16 : *Quercus incana*   (see also Colour Plate III)

Poisoning in cattle occurs mainly by ingestion of huge quantities of oak buds, leaves and twigs in spring or ingestion of large quantities of acorns in autumn, especially during periods of food shortage or drought. Poisoning is extremely frequent and particularly serious in young animals in certain regions. At lower levels of ingestion or feeding (30-40 % of the cattle diet) it is not much toxic. Most of the poisonings in Britain mainly occur due to acorns. Unripe acorns are most dangerous when large quantities of acorns come down due to gale.

Natural cases of oak poisoning have mainly been reported from U.S.A., Europe, China and India. Such poisoning in U.S.A. are mainly due to *Quercus haverdii* (sand shin oak), *Q. breviloba* and *Q. gambelli*, Europe due to *Q. robur*, India due to *Q. incana* and China due to six different species of *Quercus*. In North-West Himalaya regions of India, commonly available species of oak are *Q. incana*, *Q. semecarpifolia*, *Q. dilata*, and *Q. itex*, out of which *Q. incana* is most toxic.

Active principle(s) : Leaves and acorns contain large quantities of tannins (both hydrolysable and condensed tannins) and gallotannins in addition to some other principles like simple phenol etc. Tannins, particularly hydrolysable tannin, and their degradation products (metabolites) such as gallic acid and pyrogallol are the most likely toxic principles.

Tannins are present in highest concentration in young immature leaves. The levels of total phenols and protein precipitation capacity are higher in young and immature leaves compared to mature leaves while that of condensed tannins are higher in mature leaves. Acorns contain 7-9% pyrogallic acid, although the quantity varies according to age of the acorn. Unripe green acorns contain the highest levels of pyrogallic acids. Similarly depending on age of the tree, youngest trees have the highest / greatest levels of pyrogallic acids and tannins.

Mechanism of Toxicity : Exact mechanism of action of oak poisons is not known. But precipitation of proteins and binding to sulfhydryl groups of enzymes has been suggested to be the possible action. Depending on the quantity and the variety of oak ingested/consumed and the content(s) of tannins and phenols, signs of poisoning appear. It appears that hydrolysable tannins produce toxic effects by breaking down the alimentary canal tissues or after absorption into circulation produce pathological changes in liver, kidneys and heart when their concentration in blood exceeds the detoxifying capacity of liver.

Clinical signs : Onset of clinical signs and mortality, if any, depends on the quantity of oak leaves (buds) or acorns consumed. Signs of oak poisoning are chiefly alimentary and urinary in nature, however gastrointestinal effects are more severe and persistent and characterised by anorexia, lack of rumination, tucked up abdomen indicating abdominal pain, thirst, ruminal atony, severe constipation, black tarry, dry and pelleted faeces with shiny coat of mucus and some streaks of blood followed by dark foetid diarrhoea, often haemorrhagic. Other signs are subnormal body temperatures, extremities are cold, muzzle is dry, there may be watery discharge from the eyes, nose and mouth. Mucus membranes are pale in colour. Most of the affected animals tend to remain close to water and do not follow the herd indicating signs of polydipsia and polyuria. In severe cases, subcutaneous oedema of the neck, brisket and perineum and limbs may be observed. Ventral oedema of renal origin is characteristic especially in sheep.

Neurological effects like tremors, grinding of teeth, ataxia, depression and weakness in the hind quarters are infrequent and appear towards the terminal stage of poisoning. Haematuria and haemoglobinuria are rare. If course of poisoning is prolonged, animals become dehydrated and emaciated and blood stained nasal exudation is a frequent sign and finally the animals may collapse . Dysuria, cessation of lactation, loss of weight and body fat are also observed towards the end stage of affected animals. Goats particularly show signs of abdominal distress alongwith bleating. Some of the animals even die without showing any symptoms of toxicity and sometimes mortality rate are very high, in such cases, history and necropsy findings assist in establishing the cause of death.

Post mortem lesions:

(i)   Kidneys are pale and swollen with or without subcapsular uniformly sprinkled petechial haemorrhages. There is extensive coagulative necrosis of proximal convoluted tubules (PCT) with an early loss of nuclear and cell outlines and presence of haemoglobin and other types of casts in the lumen of damaged PCT's particularly dense casts of albumin. The necrotic epithelial lining cells are usually so intimately mixed with the proteinaceous contents of the lumen that the whole thing forms a dense, homogenous mass limited by the basement membrane and interstitial tissue. Microscopic renal lesions are nearly pathognomonic of oak poisoning. The glomeruli show little change and medulla is almost normal except for mild congestion.

(ii)   Haemorrhagic gastroenteritis with necrosis of muscular and serosal layers. Gall bladder is distended with viscid, brownish bile.

(iii)   Ascites and accumulation of excessive fluids in pericardial and pleural cavities.

(iv)   Oedema of the gastrointestinal wall and mesentery.

(v)   General congestion of the carcass with an odour of stale urine.

(vi)   Oedema of mesenteric lymph nodes.

(vii)   Abdominal fat is yellowish, soft and gives odour of urine.

(viii) Degeneration of liver and hypertrophy of gall bladder.

Diagnosis :

(i)   History of onset of the condition after feeding/ingestion of oak leaves or acorns and sudden mortality.

(ii)   Presence of buds/acorns on the pasture, evidence of ingestion with clinical signs.

(iii)   Laboratory investigation of blood indicating significant elevation in the serum bilirubin, aspartate amino transferase (AST) and lactate dehydrogenase (LDH), blood urea nitrogen and creatinine and significant decrease in haemoglobin levels. Urine examination reveals that urine is generally of low specific gravity and there is mild to moderate degree of proteinuria, haemoglobinuria alongwith occult blood, bile pigments and granular and cellular casts.

(iv)   Post mortem findings.

Treatment : Success of the treatment depends on the early diagnosis and onset of treatment though it is largely symptomatic and supportive as no specific antidote is available.

(i)   Remove the cause / source of intoxication or remove the animals from pasture.

(ii)   Administer gastrointestinal demulcents, liquid paraffin, mild oleoginous or mucilagenous laxatives .Use of saline purgatives (magnesium sulphate) is contraindicated.

(iii)   Give hepatoprotective agents.

(iv)   Administer mineralocorticoids and large quantities of water to restore fluids and electrolytes balance.

(v)   Give easily digestible feed.

(vi)   Provide supplemental feed and fluid therapy for emaciated animals.

(vii)   Thoracocentesis to remove excessive pericardial or peripleural fluids in severely dyspnoeic animals.

Control measures :

(i)   Prevent access of the animals to oak acorns or trees.

(ii)   Prune the oak trees in grazing areas.

(iii)   Isolate the animals with signs of oak poisoning. Supplement the feed/ration of animals with calcium hydroxide (15% of the ration per day) to prevent the absorption of tannins.

4.14 *EUPATORIUM*

Satish K. Garg

The weeds of *Eupatorium* species are commonly prevalent in high rainfall areas on the damp hill sides and other sheltered moist places and cover a vast expanse of cultivable land. It is a perennial weed and widely prevalent in Kangra valley of Himachal Pradesh . Ingestion or inhalation of two species of *Eupatorium* i.e. *Eupatorium riparium* (mist flower) and *E. adenophorum* (crofton weed) have been reported to be primarily responsible for pulmonary disease in horses in Australia, however, cattle, goats, sheep and swine are also affected. Milk sickness is another poisoning syndrome associated with *E. rugosum* (snake root). Nancy Hanks Lincoln, mother of Abraham Lincoln died of milk sickness in 1818. Poisoning of horses with snake root was reported for the first time in 1940. Since the toxic principle(s) of the snake root are secreted in milk, consumption of such milk has been reported to cause milk sickness in human beings and young ones. Flowering stage of plants is more toxic than the non flowering stage.

Toxic principle of some of the *Eupatorium* sp. e.g. *E. adenophorum*, *E. riparium* and *E. rugosum* is a mixture of higher alcohol trematol which needs microsomal bioactivation to form the toxic metabolite tremetone. Some members of this genus e.g. *E. semiserratum* contain flavonoides, namely - eupatorin and 5-hydroxy 3', 4', 6, 7 tetramethoxyflavone while *E. cannabinum* contains sesquiterpene lactones or pyrrolizidine alkaloids. In addition, *E. altissimum* has the antimicrobial activity. *E. altissimum*, *E. canabinum* and *E. rugosum* have been found to possess some cytotoxic and antitumour activities.

Clinical signs : Animals poisoned with *E. adenophorum* or *E. riparium* exhibit signs of coughing, rapid respiration, laboured breathing, develop heaves and have decreased exercise tolerance. Other signs are stiffness of limbs, particularly fore limbs, incoordination of gait, severe tremors, salivation, depression, recumbency, coma and death. In *E. rugosum* poisoning, also called trembling disease, there is trembling, particularly of the musculature of nose, legs and shoulders especially after exercise. The animals stand in a humped-up posture.

Other signs are constipation, vomition, regurgitation of food through nostrils, incontinence of urine, partial paralysis of throat, profuse perspiration particularly in horses, severe depression, coma and death.

Post mortem lesions :
(i)   Oedema of the lungs;
(ii)   Acute haemorrhages in lungs;
(iii)   Small nodes in the lung parenchyma dorsally near the hilus and cut surface is pale and firm;
(iv)   Microscopically, severe haemorrhages, peribronchial atelactesis and peripheral emphysema. Alveolar cells are thickened and hypertrophied, eosinophilic bronchiolitis and peribronchial fibrosis. Kidneys revealed thickening of the parietal layer of Bowman's capsule, degenerative changes in convoluted

tubules. In spleen, proliferation of reticuloendothelial cells and development of giant cells is quite conspicuous. Liver exhibits hypertrophy and hyperplasia of the hepatic cells. Nuclei are hypertrophied and accentric. Reticulo-endothelial cells migrate into the lumen of sinusoids.

Diagnosis : It is comparatively difficult, however, may be made based on :
(i)   History of exposure to *Eupatorium* through ingestion or inhaltion.
(ii)   Clinical signs.
(iii)   Post mortem findings.

Treatment : No specific antidote is available. Only symptomatic treatment need to be given

4.15 FERNS

R.K. Dawra

Ferns are pteridophytes and mostly distributed in hilly areas of the world. Types of fern vary according to the altitude, rainfall, soil type and microclimate of the place. In the mid-hills, where rainfall is adequate there is extensive growth of these plants. Most of the ferns are either ground growing (terrestrial) or lithophytes (growing on rocks) but several of them grow as epiphytes (growing on trees) as well. Usually epiphytes grow on the angiospermic trees and few on the conifers. They mainly belong to fern family *Polypodiaceae*. Members of family *Aspidiaceae*, *Thelypteridaceae* and *Athyriaceae* are largely terrestrial. These ferns usually grow in open places, humus rich forest floors, forest margins, road sides and way-sides forming conspicuous vegetation.

From animal toxicity point of view, the terrestrial species especially, those which grow either intermixed in the grasses or around the pastures where grazing animals have easy access are important. Some ferns can grow in close association with pasture grasses and grazing animals may consume significant amounts of these over several years. Some of the common terrestrial ferns are *Pteridium aquilinum* (Bracken fern), *Equisetum* sp. (Horse tail), *Pteris cretica*, *Pteris pseudoquadriaurita*, *Onychium contiguum*, *Adiantum venstum*, *Athyrium sp.*, *Dryopteris sp.*, *Polystichum sp.* etc. Out of these, *Dryopteris* and *Polystichum* are lithophytes and tend to grow terrestrially along roadsides, waysides, forest floors and forest margins. *Equisetum* grows on the moist fields and meadows, sandy banks of roads or railroad right of ways. From April to June nearly all ferns bear new fronds and from August to September these are fertile and bear spores. From November to March ferns are dormant. During winter months, the aerial parts of almost all the terrestrial ferns are killed completely or partially by snow and frost above 1500 m. But ferns growing on walls and other protected situations remain green throughout the winter. With the advent of spring together with summer heat and pre-monsoon showers the dense growth of vegetation is resumed. Some of the ferns like *Polystichum*, *Pteridium*, *Diplazium*, *Adiantum*, *Pteris* being hardy remain green throughout the year, although like other ferns, these bear new fronds only after spring.

Like other toxic plants, ferns are not the preferred vegetation by the grazing animals, in case, other grasses are available. Exposure to ferns takes place under scarcity conditions or when ferns grow in close association with grasses. Earlier recorded cases of fern toxicity were in the year of droughts indicating fern consumption by livestock under scarcity conditions. Acute toxicity to fern results when a large amount of these plants are consumed over several weeks. The general condition of the animal deteriorates and after the appearance of first symptoms, if proper treatment is not given, animal dies because of toxicity. Some fern species investigated for animal toxicity are listed in Table 4.11. The information about the other species suspected to be toxic is not available. Intermittent exposure in low doses over prolonged period causes chronic toxicity which does not respond to treatment.

Toxicity of bracken and other ferns in livestock : Toxicity of bracken (Fig. 4.17) and other ferns has been reported in a number of animal species. Toxicity symptoms are different in nonruminants and ruminants and therefore, would be discussed separately.

Table 4.11 : Some of the ferns which are toxic to livestock

Fern	Animal Species
Pteridium aquilinum	Horses, Mules, Pigs, Cattle, Rats, Sheep
Equisetum sp.	Horses and Cattle
Dryopteris filix-mas	Horses and Cattle
Cheilanthus sieberi	Cattle and Sheep
Cheilanthus tenuifolia *Cheilanthus distans* *Lindsaea linearis* *Platysome microphyllum*	These ferns have been reported to have thiaminase activity but no case of poisoning has been reported
Marsilea drummondi	Sheep
Diplazium esculentum	Cattle
Dryopteris boreii	Cattle

Fig. 4.17 : Bracken fern *(Pteridium aquilinum* (see also Colour Plate IV)

Bracken fern toxicity in nonruminants : Among the nonruminants, the animals which have been reported to be natural victims of bracken fern poisoning are horses, mules and pigs. In nonruminants the effect of bracken poisoning is essentially that of vitamins B_1 (thiamine) deficiency (Table 4.12). Experimentally, bracken toxicity in horses was induced by inclusion of sun dried bracken in the diet in 1951. Ingestion of hay containing more than 20% bracken produces toxicity signs in about one month. The symptoms are those of avitaminosis B_1. The clinical signs of bracken poisoning in horses and mules are listed in Table 4.12. Death is preceded by muscular spasms and back inflection of neck. During the onset of incoordination symptoms, blood analysis reveals leukocytosis, thrombocytopenia, increase in pyruvate levels and decrease in blood thiamine levels. Bracken poisoning in mules (field cases) with similar symptoms has been reported from Brazil. Change of feed and vitamin B_1 therapy help, even the severely ataxic horse to recover within 2-4 days.

Accidental ingestion of bracken fronds by pigs resulted in poisoning in pigs. The symptoms of poisoning (listed in Table 4.12) developed within three days. These animals recovered after injection of antibiotics and multiple vitamins preparation (containing 350 mg thiamine). Bracken poisoning in pigs can be induced by feeding dry powdered rhizomes (25 to 33% of the diet).

Table 4.12 : Toxicity symptoms of fern poisoning in different species of animals

Species of animal	Main toxicity symptoms
Horses and Mules	* Incoordination. * Staggering. * Muscular tremor, generalised congestion, pulmonary oedema and serosal and mucosal haemorrhages. * Response to B_1 therapy if started early.
Pigs	* Loss of appetite. * Vomition and constipation. * Death due to damage to heart. * Response to vitamin B_1 therapy in early stages.
Rats	* Symptoms of vitamin B_1 deficiency. * Animal respond to B_1 therapy.
Calves	* Loss of appetite. * Watery discharge from eyes and nostrils. * Swelling of throat, difficult breathing. * High temperature and death within three to five days.
Cattle	* Lack of appetite. * Slight watery discharge from eyes and nostrils. * Rumination ceases, weak pulse, laboured respiration, high temperature and constipation. * Blood in faeces, dark coloured urine, sometime containing blood. * Animals do not respond to vitamin B_1 administration.

Mechanism of bracken fern toxicity in nonruminants : Toxicity of bracken fern to nonruminants has been ascribed to presence of thiaminase (EC 3.5 99.2) in the fern which destroys vitamin B_1 (Fig. 4.18) resulting into deficiency of vitamin and symptoms which respond to vitamin B_1 therapy if started at an early stage. Thiaminase is not completely destroyed *in situ* by steaming and the entire plant is toxic.

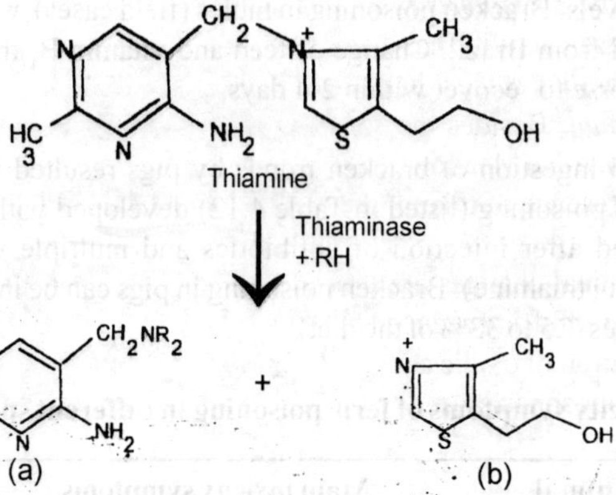

RH- is a nucleophilic agent like aromatic amines, heterocyclic amines or sulphur containing amino acids.

Fig. 4.18 : Action of thiaminase of fern origin on thiamine (vitamin B_1). Thiamine gets degraded to (a) pyrimidine and (b) thiazol, resulting in vitamin B_1 deficiency

Metabolically thiamine is involved as a cofactor in decarboxylation reactions. These include conversion of pyruvate to acetyl CoA and oxidation of α-ketoglutaric acid to succinyl CoA in citric acid cycle. So, thiamine deficiency results in impaired pyruvate utilisation. Therefore, pyruvic acid formed via glycolysis accumulates and the blood pyruvate level rises. The animal suffers from impaired energy metabolism and cellular shortage of ATP. The elevated pyruvate may affect central nervous system functions. Blood thiamine concentrations drop to 2-3 µg/dl. Response to parenteral thiamine administration is drammatic.

Bracken is poisonous while green and remains poisonous if cut in green stage, dried and stacked. So, it is not customary to harvest it for bedding until late in the season when sori are ripe and turn brown. Thiaminase activity is highest in rhizomes during summer, while in fronds it is high during early growth and decreases progressively during growing season. In bracken, in addition to thiaminase a thermostable antithiamine factor has been reported which has not been characterized so far. Using this heat stable factor, bracken poisoning was produced in equines, however, this type of heat stable factor has not been investigated in other ferns.

Toxicity of other ferns in nonruminants : Horse tail (*Equisetum arvense*) is a common thiaminase containing weed in moist areas of U.S.A. and Canada. Field cases of fern poisoning of horses have been documented from North America. Hay containing 20% or more horsetail may produce symptoms of thiamine deficiency in 2-5 weeks. *Equisetum sp* (horse tail) and *Cheilanthus sieberi* contain thiaminase activity and in horses poisoning symptoms are similar to those of bracken poisoning. These symptoms go unnoticed in early stages. There is progressive loss of condition, slow pulse and unthrifty appearance. However, appetite remains normal. During the late stages horses loose their round bellies, the muscles of hindquarters begin to waste, gait becomes unsteady and the animals may have difficulty in turning. In extreme cases, horses lie down and have difficulty in rising without assistance. Animals respond to vitamin B_1 therapy and good diet if started early. *Dryopteris filix-mas* (male fern) has been reported to be toxic to horses. Thiaminase activity has been reported in *Dryopteris filix-mas*. Thiaminase activity in *Cheilanthus sieberi* has been reported to be many fold higher than that of *Pteridum aquilinum*. Besides the species of ferns mentioned above, other ferns viz. *Equisetum palustre* L (Meadow dropwort) appears to be toxic and toxicity of it is ascribed to the presence of thiaminase and not alkaloids.

Biochemical changes are those of thiamine deficiency with hypoglycaemia during the latent period and reduced tolerance to carbohydrates. There is an increase in plasma concentration of oxalic acid, phosphate and potassium and increased activity of plasma alkaline phosphatase and cholinesterase. All biochemical changes respond to treatment with vitamin B_1. Pathological changes include atrophic degeneration of neurons in the cortex, caudate nucleus and corpora quadrigemina and loss of Purkinje cells in the cerebellum with proliferation of glial cells and haemorrhages. There is also hepatic and myocardial degeneration.

Toxicity of bracken and other ferns in ruminants : Unlike nonruminants, toxicity symptoms in ruminants are not due to deficiency of vitamin B_1. Toxicity is accompanied by acute haemorrhage syndrome. The main symptom is a progressive failure of blood forming bone marrow tissue. There is thrombocytopenia resulting in increase in blood coagulation time. A few early reports of bracken poisoning in cattle could not be confirmed because of confusion of symptoms with that of the symptoms of anthrax. Fern toxicity in ruminants was not well understood till systematic investigations were conducted. Unlike in nonruminants, it is not simple B_1 deficiency but something much more complex, which does not respond to thaimine treatment. Normally, vitamin B_1 deficiency does not develop in ruminants. *Pteridium aquilinum* generally depresses haematopoiesis in cattle. Rumen microbes are able to synthesize adequate quantity of vitamin B_1. When there is very high concentration of thiaminase as in case of *Marsilea drummondii* (Nardoo), the exposed ruminants develop vitamin B_1 deficiency and condition is called polioencephalomalacia. Nardoo is an Australian fern that grows in damp areas such as water courses. Extensive losses of sheep with typical symptoms of thiamine deficiency have been reported. Thiamine deficiency in ruminants is characterised by depression, incoordination, convulsions and cerebrocortical necrosis. Nardoo contains thiaminase levels upto 100 times those of bracken fern. Thiamine deficiency and polioencephalomalacia due to consumption of Nardoo are widespread in Australia.

Bracken toxicity in ruminants is noticed sporadically, mainly in young stock. Occasionally, milch cows kept on dried bracken bedding may become affected during winter. The ingestion of bracken results in suppression of bone marrow, aplastic anaemia and subsequently blood clotting defects. There is also serious reduction in white blood corpuscles i.e. leucopenia, granulocytopenia and thrombocytopenia.

During the late stages, there is high temperature and invasion of blood by bacteria. Urinary bladder neoplasms are also reported in cattle consuming bracken fern for long periods of time.

Mechanism of toxicity in ruminants : Cattle exposed to low doses of bracken fern over prolonged period develop tumours in urinary bladder and chronic intermittent haematuria. The compound responsible for carcinogenicity of bracken fern has been isolated. This was named as ptaquiloside by the Japanese and aquiloside A by the Dutch (Fig.4.19). This compound has also been associated with toxicity of fern in ruminants.

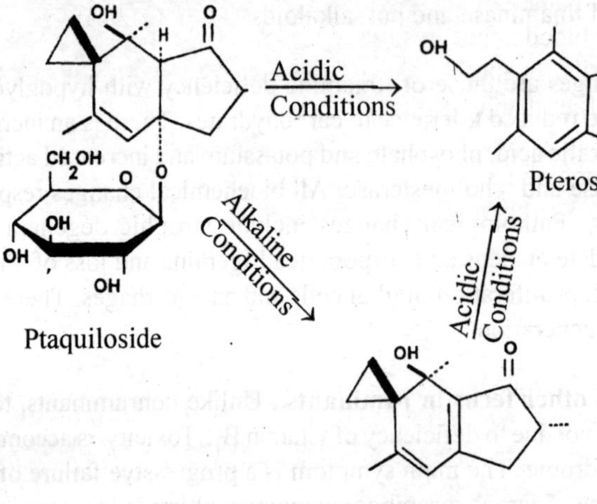

Fig. 4.19 : Ptaquiloside, the carcinogenic glycoside in bracken fern and its activation under alkaline conditions to the active carcinogen dienone. Under acidic conditions it gets converted to pterosin B.

In a study, ptaquiloside dissolved in saline when administered at the rate of 400 mg /day for 24 days, 800mg/ day for 14 days and 1600mg /day for 4 days to 6 months old calf, a decrease in neutrophilic granulocytes was observed and the changes were greatly marked by 50 days after the start of administration and granulocytopenia continued for further 35 days until slaughter despite discontinuance of ptaquiloside administration. Thrombocyte showed a relatively slow depression and reached 1×10^5 / mm^3 at the lowest level. The calf was killed 86 days after the start of experiment. Sternal bone marrow was found to be replaced with fat marrow and only small foci of erythropoietic cells and a few megakaryocytes remained. Total amount of ptaquiloside used in this experiment was 27.2 g which was extracted from about 137 kg of dry bracken fern. The ptaquiloside has been considered to be a causative factor of bracken poisoning in cattle. No gross haemorrhages have been observed during bracken poisoning due to low doses of ptaquiloside. Some workers have disputed the argument in favour of ptaquiloside as the agent responsible for fern toxicity in ruminants.

The field reports of bracken and other ferns toxicity are sporadic. In India, toxicity due to bracken or other ferns has not been reported so far. Enzootic bovine haematuria associated with chronic low dose exposure to bracken has been reported to be present in a number of localities in the hilly areas of the country.

Clinical signs : The early symptoms of bracken toxicity are loss of appetite and watery discharge from eyes and nostrils. There is swelling of throat, which causes difficulty in breathing, as a result of this an affected calf makes a roaring sound. There is high temperature and death occurs in three to five days after onset of symptoms. In older cattle early symptoms of poisoning are depression, weakness, anorexia, watery discharge from eyes and nostrils. Rumination ceases and if outdoor, the affected animals stand or lie apart from others. The pulse is weak and temperature rises slightly. Later, the discharge becomes blood tinged, respiration is laboured and muzzle becomes dry and scaly around nostrils. In the beginning of toxicity, faeces are small, dark coloured, firm and slightly drier than usual but as the condition deteriorates, enteritis supervenes and blood is passed in clots or mixed with faeces (haemorrhagic enteritis). There is haematuria and haemoglobinuria. Passage of urine ceases altogether in terminal stages of the disease coupled with sharp rise in temperature (up to 107 -108° F) due to bacteraemia. The affected animals may die within 4-10 days.

Bracken poisoning has also been observed in sheep grazing in areas infested with *Pteridium esculentum*. Affected animals exhibit loss of weight, lethargy, exercise intolerance and anaemia and diarrohea in a small percentage of the affected sheep.

Post-mortem lesions in calves :
(i)   Excess of fluid in the region of throat and neck and occasionally between the muscles of the legs.
(ii)   Petechial and massive haemorrhages under the skin (haematoma), under both visceral and parietal peritoneum, particularly on the outer wall of the rumen and abdominal surface of the diaphragm under the pleura in the heart muscles, other muscles and in the kidneys .
(iii)   Ulcers of varying size, from which severe haemorrhages have occurred are usually found on the inner walls of the intestine often in the abomasum and intestine.

Post-mortem lesions in sheep :
(i)   Ecchymotic and petechial haemorrhages throughout the carcasse.
(ii)   Selective degeneration of myeloid series cells, numerous degenerating blast cells and absence of band cells in the rib marrow.
(iii)   Femoral marrow consisted entirely of adipose tissue and haemorrhages.
(iv)   Bright blindness characterized by stenosis (narrowing) of blood vessels in the eye and progressive retinal atrophy, occurs in sheep consuming bracken fern. The causative agent has not been identified. The syndrome has been experimentally produced by feeding a diet of 50% bracken fern to sheep for 63 weeks.

Equisetum sp. has also been reported to be toxic to sheep and cattle and *Dryopteris filix-mas* and *Dryopteris borreii* in cows. The most marked symptoms of

Equisetum toxicity are a rapid and spectacular fall in milk yield and serious loss of condition. Diarrhoea is common but not always present. But *Dryopteris* sp. affected animals show anorexia and are found standing and later lying blind and comatose. Post-mortem lesions were petechiation on epicardium and major blood vessels. Infarcts were seen in mouth and ulcers in oesophagus, omasum, caecum and colon. There was zebra crossing in small intestine and anterior colon in addition to focal retinal haemorrhages and scleral congestion.

The ferns *Cheilanthus distans, C. sieberi, C tenuifloia, Lindsaea linearis, Marsilea durmmondi, Pteridium aquilinum* and *Platysoma microphyllum* have also been found to be toxic to animals. These ferns contain thiaminase and may contain other toxic constituents as well but toxicity as observed in ruminants is not due to thiaminase. *Diplazium esculentum* feeding to male calves at the rate of 5kg /day induced inappetence after about 30 days of feeding. Debility, ataxia and complete anorexia were noticed in terminal stages. There was decrease in haemoglobin, blood glucose and serum thiamine. All calves died between 41 to 150 days of experiments. Tuberculoid serous and fibrinous, noninflammatory infiltration (5-10 mm diameter) in and below the mucosa of small and large intestines have been noticed in cattle grazing on pastures where male fern *Dryopteris filix-mas* was growing.

Diagnosis :
(i)   History of the presence of the plant in hay or on the pasture and also the access of the animals to the same.
(ii)   Clinical signs- haemorrhages, anaemia, haematuria, ventral oedema after 4-6 weeks of continuous exposure to *Pteridium* sp.
(iii)   Post-mortem lesions of anaemia, pale bone marrow, petechial and ecchymotic haemorrhages on the mucous membranes.
(iv)   Decreased thiamine levels in serum.
(v)   Polioencephalomalacia of brain.
(vi)   Response to thiamine treatment.

Treatment : No specific antidote is available. Prognosis in advanced cases is very poor.
(i)   Remove the animals from the contaminated area.
(ii)   In cattle, blood transfusion is the most effective remedy. For calves, 0.4 litres and for adult cattle, 1-4 litres of blood with 3% sodium citrate, 100 ml for every 400 ml of blood together with a single intravenous injection of 10 ml of 1% protamine sulfate to counteract the effect of released heparin.
(iii)   Toluidine blue (1 gram/litre serum) at a rate of 1-2 litres/24 hours intravenously, but administer with caution.
(iv)   Batyl alcohol (2 gram), Tween 80 (5 gram) and 1% sodium chloride (100 ml) by slow intravenous injection @ 20-50 ml/day. Batyl alcohol is known to stimulate the bone marrow.
(v)   Supplement the above therapy with antihistaminics, antibiotics, vitamin B complex and oral ruminotorics to stimulate appetite.

4.16 ARGEMONE (MEXICAN POPPY, PRICKLY POPPY)

M.A. Ayub Shah

Argemone mexicana (Mexican poppy, prickly poppy or yellow poppy; Hindi- *Bharband* or *Brahmadundi;* Fig. 4.20) which belongs to the family *Papaveraceae* is an American plant which is, at present, widely occurring in India and grows wildly throughout the country in wastelands and along roadsides. Argemone oil (known as "katkar" in some parts of India) is a cause of concern because this oil has been used as an adulterant in mustard oil industries. In July-September, 1998 there was an outbreak of epidemic dropsy in New Delhi and many other cities of India due to consumption of argemone oil adulterated mustard oil causing deaths of several human beings.

Fig. 4.20 : *Argemone mexicana*   (see also Colour Plate IV)

Toxic principles : Argemone oil contains an isoquinoline alkaloid- sanguinarine, ingestion of which leads to the development of epidemic dropsy in human beings. The plant devoid of seeds contains the alkaloid berberine and protopine.

Mechanism of toxicity and clinical signs : The exact mechanisms by which sanguinarine induces toxicity is not fully elucidated. The pathophysiological changes observed in argemone-induced epidemic dropsy in man are oedema of the extremities, gastrointestinal disturbances, cardiopathies (particularly cardiac hypertrophy) and open angle glaucoma.

Feeding of feeds contaminated with argemone seeds in poultry cause a drop in egg production, oedema, depression, ataxia and cyanosis of the comb, haemorrhagic enteritis and death.

Diagnosis :

(i)   History of ingestion of argemone seeds or oil contaminated feeds.

(II)   Clinical manifestations.

Treatment : There is no specific antidote for argemone poisoning. Supportive and symptomatic treatment may be provided.

4.17 MISCELLANEOUS TOXIC PLANTS

M.A. Ayub Shah and H.S. Panwar

As thousands of toxic plants exist all over the world, it is not possible to describe each and every plant in details in this book. Further, toxic plants which are occurring in America, Australia or Africa may not have much relevance in India or the Indian subcontinent. Most of the toxic plants which are abundantly available in India and are responsible for inducing poisoning in different species of animals have already been discussed in details in this chapter. However, some of the plants which have not been discussed in details, but possess the potential to produce toxicity, if ingested by the animals, are briefly summarized below (Table 4.13)

Fig. 4.21 : *Calotropis gigantea*   (see also Colour Plate IV)

Table 4.13 Miscellaneous toxic plants, their toxicity and treatment

Plant species	Toxic principles	Toxic part	Clinical signs	Treatment
Aconitum napellum (Aconite, Monkshood wolf's bane)	Aconitine	All parts especially root	Salivation, nausea, vomition, colic, slowing of heart rate and respiration, muscular weakness, paralysis and dilated pupils. Death from asphyxia.	Heart stimulants, warming, rubbing the animal to increase circulation.
Aesculus glabra (Horse chestnut)	Aesculin (saponin glycosides)	Nuts, barks and leaves	Neurological and gastrointestinal signs.	Supportive treatment fluids and electrolytes, demulcents.
Allium sp. (Onions and garlic) *A. validum A. cepa* (onion)	N-propyl disulphide	All parts	Haemoglobinuria, anaemia and icterus Intravascular haemolysis, cyanosis.	No specific treatment, give whole blood transfusion and fluids.
Asclepias sp. (milk weed)	Steroid glycosides, toxic resinous substance	—	Depression, anorexia, weakness, ataxia, dyspnoea, fever, tetanic seizures, coma and death. If survived, GI signs within 1-2 days.	Supportive treatment, anticonvulsants, calcium containing preparations are contraindicated.
Brassica sp. (mustards) *B. sinapsis* (wild mustard) *B. alba* (white mustard) *B. nigra* (black mustard)	Glucosinolates -isothiocyanates, thiocyanates, Sinalsin Sinigrin	All parts but more concentrated in seeds	Acute gastroenteritis, colic, frothing around mouth and nose, respiratory distress.	Supportive therapy to check diarrhoea and correct electrolyte imbalance.

(*Continued*)

Plant species	Toxic principles	Toxic part	Clinical signs	Treatment
Brassica sp. (Rape, brussels sprouts etc.) B. oleracea B. napus (rape) B. napobrassica (turnip)	S-methylcysteine sulfoxide and L-5-vinyl-2-thiooxa zolidone (a goitrogen)	All parts	Anorexia, bloat, anaemia, dark reddish to brown discolouration of urine (haemoglobinuria), diarrhoea/constipation and yellowish mucous membranes, weight loss and drop in milk yield.	No specific treatment, fluids and electrolytes, blood transfusion, if necessary.
Calotropis gigantea (milk weed) Fig.4.21	Calotroxin, calactin, uscharin, gigantin	All parts	Irritation to skin and mucous membranes, acute gastroenteritis and cardiotoxicity.	Symptomatic and supportive.
Cannabis indica (Indian hemp)	Cannabinol		Dullness, sleepiness, depressed pulse and respiration rate.	Symptomatic.
Cicuta maculata, C. virosa-water Hemlock C. douglasii	Cicutoxin	Root, stems	Nausea, vomition, dilatation of pupil, hyperaesthesia, tremors, opisthotonus and violent spasmodic convulsions. Death within 60 minutes. Good prognosis if survived for two hours. Pigs are most resistant compared to most of the species.	Remove the toxicant, no specific treatment.
Colchicum autumnale (meadow saffron autumm crocus)	Colchicine colchiceine	All parts	Abdominal pain, violent purgation with passage of foetid, green or black faeces, thirst, death due to respiratory failure.	Demulcents, stimulants, atropine.

(Continued)

Plant species	Toxic principles	Toxic part	Clinical signs	Treatment
Conium maculatum (hemlock)	Coniine and pyridine type alkaloids	Root, young plants and seeds	Acute toxicity-sudden death, salivation, emesis and diarrhoea, neurological signs-muscle tremors, muscular weakness, poor vision, convulsions, coma, conciousness is usually not lost. Mousy odour of urine and exhaled air is diagnostic. Death is due to respiratory failure.	Gastric evacuation, activated charcoal, tannic acid, purgatives, artificial respiration, supportive treatment.
Euphorbia sp.	Diterpene or phorbol esters	—	Severe irritation and blistering of mouth and GIT, salivation diarrhoea or haemorrhages, dermatitis.	Apomorphine, charcoal, saline purgatives or mineral oil.
Gossypium sp. (Cotton plants)	Gossypol	Seed cake	Anorexia, weakness, pulmonary oedema and dyspnoea in pigs and horses due to myocardial degeneration and subsequent heart failure. Cardiac degeneration particularly in pre-ruminant calves. Cattle exhibit anaemia, haemolysis, dyspnoea, oedema of brisket, depression and death.	Remove the source at once. Antiarrythmic drugs may be tried. Supplement the feed with iron.
Lathyrus odoratus (sweet pea) *L. sativa* (Indian pea) *L. hirsutus* (wild winter pea) *L. pusillus* (singletary pea)	*Beta*-amino propionitrile (BAP) β-(γ-L-glutamyl) amino-propionitrile, L-α-γ-diamino butyric acid	All parts	Neurolathyrism and osteolathyrism in cattle, sweating, rapid and irregular pulse, respiratory distress, laryngeal vibration over the region of larynx, degeneration of laryngeal muscles and recurrent laryngeal nerve.	General and supportive treatment.

(Continued)

Plant species	Toxic principles	Toxic part	Clinical signs	Treatment
Lobelia sp. (Indian tobacco) *Lobelia berlandieri* *L. purpurascens* *L. urens*	Pyridine alkaloid - lobeline	All parts	Nausea, emesis, mydriasis, muscle tremors, weakness, paralysis and death.	GIT detoxification, other supportive therapy.
Lupinus sp.	Quinolizidine alkaloid-lupinine	Leaves, seeds, fruits	Acute toxicity: Salivation, ataxia, seizures, dyspnoea in sheep, head pressing and excitement, convulsions and death due to respiratory failure. Chronic toxicity : It is known as lupinosis and characterised by dullness, anorexia and jaundice. It also induces teratogenesis (crooked calf disease).	Oral detoxification, central nervous system depressants.
Melilotus officinalis *Melilotus alba* (spoiled or moldy sweet clover)	Dicoumarol	—	Coagulopathies - haemorrhages from gingiva and nose, massive internal bleeding and hematomas below the skin, joints and buttock, anaemia, haemorrhagic diarrhoea and abortion.	Vitamin K is the antidote.
Paspalum scrobiculatum (kodo)	Polycyclic hydrocarbon	Fodder and husk of grain is toxic	Tremors, rigidity, inability to sit, stand or walk, eat and drink, intermittent clonic and tonic convulsions, respiratory distress and death.	Pentobarbitone, chlorallhydras and magnesium sulphate combination, atropine.

(Continued)

Plant species	Toxic principles	Toxic part	Clinical signs	Treatment
Rhododendron sp.	Cardiac glycosides	–	Sudden death, occasional seizures, mild GIT distrubances, decreased atrioventricular conduction followed by AV block.	Symptomatic and supportive.
Solanum sp. S. *tuberosum* (potato) S. *nigrum* (garden nightshade) S. *pseudocapsicum* S. *melongena* S. *lycopersicum* (tomato)	Tropane alkaloids, glycoalkaloids (solanine and solanidine)	Sprouts, vines, green skin, green berries potato tubers	Colic, diarrhoea, salivation, ataxia, weakness, haemolysis of RBCs, bradycardia, hypotension and paralysis. Potato poisoning occurs in three forms : i) Nervous form :-Stupor and depression. It is more predominant in horses, cattle, pigs and poultry. ii) Gastric form : Salivation and diarrhoea. iii) Exanthematous form : Ulcerative stomatitis, conjunctivitis and vesicular and scurfy eczema of legs.	GIT detoxification, cardiac monitoring for the effect of atropine.
Taxus baccata (European yew) T. *cuspidata* (Japanese yew)	Taxine alkaloides A and B	All parts	Trembling, dyspnoea, collapse, death. In cattle, cessation of milk yield, incoordination of movement, slight tympany.	Removal of stomach/rumen contents and replacement therapy.
Veratrum californicum	Cyclopamine, jervine, veratrosine	All parts	Teratogenic effect-single or double fused occular orbit, brachygnathia, large limbs as a result of prolonged gestation.	No treatment.

CHAPTER 5

PHOTOSENSITIZATION

Satish K. Garg

Photosensitization is a syndrome of abnormal sensitivity of the superficial layers of unpigmented or light skinned areas of the body, namely-nose, face, back, udder, testes, teats, mucosa, cornea etc. to ultraviolet (UV) and visible light probably due to the presence of some abnormal substance, termed as photodynamic agent, in the peripheral circulation. The reaction is limited to body areas which receive direct sunlight e.g. teats and udder in cows; head, face, ears, mandibular area and cornea in sheep and goats etc. Severity of the condition depends on the amount of photodynamic substance consumed, duration of ingestion and time of exposure to sunlight. This syndrome may be observed after a few hours to few days of ingestion on exposure to strong sunlight. Thus, photosensitization generally involves a triad of hepato-toxic plant, fresh-green feed in the diet and exposure to bright sunlight.

Depending on the circumstances how photodynamic substance accumulates in the body by ingestion of the preformed photosensitizing agents, abnormal metabolism and/or reduced excretion of phylloerythrin, a product of chlorophyll breakdown, due to hepatic damage and metabolic defects in porphyrin metabolism, three types of photosensitization reactions, namely-primary, secondary and congenital photosensitization have been identified. Table 5.1 enumerates various agents responsible for primary and secondary photosensitization. Almost all species of animals, namely bovines, ovines, caprines, equines and birds are affected. Canines and felines are rarely affected.

Primary photosensitization : It is primarily due to ingestion of exogenous agents like drugs/chemicals - phenothiazine (an anthelmintic), rose Bengal and acridine dyes or photodynamic agents present in certain plants e.g. hypericin in *Hypericum perforatum*, fagopyrin in *Fagopyrum esculentum*, *F. sagitatum* and *Polygonum fagopyrum*, furanocoumarins in *Ammi visnaga*, *Ammi majus*, *Thamnosma texana*, *Cymopterus watsoni* and *C. longipes* or perloline in *Lolium perenne* (Table 5.1). On consumption of these plants, active principle is released in the stomach, comes to peripheral circulation in the capillaries under skin and on exposure to sunlight, these sensitize skin similar to that by porphyrins. All species of animals are affected though susceptibility varies between species and even between animals within a species.

Mechanism of action: Primary photosensitizers (photodynamic agents) interact directly with cellular constituents particularly pyrimidine bases and nucleic acids and inhibit DNA synthesis and produce cellular damage.

Table 5.1: Agents producing primary photosensitization

Plants :	*Ammi majus* (Bishop's weed)
	Ammi visnaga
	Cooperia pedunculata (Rain lily)
	Cymopterus longipes
	Cymopterus watsoni (Spring parsley)
	Fagopyrum esculentum (Beech wheat)
	F. sagittatum (Buck wheat)
	Hypericum perforatum (St. John's Wort)
	Lolium perenne
	Parthenium hysterophorus
	Polygonum fagopyrum
	Thamnosma texana (Dutchman's breeches)
Dyes:	Acridine
	Rose Bengal
Drug metabolite	Phenothiazine sulfoxide

Clinical signs : Toxic doses are not known. There is a latent period of several days before clinical signs become apparent. Clinical signs appear within a few minutes after exposure of the animal to sunlight. In young birds, acute inflammatory effects are characterised by erythema and blistering on beaks and exposed skin followed by thick crusts and scabs, keratoconjunctivitis, closure of eyes with serous fluid which dries and seals the eyelids. In chronic conditions in chicken, there is deformation of upper beak, feet, legs, sloughing of comb and wattles and scarring of ocular opening. The ocular lesions are more intense in ducks and geese.

In sheep and cattle, acute effects are marked by congestion, hyperemia, oedema, and blistering of the nonpigmented areas of muzzle, ears, vulva and mammary glands, nasal discharge, watering eyes, keratoconjunctivitis and cloudy cornea leading to blindness in some of the cases. Chronic effects in these species are almost the same alongwith permanent damage to tips of ears. The animals seek shade and turn their backs to sun. Neurological signs like ataxia, dizziness and convulsions may be present. In severe cases, death of the affected animals may occur.

Secondary photosensitization or hepatogenous photosensitization : Some substances of plant or some other origin on ingestion result in hepatic damage and obstruction of bile duct. As a result, phylloerythrin, the normal end product of chlorophyll metabolism in the stomach or rumen is not excreted in bile due to hepatitis or biliary obstruction. Hepatotoxic photosensitization is more common in animals grazing on pasture, though it can occur in animals fed on hay or other stored feeds or concentrates. Hepatic damage is of serious consequence and sometimes, the animals may die even without exhibiting any signs of photosensitization if the animals are kept in dark or maintained on chlorophyll free diet.

Not only the hepato-toxic plants, certain mycotoxins, blue green algae and chemicals (Table 5.2) but also hepatic fascioliasis and rift valley fever may result in secondary photosensitization.

Table 5.2: Agents producing secondary photosensitization

Plants	Agave lecheguilla
	Brachiaria brizantha
	Crotolaria retusa
	Holocalyx glaziovii
	Kochia scoparia
	Lantana camara
	Lippia rehmanni
	Lupinus angustifolius
	Myoporum laetum
	Nolina texana
	Narthecium ossifragum
	Panicum miliaceum
	Senecio jacobaea
	Sphenosciadium sp.
	Tribulus canescens
	Tribulus terrestris
	Tetradymia glabrata
Mycotoxin	Pithomyces chartarum
Blue Green Algae	Microcystis flos aquae
Chemicals	Carbontetrachloride
	Corticosteroids
	Phenanthridium
	p-dehydroxybenzene

Mechanism of action : Secondary photosensitization is not primarily because of the photodynamic agent present in the feed/fodder, rather it is the sequelae of extensive liver damage and obstruction of bile duct. Phylloerythrin- the photodynamic agent, a breakdown product of chlorophyll, not excreted out results in its accumulation in the body. Through peripheral circulation, it comes to capillaries of skin and on exposure to bright sunlight photosensitization is produced.

Congenital photosensitization: It is often hereditary in origin due to certain enzymatic abnormalities or inadequate production of enzymes like catalase, glucose-6-phosphate dehydrogenase etc. in RBCs and disruption of haeme biosynthetic pathway. As a result, porphyrin or its derivatives accumulate in the body including skin. These substances in skin when get exposed to bright sunlight get excited, interact with cellular macromolecules or molecular oxygen to generate toxic free radicals.

Congenital porphyria is a metabolic defect due to abnormality in haeme synthesis; ferroprotoporphyrin has been observed in cattle, swine and cats. As a result, protoporphyrin, uroporphyrin II and coproporphyrin III accumulate in higher concentrations in serum and tissues. Accumulation in dentine and bones give rise to red colour and results in pink tooth or osteohaemochondrosis. There is increased excretion of porphyrins in urine (porphyrinuria) which is reddish brown in colour. Kidneys and lungs of affected animals also fluorosce in U V light. Sometimes anaemia due to inadequate biosynthesis of haemoglobin and reduced life span of erythrocytes is also observed.

Pathogenesis : The mode of photosensitization or pathogenesis is the same irrespective of the type as it depends on the level of photodynamic agent in the capillaries under skin. On exposure to light of a particular wavelength, dermatitis lesions occur on unpigmented skin areas particularly when not covered with thick coat of hair or wool. Thus, the lesions are more severe on dorsal parts of the body. Penetration of light rays to sensitized tissues induces release of histamine which is responsible for local cell reactions - oedema and local cell death followed by intense itching and sloughing of tissues and skin. In certain cases, particularly hepatotoxic-photosensitization, nervous symptoms may also appear.

Clinical signs : Clinical signs of photosensitization are similar irrespective of the type of syndrome except in hepatogenous photosensitization where anorexia, depression, diarrhoea, constipation and icterus are the first to be observed due to liver damage. Hyperemia is the first change, followed by oedema and swelling. Serous fluid oozes out to the surface of skin followed by necrosis and exfoliation of the superficial layer of skin. In sheep and goats ears, nose, muzzle, eyes, intermandibular region are affected while in cattle udder and teat are worst affected. Tips of ear are the first to show the lesions and bend downward and get dropped. Mandibular area is so distended and bulged with oedematous fluid that it gives a bottle jaw appearance and head becomes big, thus syndrome is also referred as "*Big head in sheep*". Compared to sheep, conjunctivitis is more pronounced while oedema is less marked in cattle. Cutaneous lesions heal in days to weeks. However, in severe cases, infection and gangrenous changes occur in the ulcerated areas where sloughing off of large areas take place and some of the animals even die due to shock.

There is increased lacrimation, initially watery which later becomes thick, swelling and closure of eyelids and nostrils. In sheep, cornea becomes clouded leading to blindness. Break of wool and high mortality is also characteristic in sheep. The symptoms in equines are similar to that of cattle and sheep except colic, inflammation of the mucous membrane of mouth and brain.

Post mortem lesions :
(i)   Presence of photosensitization-inducing plants in the rumen/stomach.
. (ii)   Liver is enlarged, granular and margins may be thin and gall bladder is distended in hepatogenous photosensitization.
(iii)   Body tissues are yellowish in colour in icterogenic photosensitization.
(iv)   Pink - brown discoloration of teeth and bones in congenital porphyria.

(v) Degenerative changes in liver and kidneys.

(vi) Extensive subcutaneous oedema.

(vii) Enlargement of local lymph nodes.

Diagnosis :

(i) History and evidence of ingestion of some photodynamic agent.

(ii) Clinical signs.

(iii) Icteric mucous membrane.

(iv) Photosensitization lesions on unpigmented or light skinned areas.

(v) Post mortem lesions.

(vi) Aggravation of skin lesions on exposure to bright sunlight and relief in shaded areas.

(vii) Laboratory investigations indicating elevated liver enzymes (*gamma* -glutamyl transferase, sorbitrate dehydrogenase, aspartate aminotransferase).

(viii) Increased serum phylloerythrin concentration.

Differential diagnosis :

(i) Sun burns.

(ii) Mycotic dermatitis.

(iii) Big head of rams due to *Clostridium novyii*.

(iv) Haemoglobinuria.

(v) Myoglobinuria.

Treatment : No specific antidote is available.

(i) Prevent further ingestion of the responsible agent i.e. remove the cause.

(ii) Remove the animal(s) from sunlight. Keep the animals in dark and shaded areas until the toxin has been excreted (5-7 days).

(iii) Topical treatment of the dermatitis lesions with demulcents, antibiotics and corticosteroids ointments.

(iv) Systematically give antihistaminics and antibiotics.

(v) Keep flies and other ectoparasites away from the lesions.

(vi) Administer laxatives or saline purgatives to remove the offending/obnoxious ingesta from the stomach/rumen.

(vii) Administer liver tonics and stimulants.

(viii) In severe cases, dextrose saline if lesions are extensive.

CHAPTER 6

TOXICOLOGY OF AGROCHEMICALS

6.1 GENERAL INTRODUCTION

M.A.Ayub Shah and A.K.Srivastava

Pesticides are one of the most widely used agrochemicals of toxicological importance. With the advent of green revolution, the use of pesticides has increased many folds. Some of the pesticides are also used extensively as acaricides/ ectoparasiticides in veterinary medicine to control insect pests of both the animals and birds. In addition, pesticides are also widely used to control insect vectors of public health importance.

The United States Environmental Protection Agency (USEPA) defines pesticide as "any substance or mixture of substances intended for preventing, destroying, repelling or mitigating any pest". In other words, pesticides are any chemical, physical or biological agents that would kill or destroy unwanted plant or animal pests.

Pesticides consist of a large variety of chemical agents having diverse chemical structures and biological activities which form the largest group of potentially toxic agents which are intentionally introduced into the environment for protection against the pests of plant, animal and human importance. In addition to the availability of better quality and varieties of seeds and fertilizers, use of pesticides has helped in bringing about the green revolution and making the country self sufficient in foodgrains, pulses, vegetables, fruits etc. by checking substantial losses caused by the plant and rodent pests. Also their use in animal husbandry and public health improves the health and productivity of animals and birds and the health of mankind.

People often confuse the term "pesticide" with "insecticide" and use these interchangeably, however, it is erroneous since pesticide is a general term and insecticide is a class of pesticides. For convenience, pesticides may be categorised as summarised in table 6.1.

(A) Insecticides:

Insecticides are a heterogeneous group of chemicals whose desired activity is killing of insects in a very selective and specific manner. However, most of the chemicals that are used as insecticides are not highly selective and result in poisoning in many nontarget species including man and domestic animals. The increased use of insecticides since the end of second world war has introduced a great health hazard to livestock and human beings. It is extremely important to be aware of the potential danger of these chemicals to man and domestic animals due to their indiscriminate and improper use.

Table 6.1: Classification of the pesticides

Class	Chemical nature	Examples
A. Insecticides	a) <u>Organochlorines</u>	
	i) Dichlorodiphenylethanes	DDT, dicofol
	ii) Chlorinated cyclodienes	Aldrin, dieldrin, endosulfan
	iii) Hexachlorocyclohexanes	Lindane
	b) <u>Organophosphates</u>	
	i) Phosphate	
	ii) Phosphonate	
	iii) Phosphorothionate	
	iv) Phosphorothiolate	
	v) Phosphorodithioate	
	vi) Phosphoramidate	
	c) <u>Carbamates</u>	Carbaryl, aldicarb
	d) <u>Pyrethrins and pyrethroides</u>	
	i) Natural pyrethrins	Pyrethrum extract
	ii) Synthetic pyrethroides	Allethrin, permethrin, flumethrin, cypermethrin
B. Herbicides or weedicides		
	i) Dinitro compounds	Dinitrophenol
	ii) Phenoxyacetic acids	2,4-D, 2,4,5-T
	iii) Bipyridinium compounds	Paraquat, diquat
	iv) Triazenes	Atrazine, propazine
	v) Chloroalipathic acids	Dalapan
	vi) Substituted ureas	Diuron, isoproturon
	vii) Substituted dinitroanilines	Pendimethalin
C. Rodenticides	i) Inorganic agents	Zinc phophide, fluoroacetate
	ii) Dicoumarol derivatives	Warfarin
	iii) Glycosides	Red squill

Insect pests are responsible for destruction and damage of one third of the world's crop. In India, the use of chemicals to control insect pests started in 1947 and their use has gradually increased. During the last few decades, there has been tremendous increase in the synthesis, manufacture and application of new chemicals for the control of insect pests. Increased awareness and concern for ecological implications of the use of insecticides is gaining importance to safeguard the health of man, livestock and wild animals. Insecticides are among the most frequently encountered chemicals in daily life. They are commonly used : (1) to increase the production and quality of agricultural products, (ii) to minimize the damage caused by insects during storage of food grains, (iii) to control ectoparasites of domestic animals, (iv) to control certain vector borne diseases, (v) to repel household pests and (vi) as anthelmintics in livestock. Inspite of their persistent and ever increasing use in agriculture, insecticides remain a major public health hazard as well. The residues of most of the commonly used insecticides can be detected in agricultural produce (like foodgrains, pulses, vegetables, fruits etc.), dairy products (like milk, butter, ghee etc.), poultry products (eggs and poultry meat) and meat of other food animals. Further, these compounds give a serious alarm with regard to environmental pollution and accidental poisoning in man, livestock, poultry and wild animals.

There is increased concern over the possible toxicological hazards posed by indiscriminate use of insecticides in agriculture. Concern about the effect of insecticides on health has been raised in a number of WHO reports. The toxicity potential of different insecticides varies greatly in man and domestic animals and the adverse effects on health may be reduced considerably by selecting the least toxic compound(s). Most insecticides affect the nervous system of insects but they also possess some activity against the mammalian nervous system, their neurotoxic effects in man and animals are often prominent.

In agricultural practice, insecticides are usually employed as sprays- applied either from ground or air. The hazard to man and animals occur due to percutaneous absorption or by ingestion of their immediate residues.

On the basis of their chemical nature , insecticides may be further categorised as:

a) Organochlorine or chlorinated hydrocarbon insecticides : e.g. DDT, endrin, heptachlor etc.

b) Organophosphorus insecticides : e.g. parathion, sumithion, malathion etc.

c) Carbamate insecticides : e.g. carbaryl, aldicarb etc.

d) Synthetic pyrethroid insecticides: e.g. allethrin, permethrin, cypermethrin, fenvalerate, fluvalinate, flumethrin etc.

(B) Herbicides or weedicides : Herbicides/weedicides are agents which are used to destroy undesirable plants/weeds e.g. dinitro compounds, phenoxyacetic acid (2,4-D, 2,4,5-T), chloroalipathic acids (dalapon), triazenes (atrazine, simazine), bipyridinium compounds (paraquat, diquat), substituted ureas (monouron, diuron, isoproturon) and substituted dinitroaniline compounds (pendimethalin).

(C) Rodenticides : Rodenticides are agents which are used to destroy the rodent pests e.g. warfarins, zinc phosphide, fluoroacetate etc.

Common sources of pesticide poisoning: The major sources of pesticide-poisoning are:

a) Accidental exposure : Accidental exposure is the most common source of pesticide poisoning and can result from:

i) Ingestion of pesticide sprayed crops, drinking of water from paddy fields treated with pesticide may cause death of livestock. Similarly, animals may get poisoned following ingestion of paddy straw or other fodders sprayed with insecticides. Such fatalities have been most frequently encountered with endrin.

ii) Ingestion of contaminated concentrates particularly when the pesticide sprayed grains unfit for human consumption are used in the formulation of livestock concentrate feeds.

iii) Feeding or watering of animals in insecticide contaminated containers.

iv) Contamination of fodder or concentrates due to improper disposal of empty insecticide containers.

v) Spraying of walls and interior of animal house/stores with insecticides. There is danger of licking the walls and pillars by the animals.

vi) Improper storage e.g. keeping the insecticide in the animal house within the reach of the animals or in stores where animal feeds are also stored.

vii) Improper and indiscriminate use of insecticides.

viii) Contamination of pastures during aerial spraying of insecticides on the field crops.

ix) Contamination of ponds, lakes, rivers etc. due to aerial spray of insecticides or discharge of effluents from the pesticide industries to nearby water reservoirs.

b. Intentional exposure :

i) Residues of pesticides in agricultural produce e.g. grains, pulses, vegetables, fruits etc.

ii) Residues of insecticides in animal products like meat, milk and eggs.

iii) Consumption of insecticide for committing suicide in human beings.

C. Occupational exposure:

i) Workers of pesticide manufacturing industries are constantly exposed to the hazards of pesticides.

ii) Faulty spraying of pesticides by unskilled workers or inefficient equipments.

6.2 ORGANOCHLORINE INSECTICIDES

M.A. Ayub Shah

The organochlorines are the first generation insecticides. They are also known as chlorinated hydrocarbons. These compounds are mainly used as contact insecticides and ectoparasiticides.

The chlorinated hydrocarbons are divided into four groups:

i) Dichlorodiphenylethanes e.g. DDT, dicofol, perthane, methoxychlor etc.
ii) Chlorinated cyclodienes e.g. aldrin, dieldrin, endrin, chlordane, heptachlor, endosulfan etc.
iii) Hexachlorocyclohexanes e.g. lindane.
iv) Miscellaneous group e.g. kepone-solely chlordecone and mirex.

The toxic and lethal doses of some of the commonly employed chlorinated hydrocarbon insecticides in different species of animals and birds are summarized in table 6.2.

Table 6.2: LD$_{50}$ values and toxic and lethal doses (mg/kg) of some of the chlorinated hydrocarbons in different species of animals

Compound	Species	LD$_{50}$	Toxic Dose	Lethal Dose
DDT	Rats	87-300	—	—
	Rabbits	250	—	—
	Dogs	500-750	300	—
	Cats	400	250	—
	Sheep & goats	>1000	250	—
	Ducks	2240	—	—
	Cattle (adult)	—	—	500
	Calves	—	—	250
Dicofol	Rats	575-1000	—	—
	Rabbits	1870	—	—
	Guinea pigs	1810	—	—
	Dogs	>4000	—	—
Aldrin	Rats	39	—	—
	Rabbits	50	—	—
	Dogs	5	—	—
	Cats	10-15	—	—
	Ducks	520	—	—
	Cattle (adult)	—	25	40-50

(Continued)

	Calves	—	25	—
	Sheep	—	15	40-50
	Goats	—	15	—
Dieldrin	Rats	40		
	Rabbits	45		
	Dogs	65		
	Cats	300-500		
	Chickens	20		
	Goats	100		
	Cattle (adult)	—	25	60
	Calves	—	10	
	Sheep	—	25	50-75
Endosulfan	Rats	50-110	—	
	Cats	2	—	
	Ducks	200-300	—	
	Dogs	—	200	
Endrin	Rats	3	—	
	Rabbits	7	—	
	Cats	3-6	—	
	Chickens	2-4	—	
	Ducks	6	—	
	Goats	2.5	—	
Chlordane	Rats	283	—	—
	Rabbits	300	----	----
	Dogs	200-700	—	—
	Cattle (adult)	—	90	—
	Calves	—	25	
	Sheep	—	50	
Chlordecone	Rats	114-140	—	—
	Rabbits	71	—	—
	Dogs	250	—	—
Lindane	Rats	76-225	—	—
	Rabbits	40-75	—	—
	Cats	35	—	
	Dogs	30-200	—	—
	Hens	70	—	
	Ducks	>2000	—	—
	Calves	—	5	5-15
	Cattle (adult)	—	20	25
	Sheep and goats	—	25-50	100
	Horses	—	—	30

i) Dichlorodiphenylethanes: Major compounds under this chemical group are dicofol, methoxychlor and dichlorodiphenyltrichloroethane (DDT; Fig. 6.1). DDT was synthesized in 1874; however, its insecticidal property was discovered only in 1939 by Paul Muller, a Swiss Chemist who was awarded Nobel Prize in 1948 for his contribution. However, the euphoria of its effectiveness in controlling pests was shortlived due to its persistence in the environment for a prolonged time. The residues of DDT can be detected in soil long after the discontinuance of its use . Further, DDT is also stored in the body fats of animals which is eliminated at extremely slow rate and has biomagnification potential. One of the most adverse effects of DDT due to biomagnification is the decline in the population of certain birds due to thinning of their egg shells. Because of its persistence in the biosphere, the use of DDT has been banned by almost all the countries including India. However, dicofol and methoxychlor are still in use.

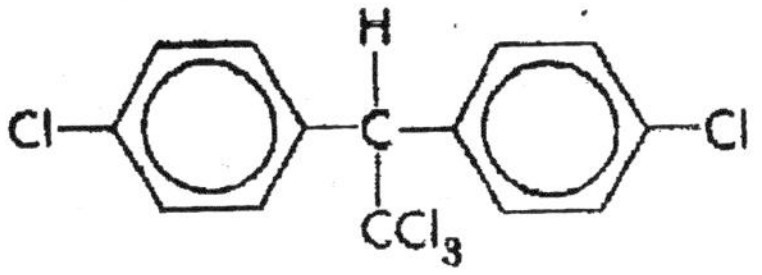

Fig. 6.1 : Chemical structure of DDT

Amongst the dichlorodiphenylethanes, DDT is the most widely studied compound. All these agents share common mechanism of toxicity; however, dicofol and methoxychlor are comparatively less toxic and less persistent in the biosphere and are rapidly eliminated from the body.

DDT has been reported to induce hepatic tumour in mice and also has some oestrogenic activity which can promote the growth of oestrogen-dependent tumors in laboratory animals.

ii) Cyclodienes: The chlorinated cyclodienes e.g. aldrin, dieldrin, endrin, heptachlor, chlordane, endosulfan etc. (Fig. 6.2) are widely used and comparatively more toxic than the dichlorodiphenylethane insecticides. These compounds are well absorbed from the skin and are reported to cause toxicity through incidental contacts. Some of the cyclodiene

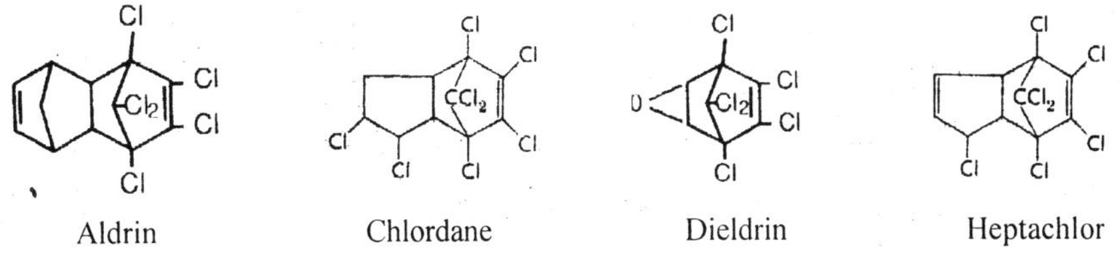

Aldrin   Chlordane   Dieldrin   Heptachlor

Fig. 6.2 : Chemical structures of cyclodiene compounds

compounds are carcinogenic which generally cause hepatic tumours in mice and non-cancerous lesions characterized by necrosis or hepatocellular swelling in rats. Heptachlor, chlordane and aldrin are carcinogenic to mice. However, endrin an *isomer* of dieldrin is not carcinogenic in mice. Like DDT and other dichlorodiphenylethanes, chlorinated cyclodienes are also stored in the adipose tissue and are slowly eliminated from the body and residues of these agents have been detected in tissues.

iii) Hexachlorocyclohexanes: Hexachlorocyclohexanes (HCH) or benzene hexachloride (BHC) is a mixture of eight *isomers* (Fig.6.3). Out of the eight isomers, the *gamma*-isomer-lindane is the most toxic and the insecticidal activities of HCH reside with this isomer. HCH is not well absorbed from the skin compared to cyclodienes; however, lindane has been associated with poisoning following dermal exposure in man. Risk of poisoning following dermal exposure appears to be highest if there are abrasions of the skin. Lindane induces hepatic tumor in mice following chronic administration.

BHC   Lindane

Fig. 6.3 : Chemical structures of hexachlorocyclohexanes

iv) Miscellaneous organochlorines: Mirex and kepone (chlordecone) are structural analogs (Fig.6.4). As both these agents cause identical effects, they may be having a common mechanism of action.

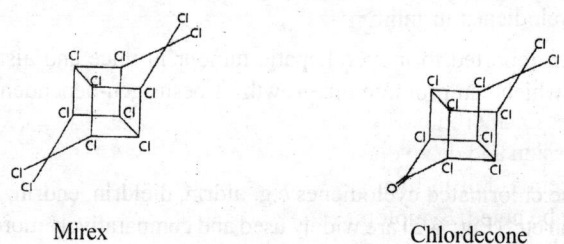

Mirex   Chlordecone

Fig. 6.4 : Chemical structures of miscellaneous organochlorines

Sources of poisoning:
i)   Ingestion of organochlorine contaminated feeds and water by the animals.
ii)   Inhalation or absorption from the skin during topical application as ectoparasiticides.

Absorption and fate of chlorinated hydrocarbons: Organochlorine insecticides are water insoluble but soluble in oil and organic solvents. These are rapidly absorbed from the oily preparations and are capable of penetrating the intact skin when applied in oily solution or emulsion. Dieldrin is an exception which may be absorbed even from dry powder form. However, all the compounds, in powder form, can easily penetrate the cuticle of insects compared to mammalian skin and intestinal mucosa which explains its greater toxicity to insects than in mammals.

Biotransformation of DDT in animals and plants involves dehydrochlorination catalysed by DDT dehydrochlorinase which converts DDT into non-polar and persistent metabolite DDE and the resultant conjugates are the major metabolites which are excreted in urine. The major metabolic pathway for methoxychlor is o-demethylation and subsequent conjugation and excretion. In mammals, methoxychlor is rapidly degraded by liver and the non-toxic metabolite(s) come to the intestines and get excreted through faeces. Its low toxicity and tissue accumulation is due to rapid detoxification and slow gastrointestinal absorption. Aldrin is metabolised by microsomal enzymes to dieldrin which are more toxic than the parent compound. Except methoxychlor, other organochlorine insecticides are stored in the body fat. However, none of these agents are known to accumulate in other vital organs.

Mechanism of toxicity: The chlorinated hydrocarbons are neuro-poisons. By virtue of their high lipid solubility, these agents can enter the neural membrane with ease and interfere with normal functioning of the nerve membrane sodium channels. DDT acts by (1) reducing the potassium transport through pores; (2) inactivating sodium channel closure; (3) inhibiting Na^+-K^+ and Ca^{2+}-Mg^{2+} ATPases and (4) inhibiting calmodulin-Ca^{2+} binding with release of neurotransmitter. The cyclodiene compounds act on the chloride ion (Cl^-) transport by antagonizing the *gamma* amino butyric acid (GABA) receptors in the Cl^- channels and also inhibit the Ca^{2+}-Mg^{2+} ATPase. However, the relative importance of these two mechanisms (GABA blockade and inhibition of Ca^{2+}-Mg^{2+}ATPase) in chlorinated cyclodiene neurotoxicity is yet to be understood. Lindane, an isomer of HCH binds to the GABA receptors thereby producing an inhibition of GABA -dependent chloride influx into the neuron. However, lindane is less potent than the cyclodienes in inhibiting the GABA-dependent Cl^- influx and is less toxic.

Clinical symptoms: Dichlorodiphenylethanes poisoning cause initial stimulation of CNS followed by depression and death due to respiratory failure. In chronic poisoning, liver damage, hypoglycemia, fall in liver glycogen concentration, blood lactacidemia and hyperkalemia may be noted. Symptoms of cyclodiene compound poisoning are similar to DDT poisoning but more severe in nature and are characterized by grinding of teeth, difficult respiration, snapping of the eyelids and frequent urination. Other signs include walking backwards, wall climbing, aimless jumping and violent frenzied behaviour.

Salient clinical features observed in chlorinated hydrocarbon poisoning may be categorized as:

1. Behavioural changes:
· initial anxiety,
· aggressiveness,
· abnormal posturing,
· jumping over unseen objects,
· wall climbing and
· madness syndrome.

2. Neurological symptoms:

- · hypersensitivity to external stimuli,
- · fasciculation and twitching of the facial and eyelid muscles,
- · spasm and twitching of the fore- and hind quarter muscles,
- · champing of the jaw and
- · hyperthermia.

If death does not take place at this stage, the animals may go into coma state.

3. Cholinergic manifestations:

- · vomiting,
- · marked salivation,
- · mydriasis,
- · diarrhoea and
- · micturation may also be observed .

Post-mortem lesions : There are no specific lesions in the nervous system. However, acute aldrin poisoning may cause hepatitis and acute tubular nephrosis. Chronic DDT and methoxychlor toxicoses may produce focal centrilobular necrosis of the liver.

Diagnosis : Diagnosis of organochlorine insecticides poisoning may be made based on:

i)   History of exposure to the insecticide.
ii)   Clinical symptoms and post-mortem lesions.
iii)   Analysis of feeds and/or biological samples like liver and kidneys in dead animals and blood and milk samples in living animals for the presence of organochlorine compounds.

Differential diagnosis : Organochlorines poisoning should be differentiated from the following poisonings:

i)   Salt poiscning: history and absence of hyperthermia.
ii)   Strychnine poisoning: convulsions are tonic and absence of behavioural abberations and locomotor disturbances.
iii)   Fluoroacetate poisoning: convulsions not elicited by external stimuli.
iv)   Nicotine poisoning: only cholinergic signs are exhibited.
v)   Anticholinesterase insecticide poisoning: only parasympathetic signs, no behavioural changes or hyperthermia.
vi)   Lead poisoning: no abnormal posturing.

Treatment : No specific antidote is available. Treatment is only symptomatic and supportive.
i)   Remove the source of poisoning at once.
ii)   Administration of non-oily purgatives.

iii)   Control convulsions by administering barbiturates (pentobarbital sodium) or benzodiazepines in dogs and cats; and chloralhydrate or pentobarbital sodium in ruminants. CNS depressants/anaesthetics are contraindicated if the animal is already depressed.

iv)   A small dose of atropine sulfate may be given to control the parasympathetic signs.

v)   Intravenous administration of calcium borogluconate is recommended to prevent liver damage and nullify the effect of preconvulsive increase in K^+-ion concentrations.

vi)   Activated charcoal (2 kg stat followed by 1 kg daily for 2 weeks) may adsorb the insecticide that is excreted into the intestine through bile and suppresses the recycling or enterohepatic circulation of insecticide and promotes its excretion through faeces.

vii)   Phenobarbital (10 mg/kg /day) may be tried to induce hepatic microsomal enzymes and to promote faster metabolism and excretion.

6.3 ORGANOPHOSPHOROUS INSECTICIDES

A.K. Srivastava and V. K. Dumka

Of the various groups of insecticides used in agriculture, veterinary and public health practices, organophosphorous insecticides (OPIs) constitute the bulk. These compounds are preferred due to their high selectivity, low toxicity in mammals and their rapid degradation in animal body and ecosystem. With indiscriminate use of these compounds, several cases of poisoning frequently occur in man, domestic animals and wild life. Although, OPIs have comparatively low toxicity in mammals than in insects, much higher acute toxicity has been observed in animals from careless use or following accidental ingestion of these chemicals. Besides their acute toxic effects, OPIs are also implicated in various disorders and diseases like cancer, reproductive disorders, peripheral neuropathy, impaired immune functions and neurobehavioural changes.

In quest to find new organophosphorous insecticides of low toxicity and high insecticidal activity, numerous compounds have been synthesized and introduced into day to day use.

The first organophosphate insecticide to be introduced was tetraethylpyrophosphate (TEPP), which was developed in Germany during second world war as a substitute for nicotine. This was followed by extremely toxic nerve gas chemical warfare agents like ethyl N-dimethyl phosphoroamidocyanidate (tabun) and isopropyl methylphosphonofluoridate (sevin). These compounds were highly toxic to mammals and were rapidly hydrolysed by moisture. This led to the discovery by Schrader in 1944 of more stable compound parathion (o, o-diethyl o - p-nitrophenyl phosphorothioate) and its oxygen analogue paraoxone (o, o-diethyl o-p-nitrophenyl phosphate). Because parathion exhibited a wide range of insecticidal activity, low volatility and sufficient stability in water and mild alkali, it became one of the most widely used organophosphorous insecticide. Diisopropyl phosphorofluoridate (DFP) is the most extensively studied organophosphate insecticide because of three reasons: (a) early development and laboratory evaluation during world war II; (b) high lipid solubility resulting in its extensive penetration through BBB; and (c) complete and irreversible inhibition of acetylcholinesterase (AChE) enzyme by alkyl phosphorylation.

The high degree of toxicity of the early preparations led to intensive research in order to find compounds that were less toxic to mammals. More than 30,000 different organophosphate compounds have been tested for insecticidal activity so far. Products like diazinon, dimethoate and malathion possess toxicity equivalent to that of organochlorine compounds in mammals. Some less toxic ones like coumaphos, crufomate, fenchlorvos and trichlorphon have been used as insecticides. Other compounds like dichlorvos, haloxon and naphthalphos are being used as anthelmintics.

Nomenclature and chemistry: The anglo-american system of nomenclature is most widely accepted for naming of compounds containing one atom of phosphorous and these constitute a majority of commercial OPIs. The term "organophosphate" is a generic term to cover all the toxic organic compounds containing phosphorous. The organophosphates are esters, amides or other simple derivatives of phosphoric and thiophosphoric acids. Based on the different atoms/chemical groups attached to the phosphorous, organophosphate compounds are grouped into different classes (Table 6.3).

Table 6.3: Classification of OPIs based on the chemical nature.

Class	Chemical group
Phosphate	
Phosphonate	
Phosphorothionate	
Phosphorothiolate	
Phosphorodithioate	
Phosphoramidate	

Most of the organophosphates are esters of alcohols with a phosphoric acid or anhydrides of phosphoric acid with some other acid. e.g. parathion is an ester of the acid $(OH)_3P=S$ with two molecules of ethanol and one of p-nitrophenol. Structurally it is called diethyl p-nitrophenyl phosphorothionate.

The general structure of most of the organophosphates is :

$(RO)_2P(A)X$; where R= methyl or ethyl, A= sulfur or oxygen and X=different variables.

Some of the commonly used OPIs are summarised in Table 6.4 and their structures are shown in Fig. 6.5.

Table 6.4 : Some of the commonly used OPIs.

Chlorfenvinphos	Methylparathion
Chlorpyriphos	Parathion
Chlorothion	Phorate
Coumaphos	Sarin
DFP	Soman
Diazinon	Sumithion
Dichlorvos	Tabun
Dimethoate	TEPP
Echothiophate	TOCP
Malathion	Trichlorfon

Further, OPIs are categorised into two broad groups on the basis of their activity :

a) Direct acting OPIs : These compounds act by directly inhibiting the cholinesterase enzyme e.g. dichlorvos, fenchlorvos, diisopropylfluorophosphate (DFP), tabun, serin, soman, tetraethylpyrophosphate (TEPP), mervinphos, disulfoton, azinphosmethyl, chlorfenvinphos, diazinon, dimethoate, ronnel, abate, trichlorfon, chlorothion, triazophos, chlorpyriphos and anilophos.

b) Indirectly acting OPIs : These insecticides as such are inactive but are biotransformed in the body to toxic metabolites which inhibit cholinesterase enzymes. Some of the common examples of such OPIs and their toxic metabolites are listed in table 6.5.

Table 6.5 : Indirectly acting OPIs and their toxic metabolites

Compound	Active metabolite
Malathion	Malaoxon
Parathion	Paraoxon
Fenthion	Fenoxon
Fentrothion	Fentroxon

Organophosphate insecticides are also classified according to the manner in which they exert their insecticidal action :

a) Contact poisons : Some organophosphate insecticides exert their insecticidal action like the chlorinated hydrocarbon insecticides and are called contact poisons e.g. parathion, malathion, paraoxon, methylparathion etc.

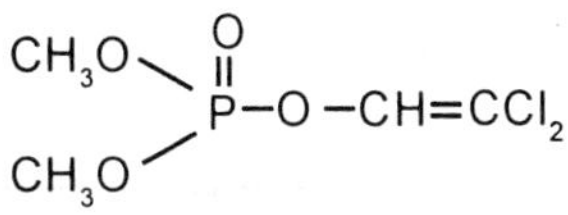

$$CH_3O \diagdown \overset{\overset{O}{\|}}{P} - O - CH = CCl_2$$
$$CH_3O \diagup$$

Dichlorvos

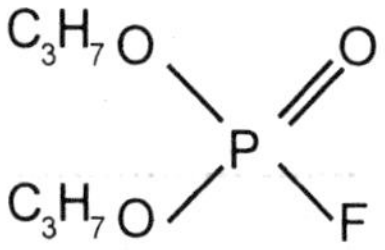

$$C_3H_7O \diagdown \overset{\overset{O}{\|}}{P} \diagdown F$$
$$C_3H_7O \diagup$$

DFP

$$C_2H_5O \diagdown \overset{\overset{O}{\|}}{P} \diagdown SCH_2CH_2\overset{+}{N}(CH_3)_3$$
$$C_2H_5O \diagup$$

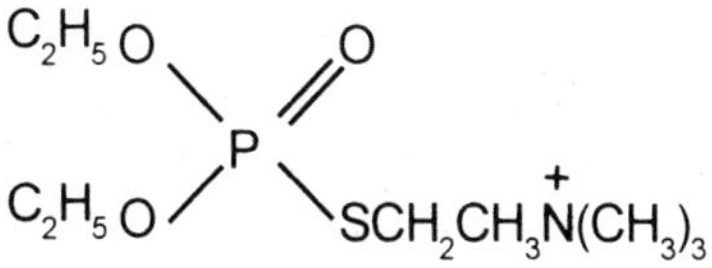

Diethoxyphosphinylthiocholine

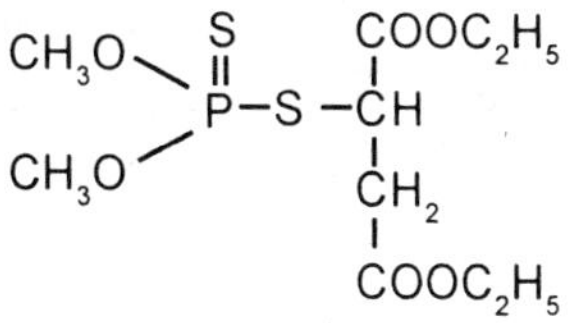

$$CH_3O \diagdown \overset{\overset{S}{\|}}{P} - S - \overset{\overset{COOC_2H_5}{|}}{\underset{\underset{COOC_2H_5}{|}}{\underset{CH_2}{|}}}CH$$
$$CH_3O \diagup$$

Malathion

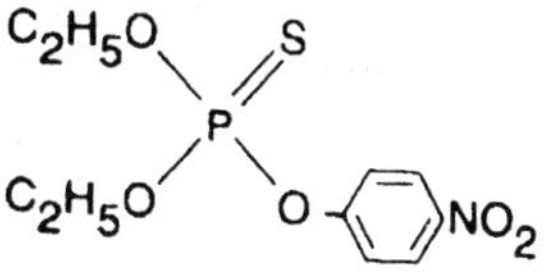

$$C_2H_5O \diagdown \overset{\overset{S}{\|}}{P} \diagdown O - \underset{}{\bigcirc} NO_2$$
$$C_2H_5O \diagup$$

Parathion

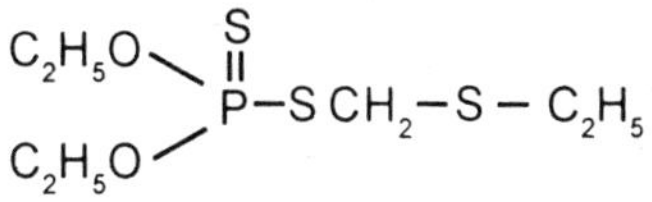

$$C_2H_5O \diagdown \overset{\overset{S}{\|}}{P} - S\,CH_2 - S - C_2H_5$$
$$C_2H_5O \diagup$$

Phorate

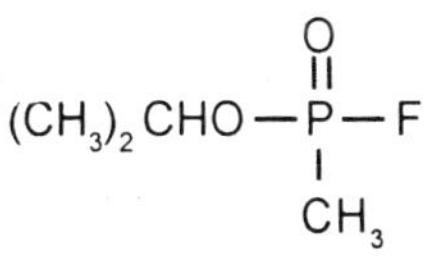

$$(CH_3)_2CHO - \overset{\overset{O}{\|}}{\underset{\underset{CH_3}{|}}{P}} - F$$

Sarin

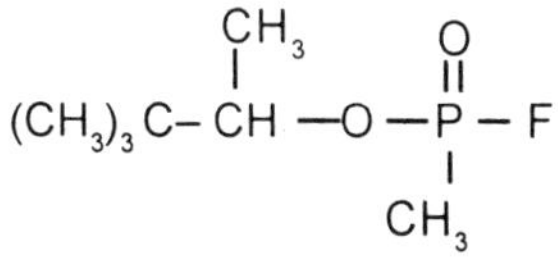

$$(CH_3)_3C - \overset{\overset{CH_3}{|}}{CH} - O - \overset{\overset{O}{\|}}{\underset{\underset{CH_3}{|}}{P}} - F$$

Soman

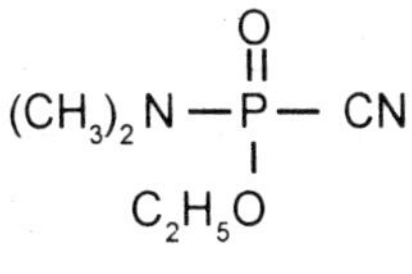

$$(CH_3)_2N - \overset{\overset{O}{\|}}{\underset{\underset{C_2H_5O}{|}}{P}} - CN$$

Tabun

$$C_2H_5O \diagdown \overset{\overset{O}{\|}}{P} \diagup O \diagdown \overset{\overset{O}{\|}}{P} \diagup OC_2H_5$$
$$C_2H_5O \diagup \qquad \diagdown OC_2H_5$$

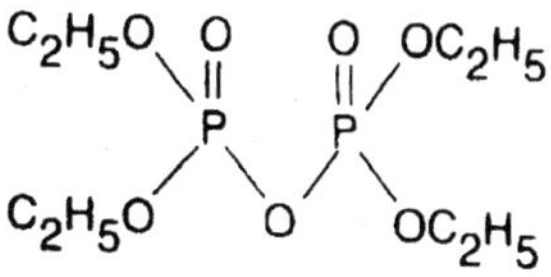

TEPP

Fig. 6.5 : Structures of some of the common OPIs.

b) Selective systemic insecticides : Such compounds are absorbed into the sap, of plants, remaining active and soluble for a reasonable period and are toxic to the plant pests but not to their predators. e.g. dimethoate, mipafox, schradan, dementon etc.

Mechanism of toxicity : The OPIs owe their toxicity to their ability to inhibit acetylcholinesterase, the enzyme which is responsible for hydrolysis of acetylcholine to acetic acid and choline, released from nerve endings.

Two types of cholineterase enzymes are present in the body. True cholinesterase, acetylcholinesterase (AChE) or specific cholinesterase is highly specific in its action and is found mainly in erythrocytes, muscles and nervous system. Pseudocholinesterase, false cholinesterase or butyrocholinesterase (BuChE) is a non specific enzyme, which is capable of hydrolysing a wide variety of esters and occurs in plasma and nervous system. Some of the organophosphorous compounds inhibit one of these enzymes more than the other, but few are entirely specific. The main reason for differentiating between AChE and BuChE is because all the toxicological effects of anticholinesterase agents are due to inhibition of AChE with the consequent accumulation of endogenous acetylcholine (ACh). Inhibition of BuChE at most of the sites produces no apparent toxic symptoms.

In general, the peripheral cholinergic fibers contain higher concentration of AChE enzyme than the noncholinergic fibers. AChE enzyme is distributed throughout the length of the neuron. The enzyme is located in the outer basal lamina (basement membrane), rather than the post synaptic membrane. At the motor end plate of skeletal muscles, most of the AChE is localized at the surface and infoldings of the post-junctional membrane or subneural apparatus. The basic structure of AChE is tetrameric arrangement with a molecular weight of 80,000 each. Usually three tetrameric units are linked through disulfide bond to a filament. Each unit of AChE enzyme has two active sites - anionic and esteratic.

The normal levels of AChE established in different species of animals are given in table 6.6 for the benefit of the readers.

Table 6.6 : Levels of acetylcholinesterase enzyme (n mol. acetylcholine hydrolysed/min/ml) in different species of animals.

Species	Acetylcholinesterase level	
	RBCs	Plasma
Buffalo	2154±51	178±22
Cattle	2096±164	127±10
Sheep	1926±102	146±5
Goat	2165±82	303±13
Dog	1030±95	84±5
Rabbit	1690±125	60±4
Chicken	463±119	119±7

Hydrosis of ACh by AChE: The active centre of AChE consists of a negative sub-site (anionic site) which attracts the quarternary group of choline. At the positive sub-site (esteratic site), the nucleophilic attack occurs on the acyl carbon of ACh. During enzymatic attack on the ester, a tetrahedral intermediate is formed which releases choline and forms acetylated enzyme. The acetylated enzyme undergoes hydrolysis with the release of acetate and active enzyme. AChE has the capacity to hydrolyse 3×10^5 ACh molecules per molelcule of enzyme per minute. The hydrolysis of ACh by AChE enzyme is shown in the Fig. 6.6.

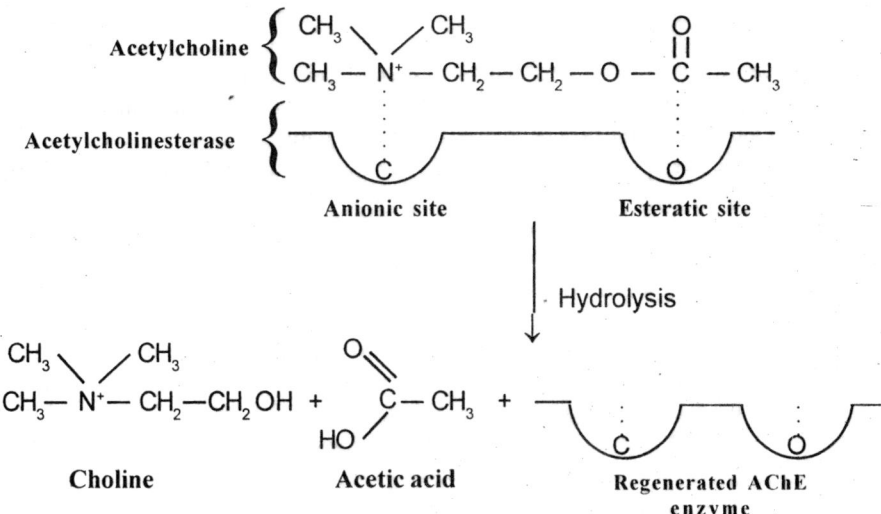

Fig. 6.6 : Hydrolysis of acetylcholine by AChE

Inactivation of cholinesterase by OPIs involves a reaction in which one substitute group is lost producing a dialkyl phosphoryl enzyme. The primary parasympathomimetic effects of organophosphorous insecticides in both pests and mammals are attributed to phosphorylation of serine residue of the active site of acetylcholinesterase with the formation of covalent bond between the enzyme and inhibitor. This inhibition leads to accumulation of acetylcholine at the nerve endings thus producing symptoms of parasympathetic stimulation. OPIs bind to the esteratic site of the enzyme as shown in Fig. 6.7.

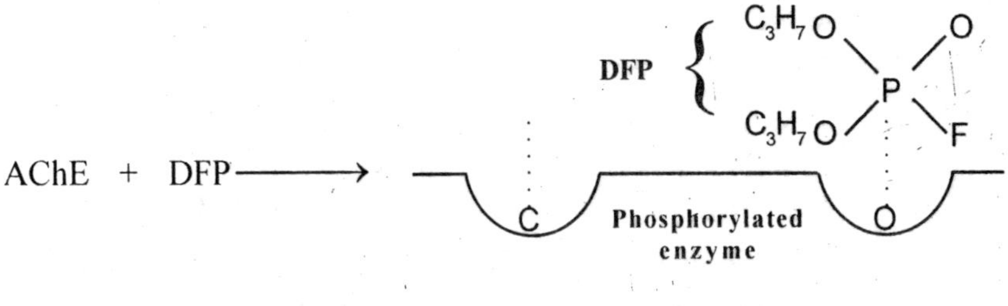

Fig. 6.7 : Inhibition of AChE by DFP.

The phosphorylated AChE is highly stable and inactive and is unable to hydrolyse acetylcholine. Poisoning from OPIs, therefore, results due to persistence of excess of acetylcholine at the nerve endings where it normally exists momentarily during impulse transmission. Besides their anticholinesterase activity, some OPIs also bring about other pathological changes. For example, DFP, mipafox and triorthocresyl phosphate (TOCP) produce demyelination of the peripheral nerves and white matter of spinal cord.

Absorption, fate and excretion: The commonly encountered OPIs barring a few exceptions like ecothiophate are highly lipid soluble and are rapidly absorbed by practically all routes including gastrointestinal tract, skin, mucous membranes and lungs. Dermal absorption is highly influenced by the solvent used. The extent of absorption is more if these are applied in organic solvents. The metabolism and excretion of OPIs is a very complex process and it is dependent on chemical nature of the compound, route of entry and the species of animals involved. The important metabolic pathways of OPIs are discussed below. On reaching the tissues, the directly acting OPIs are rapidly hydrolysed enzymatically while exerting a direct inhibitory action on cholinesterase. The indirectly acting OPIs inhibit cholinesterase in a similar manner after conversion into their active metabolites. OPIs are hydrolysed in the body by a group of enzymes called phosphoryl phosphatases. These enzymes are widely distributed in plasma and various tissues and are not inhibited by organophosphorous compounds. Malathion and other organophosphorous compounds containing carboxylesters undergo hydrolysis at these ester linkages by a reaction catalysed by plasma esterases. These esterases can be inhibited by organophosphates. Thus, the toxicity from exposure to two OPIs may be supra-additive.

Organophosphates undergo mainly two types of metabolism, either activative or degradative. Activative metabolism converts a poor anticholinesterase inhibitor to a stronger one whereas degradative metabolism is just the reverse.

Activation : It involves P=S to P=O conversion called desulfuration, hydroxylation (found only in case of phosphoramides), thioether oxidation, cyclization and microsomal activation.

Degradation : It primarily occurs by hydrolytic routes, which leave an anionic group attached to the phosphorous, thereby reducing its positivity. It involves a number of enzymes such as phosphatases, carboxylesterases and amidases. The most common hydrolysis is by phosphatases which convert insecticides to their degradation products as shown below:

Sumithion	$\longrightarrow$	Ethyl sumithion
Malathion	$\longrightarrow$	Demethyl malathion
Dichlorvos	$\longrightarrow$	Dimethyl phosphate
Mevinphos	$\Longrightarrow$	Dimethyl phosphate

Organophosphates containing carboxamide group are cleaved by amidases e.g.

Dimethoate	$\longrightarrow$	Dimethoate acid
Imidan	$\longrightarrow$	Phthalamic acid

In compounds containing carboxy ester group, cleavage occurs by carboxylesterase, e.g.

Malathion $\longrightarrow$ Malathion acid

Certain other metabolic processes may also play a role in degradation of OPIs :

Reductive degradation : These reactions result in severe lessening of the positivity of phosphorous. The commonest mechanism is hydrolysis which introduces an anion. The reduction of a nitro group to an amino group has the same effect and it has been shown to occur in rumen.

N-Dealkylation and N-hydroxylation : The N-derivatives of amino or amidic organophosphates differ considerably in their anticholinesterase activity. The removal or modification of N-substituents leads to either activation or degradation or neither of the two. N-dealkylation is catalysed by hepatic microsomes and is preceded by hydroxylation of the alkyl group.

Nonenzymatic reactions : Organophosphates undergo certain non-enzymatic reactions which alter the biological properties of these compounds in important ways. The most crucial among these reactions are hydrolysis and isomerization.

Hydrolysis : All the organophosphates can be hydrolysed at a rate directly related to the alkalinity and this reaction determines the half life of an OPI e.g. the half lives of some organophosphates at pH 8 are as follows : TEPP 73 h, DFP 226 h, thiodementon 14,200 h, paraoxon 22,200 h, methyl parathion 50,200 h, parathion 203,000 h and thionodementon 1,250,000 h.

Certain amino acids, hydroxamic acids, chlorine, inorganic phosphates, copper and molybdenum ions promote hydrolysis of organophosphates.

Isomerization : Properties of the *isomers* of certain organophosphates differ in important ways from those of the parent compound, e.g. S-ethyl parathion, an isomer of parathion and S-alkyl malathion, an isomer of malathion. These isomers are responsible for the direct anticholinesterase activity, increased susceptibility to hydrolysis and increased potency against cholinesterase.

Excretion: Many OPIs are excreted in milk, urine and bile. If insecticides are used judiciously, their traces are not found in urine 1-2 weeks after termination of treatment. Some insecticides and their metabolites can cross the placenta and inhibit the foetal AChE enzyme, producing toxicity to the foetus.

Toxicity: Organophosphorous compounds vary greatly in toxicity. TEPP is ten thousand fold more toxic than temephos. There is marked sex difference in susceptibility to OPIs. Female rats are considerably more susceptible to dichlorvos, dioxathion, EPN and parathion whereas, male rats are more susceptible to temephos, dimethoate, fenthion and schradan.

All the organophosphorous insecticides produce a cumulative effect by progressive inhibition of cholinesterase. With regard to cumulative effect, schradan is most toxic while malathion is least toxic. A number of organophosphorous compounds have been found to be active in potentiating the toxicity of malathion. e.g. EPN prevents the hydrolysis of malathion, thus, increasing its toxicity. The toxicity of OPIs is also increased by other xenobiotics such as urea, copper sulphate and some of the plant poisons. There is a wide variation in the degree of toxicity to OPIs. The oral acute LD_{50} in rats ranges from 1.0 mg/kg for the extremely toxic OPI like TEPP to 8.6 gram/kg for safer insecticides like temephos. The oral acute LD_{50} values of some of the important OPIs in different species of animals are given in Table 6.7.

Table 6.7 : LD_{50} values of some of the organophosphate insecticides in different species of animals.

Insecticide	Species	Oral LD_{50} (mg/kg)
Azinphos-methyl	Rat	10-20
Carbophenothion	Rat	30
Coumaphos	Rat	15
Malathion	Rat	1,375
Ronnel	Rat	1,250
Abate	Rat	8,000
Phosphomidon	Rat	11-27
Fenitrothion	Rat	365
	Sheep	770
	Cow	300
	Buffalo	217
	Pig	300-1,000
	Cat	142
Ethion	Rat	13-96
	Guinea pig	45
Phorate	Rat	2-4
	Calves	0.1
Dimethoate	Rat	300-350
	Mouse	60
	Guinea pig	600
	Rabbit	400
	Hen	35
Dichlorvos	Rat	56-80
	Mouse	124
	Dog	1,090
	Chicken	15

Clinical signs : Organophosphates kill the animals, both vertebrates and invertebrates, by inhibiting the acetylcholinesterase enzyme with consequent disruption of nervous activity. Acetylcholinesterase enzyme declines profoundly, with death usually occurring when the brain AChE is inhibited to the extent of 95%. Other cholinesterase may or may not reflect the brain ChE inhibition. Erythrocytes cholinesterase is a good index but is inaccurate for compounds such as ionic or amidic organophosphates that penetrate the brain poorly.

The inhibition of AChE results in accumulation of endogenous acetylcholine in the nerve tissues and effector organs with consequent signs and symptoms. Acetylcholine is the chemical neurotransmitter at post-ganglionic parasympathetic nerve endings, preganglionic nerve endings of both parasympathetic and sympathetic nerves, somatic motor nerves to skeletal muscles and certain synapses in the central nervous system. Acetylcholine in the blood and brain rises to the extent of two to three fold in organophosphate poisoning in cats, dogs and rabbits. The accumulation of ACh mimics the muscarinic, nicotinic and central nervous system actions of acetylcholine and also produces the delayed neurotoxic effects. Various types of symptoms in animals are as follows :

(i) **Muscarinic symptoms :** Muscarinic receptors for acetylcholine are found primarily in smooth muscles, heart and exocrine glands. Symptoms of OPI toxicity resulting from the stimulation of these receptors include bronchoconstriction and increased bronchial secretions, hypersalivation with drooling of watery saliva, lacrimation, excessive sweating, increased gastrointestinal tone, frequent and involuntary urination and defecation, diarrhoea, nausea, vomition, abdominal cramps, difficult breathing, hypotension, bradycardia, pupillary constriction (miosis) etc.

(ii) **Nicotinic symptoms :** Nicotinic symptoms result from accumulation of acetylcholine at skeletal motor nerve endings and autonomic ganglia. Muscular effects include muscular weakness, twitching, fasciculations, cramps, tremors and atrophy. Weakness of the muscles of respiration results into dyspnoea and cyanosis.

(iii) **CNS symptoms:** These include restlessness, convulsions, ataxia, depression of respiratory and circulatory centres leading to cardiac and respiratory arrest, tremors and coma. Death in OPI poisoning results from asphyxia due to respiratory failure.

(iv) **Delayed neuropathy :** The delayed neuropathy is produced by several phosphate triesters and most commonly by triorthocresyl phosphate (TOCP). The functional disturbance begins in the distal part of lower limbs. Motor weakness with ataxia occurs which progresses to increased weakness and flacidity of the limbs. Although, the delayed neurotoxic effects of OPIs can result from single toxic dose, the onset of neuropathy is generally delayed. The inhibition of neurotoxic esterases (NTE) has been found responsible for the organophosphorous compound-induced delayed neurotoxicity (OPIDN).

(v) **Localized effects :** Localized effects at the site of exposure may be observed in the absence of obvious symptoms of systemic absorption. Exposure to vapours, dusts or aerosols of OPIs produce local effects on the smooth muscles of eyes and respiratory tract resulting in miosis, blurred vision, spasm of accomodation, bronchoconstriction, watery nasal discharge and hyperemia. Local effects of dermal exposure include sweating and fasciculations at the site of contact. Gastrointestinal manifestations appear after oral ingestion.

(vi) Systematic effects : Irrespective of the route of administration, systemic effects in general are similar but sequence of appearance of the toxicity symptoms may differ as summarized below :

Route of administration	First appearance of symptoms
Spray and dust	Respiratory and occular
Oral	Gastrointestinal
Dermal	Localized sweating

Acute toxicity due to organophosphate insecticides is more common than chronic toxicity. In acute toxicity, the onset of symptoms after exposure to OPIs is usually rapid, within 10 minutes to two or three hours. The duration of symptoms is generally from 1-5 days. In fatal untreated poisoning, death usually occurs within 24 hours. The onset and duration of symptoms differ markedly for different copounds by virtue of their differences in metabolism, biodistribution and affinity to AChE enzyme.

Post-mortem lesions: There are no characteristic lesions of organophosphorous compounds poisoning, however, some of the common lesions are:
i)   Haemorrhagic gastroenteritis.
ii)   Pulmonary oedema.
iii)   Degenerative changes in liver and kidney, which are usually secondary to the clinical signs.

Diagnosis : Exposure to OPIs can be diagnosed on the basis of following:
(i)   History and circumstantial evidences.
(ii)   Symptomatic evidence : The important toxic symptoms are noted and differentiated from the toxicity of other group of insecticides. Although some of the toxic symptoms of OPIs and chlorinated hydrocarbon insecticides (CHI) are overlaping and confusing, yet these can be differentiated from each other based on certain clinical symptoms (Table 6.8).

Table 6.8 : Differences between the organophosphate insecticides (OPIs) and chlorinated hydrcarbon insecticides (CHIs) poisoning based on clinical signs

Symptoms	OPIs	CHIs
Salivation	Watery saliva, falling drop by drop.	Froathy saliva, sticks on nose, lips, muzzle and mouth.
Pupil size	Miosis (constricted).	Normal.
Body temperature	Hypothermia.	Hyperthermia.
Aimless running	Animal runs in straight direction.	Animal moves in circles.
Abnormal behaviour and posture	Absent.	Abnormal posture such as opisthotonus, head between fore legs and abnormal behaviour such as licking of skin till hair and epidermis are generally present.

Treatment: Treatment of OPIs poisoning in man, domestic and wild animals should be aimed at:

(i)   to abolish the muscarinic effects due to excess of acetylcholine.

(ii)   to regenerate the inactivated enzyme.

The first line of treatment consists of the administration of atropine at the dose rate of 0.2-0.5 mg/kg. It is administered as 0.15 per cent solution of atropine sulphate in physiological saline. One fourth of the above dose is given intravenously and three fourth intramuscularly or subcutaneously. The treatment may be repeated symptomatically if necessary or at 6 to 12 h intervals with half the dose. Atropine, which is a physiological antidote to OPI poisoning, is a competitive antagonist of ACh for muscarinic receptors. It controls or abolishes the muscarinic effects, but has no influence on the nicotinic symptoms.

The second line of treatment involves the use of oxime reactivators such as diacetyl monoxime (DAM), 2-pyridine aldoxime methiodide (2-PAM, pralidoxime), obidoxime, TMB 4, aldoxime, H 16 or MINA, which reactivate the phosphorylated AChE enzyme and greatly accelerate clinical recovery. It also serves to reduce the quantity of atropine required. The chemical structures of some important oxime reactivators are shown in figure 6.8.

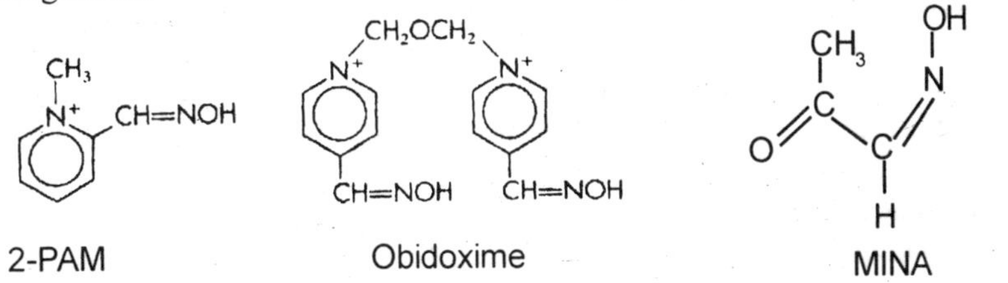

Fig. 6.8 : Chemical structures of some important oxime reactivators.

The mechanism of reactivation of inactivated acetylcholinesterase is shown in Figure 6.9

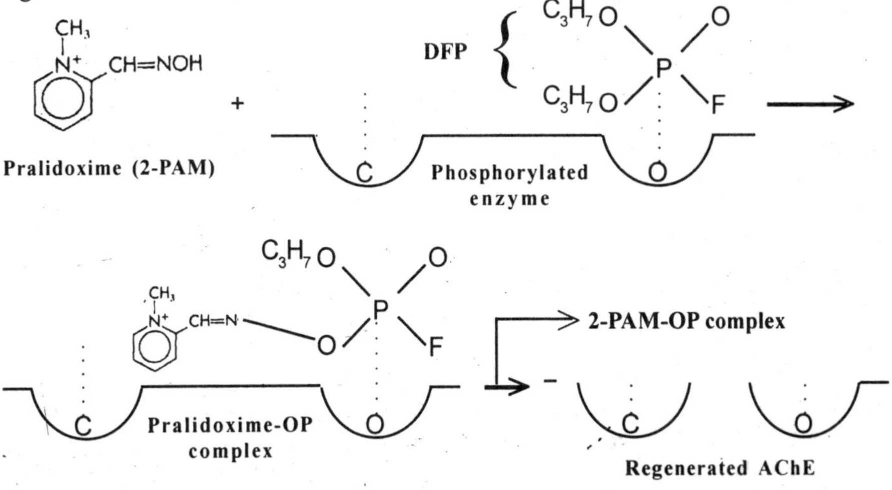

Fig. 6.9 : Reactivation of phosphorylated AChE by 2-PAM

Oxime reactivators after combining at the anionic site exert a nucleophilic attack on the phosphorous of insecticide. The oximephosphate complex is then split off leaving behind the regenerated AChE enzyme.

DAM, which is inexpensive and freely available, is preferred over 2 PAM which is very expensive. Other advantages of DAM over 2-PAM are:

(i)   Compared to 2-PAM, DAM has several fold greater penetration through BBB.
(ii)   DAM can directly combine and neutralize the OPIs in body for enhancing its excretion.
(iii) The rate of reactivation of inhibited AChE enzyme by DAM is faster than 2-PAM.
(iv)   DAM has longer elimination half life and requires less frequent administration than 2-PAM.

2-PAM and DAM are administered at the dose rate of 30 mg/kg as 6 per cent solution in normal saline intravenously or intramuscularly. 2-PAM is used as chloride, iodide or mesulate salt. DAM can be repeated at 12 h at the dose rate of 15 mg/kg intramuscularly, while 2-PAM is repeated at 4 h at the dose rate of 15 mg/kg by IM route. High dose of pralidoxime and other related acetylcholinesterase reactivators can themselves cause neuromuscular blockade and other effects including inhibition of AChE enzyme. Such actions are minimal at the doses recommended for clinical use. If oxime reactivators are injected intravenously more rapidly then mild weakness, blurred vision, dizziness, nausea and tachycardia may be observed. Further, acetylcholinesterase reactivators are not effective after "ageing" of phosphorylated enzyme. The ageing is probably due to loss of one alkyl or alkoxy group, leaving a much more stable monoalkyl or monoalkoxy phosphorylated-AChE. Phosphonates containing tertiary alkoxy group are more prone to ageing than the secondary or primary congeners. The oximes as a group are largely metabolized by liver and the metabolites are excreted through kidneys.

Further treatment includes decontaminating the skin, stomach and eyes promptly. The skin should be washed with an alkaline soap, which will not only remove, but will also help to hydrolyze the phosphate ester. Appropriate clinical procedure for evacuating the stomach and cleansing the eyes may be indicated. In very severe cases, artificial respiration should be provided, preferably by mechanical means.

General precautions : There is a great potential for accidental exposure to insecticides. Fatal poisoning can be prevented by taking certain precautions :

(i)   All insecticides should be stored after proper labelling.
(ii)   They should be kept out of the reach of children and domestic animals.
(iii) After use, all the empty containers should be completely disposed off.
(iv)   Proper attention should be paid during handling of insecticides to avoid spillage.
(v)   Insecticides should be handled only after wearing proper protective clothing, gloves, mask etc.
(vi)   Care should be taken to avoid the use of insecticide sprays on food items, fodder etc.

(vii)   Animals should not be allowed to graze on freshly sprayed pastures and similarly, sprayed vegetables, fruits or fodders should not be fed to animals.

(viii)   Proper precaution should be taken to avoid intake of insecticide residues through food, milk, eggs, meat etc.

(ix)   Preferably safer insecticides and compounds having low persistence in the environment should be used.

(x)   The user should have some basic knowledge about the toxicity and antidotes of the compounds being used.

(xi)   Availability of specific antidotes of insecticides as well as compounds used for general treatment of poisoning must be ensured before hand.

(xii)   All cases of insecticide poisoning must be considered as emergencies and immediate treatment should be provided without losing any time.

6.4 CARBAMATE INSECTICIDES

A.K. Srivastava and V.K. Dumka

Carbamates have been used in medicine for many years as parasympathetic drugs. Since long, in the West African Witchcraft trial, the calabar bean (*Physostigma venenosum*), which contains a biologically active carbamate was used as an ordeal poison. About one hundred years ago, physostigmine (eserine), the only known naturally occurring carbamate ester has been isolated. Carbamates have been synthesized by several workers and found to be active against a number of arthropods. During recent years, a number of carbamates have come into use as insecticides.

Carbamates are not broad spectrum insecticides and these show erratic patterns of selectivity to insects. They are widely used for the control of ectoparasites in both large and small animals. Because of extreme toxicity they are recommended only for limited use. Carbaryl, which has found some application in the control of ectoparasites in both large and small animals, is the most widely used carbamate.

The basic structure of carbamates is

$$R_3 - O - C(=O) - N \big< {{R_1} \atop {R_2}}$$

R_1 and R_2 = -H, -CH_3, -C_2H_5, propyl or other short chain alkyl groups.
R_3 = Phenol, naphthalene or other cyclic hydrocarbons.

Carbaryl (Sevin), the best known carbamate insecticide is a naphthyl carbamate. Other compounds of this group are heterocyclic, phenolic or oxime carbamates.

Carbaryl   Propoxur

The more frequently used carbamate insecticides fall under two broad categories- N-methyl carbamates and N-dimethyl carbamates. Some of the common carbamate compounds under these classes are listed in table 6.9.

Absorption and excretion : Carbamates are well absorbed through skin, lungs and gastrointestinal tract; and are widely distributed in tissues. They get rapidly metabolised and excreted in urine mainly as sulphate or glucuronide conjugates within 24 hours. A small amount is also eliminated via faeces and milk. Unlike the OPIs, most of the carbamates produce low dermal toxicity.

Table 6.9 : Some of the commonly used carbamate compounds.

N - methyl carbamates	N- dimethyl cabamates
Carbaryl	Dimetan
Mesurol	Isolan
Propoxur	Dimetilan
Aldicarb	Pyramat
Metacil	

Metabolism: Some tissues contain a carbamate hydrolysing enzyme but degradation of carbamates can also be catalysed by non-specific proteins. Cleavage of the C-O-N bond is most important and the metabolites are hydroxylated derivatives, hydrolysis products or conjugates of these. Metabolism of carbamates by hepatic microsomes is quite complex yielding several unidentified products. The species variation in carbaryl metabolism was not found much and the extent of breakdown and production of each metabolite was roughly the same. Various oxidation steps catalysed by mixed function oxidase also occur. Decarbamylation i.e., hydrolytic removal of the $OC(O)NHCH_3$ group, is a major feature of carbamate metabolism. For N-dimethylcarbamates, hydroxylation of one methyl group is an important process of metabolism. The enzyme system involved is found in liver microsomes which converts $P = S$ to $P = O$ groups. This system hydroxylates compounds like dimetan, isolan and pyrolan. Hydrolysis of the carbamic acid ester linkage results in metabolites that lack anticholinesterase activity. e.g. mesurol is converted to sulfoxide and sulphur and zectran is degraded to dimethyl-p-benzoquinone

In addition to metabolic conversion, carbamates can also undergo light-catalysed breakdown yielding products having anticholinesterase activity.

Toxicity: Different carbamate insecticides produce varying degree of toxicity in animals depending upon their chemical structure. The oral acute LD_{50} of some carbamate compounds in rats are presented in table 6.10.

Table 6.10 : Oral LD_{50} (µg/g) values of some of the carbamate compounds in rats

Carbamate	Oral LD_{50} (µg/g)
Carbaryl	540
Dimetan	150
Isolan	13
Mesurol	100
Pyrolan	90
Zectran	60
Carbofuran	8-14

Carbaryl is of low toxicity in warm blooded animals. The LD_{50} values of carbaryl in different species of animals are presented in Table 6.11.

Table 6.11 : Oral LD$_{50}$ values of carbaryl in different species of animals.

Species	LD$_{50}$ (mg/kg)
Rats	850
Mice	200
Ducks	3000
Fowls	2000
Chicks	500

Water based suspension of carbaryl will not produce poisoning in concentrations ranging from 1-4% in cattle, 2% in sheep and 1% in goats. Dairy cattle were fed 50 to 200 ppm of carbaryl in the total diet for 30 days without evidence of injury. Carbamates when used in combination produce synergistic effects.

Mechanism of toxicity : The mode of action of carbamates, like organophosphates is the inhibition of acetylcholinesterase. Carbamates are potent and reversible inhibitors of cholinesterase enzyme. In addition, all the carbamates inhibit aliesterase enzyme of insects, but they kill insects and mammals entirely by cholinesterase inhibition.

Mechanism of cholinesterase inhibition : Carbamates react with cholinesterase forming a complex. This is followed by carbamylation of the enzyme at the serine OH group in the active site. The third step is hydrolysis or decarbamylation to yield the original enzyme. There is leaving behind of a group e.g. 1-naphthol in carbaryl, which is released in the second step.

Carbamylation reaction is reversible from the point of view of AChE enzyme, however, it is not reversible from the point of view of carbamate insecticide which is cleaved in the process. Carbamate is steadily destroyed and after some time all the carbamate is destroyed and the enzyme will totally recover .

The mode of action of carbamates differs from that of OPIs in the following ways:

i)   Carbamates inhibit AChE by carbamylation unlike the phosphorylation caused by OPIs.
ii) AChE inhibition by carbamates is reversible but that by OPI's is irreversible.
iii) Decarbamylation of carbamates is easier than dephosphorylation of OPI's.
iv) Carbamates can detach from AChE much more easily and rapidly than OPI's.
v)   Carbamates bind both at anionic as well as esteratic site of AChE whereas, OPI's bind only at the esteratic site.
vi) Unlike in OPI's toxicity, oxime reactivators are ineffective in the treatment of carbamate poisoning.
vii) Organophosphates can virtually completely inhibit the cholinesterase while inhibition of cholinesterase is partial with carbamates.

Clinical signs : In general carbamates produce toxicity similar to that of organophosphates, but poisoning by carbamates is less severe and the effects do not last long.

Symptoms of acute poisoning are typically cholinergic with salivation, lacrimation, miosis, convulsions and death.

Chronic toxicity produces signs of neuromuscular type which are characterized by incoordination, ataxia, recumbency and prostration but there is no demyelination or damage to nerves.

The lethal dose of carbamate insecticides is many fold higher than the dose causing initial signs of poisoning when compared to organophosphate insecticides.

Post mortem lesions : The post-mortem findings in acute poisoning are usually limited to congestion and oedema of lungs, liver and kidneys and petechial haemorrhage of gastric mucosa. Neuromuscular lesions are found in chronic poisoning.

Diagnosis : Carbamate poisoning is much more difficult to diagnose than OPIs toxicity because of the following two reasons :
(i)   Carbamates are rapidly metabolized so they are not detected in tissues and blood.
(ii)   Unlike in OPI's toxicity, the level of AChE is normal in carbamate poisoning due to reversible inhibition.

Diagnosis is thus made on the basis of history and the response of the animals to atropine therapy.

Treatment : Carbamate poisoning is treated by administration of atropine sulphate in a similar manner and same dose levels as that in case of OPI toxicity. Diuretics such as hydrochlorthiazide are helpful. Cholinesterase reactivators like pralidoxime are ineffective. Administration of oxime reactivators is, in fact, contraindicated since they aggravate the toxicity of carbamates.

6.5 SYNTHETIC PYRETHROIDS

M.A. Ayub Shah

Pyrethrins are natural insecticides obtained from the flowers of *Chrysanthemum cinarariaefolium* and *C. roseum* which possess potential insecticidal properties without mammalian toxicity potential. However, these natural compounds are unstable to light and quite expensive to be used as insecticides. Therefore, modification of the chemical structures of the natural pyrethrins was attempted to develop a series of photostable synthetic pyrethroids with improved physical and chemical properties and greater biological activity. Due to rapid degradation and nonpersistence in the environment, synthetic pryrethroids are one of the most widely used insecticides today.

The pyrethroids are esters of cyclopropane carboxylic acids with alkenyl methyl cyclo-pentanolone alcohols (e.g. chrysanthemic acid and pyrethrolone in pyrethrin I). Pyrethrin -I (Fig. 6.10) contains all the basic features for good insecticidal activity.

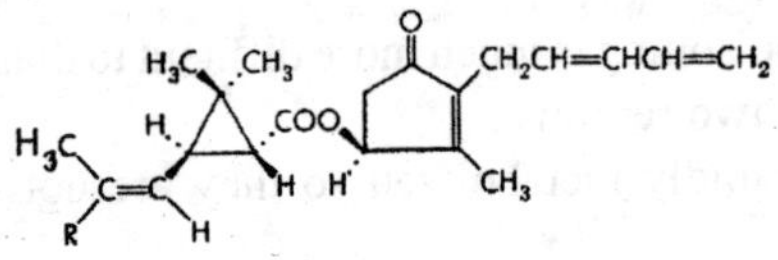

Fig. 6.10 : Chemical structure of pyrethrin

On the basis of their chemical nature, pyrethroids are categorized as:

(i) non- *alpha*-cyano containing pyrethroids (also known as Type- I pyrethroids) e.g. allethrin, permethrin, cismethrin, biorsmethrin etc.

(ii) an *alpha*-cyano containing pyrethroids (also known as Type- II pyrethroids) e.g. deltamethrin, cypermethrin, cyhalothrin, *lambda* cyhalothrin, fenvalerate, fluvalinate, flumethrin etc.

The chemical structures of some commonly employed synthetic pyrethroid insecticides are shown in Table 6.12.

Sources of poisoning:

(i) Ingestion of pesticide contaminated feeds or water.

(ii) Dermal absorption from topical application such as spray or pour-on to control external parasites.

Toxicity: Synthetic pyrethroides are safer insecticides compared to OCIs and OPIs. The LD_{50} values of some of the commonly employed synthetic pyrethroids in different species of animals are shown in Table 6.13.

Table 6.12: Chemical structures of some of the commonly employed synthetic pyrethroid insecticides

Synthetic pyrethroid	Chemical structure
Allethrin	
Cypermethrin	
Deltamethrin	
Fenvalerate	
Permethrin	

Toxicokinetics: There is no evidence till date which implicates pyrethroids causing chronic poisoning in animals or man. Long-term exposure to animals yield high "no effect levels" indicating that there is little or no accumulation of these insecticides in the body. This may be due to an efficient detoxification of these chemicals in man and other mammals.

Esters of pyrethroids are susceptible to hydrolytic degradation probably by non specific carboxylesterases associated with the microsomal fraction of tissue homogenates in many species of animals. Ester hydrolysis of pyrethroids as a mode of degradation is further supported by the fact that organophosphorus insecticides, which are known inhibitors of tissue esterases, potentiate pyrethroids toxicity in many species. In mammals and some species of insects and fishes, the microsomal monooxygenase system present in the tissues is involved extensively in the detoxification processes of pyrethroid esters.

Table 6.13: LD$_{50}$ values of some of the synthetic pyrethroids in different species of animals

Compound	Species	LD$_{50}$ (mg/kg)
Permethrin	Rats	236-1300
	Mice	500
	Rabbits	4000
	Chickens	>3000
Cypermethrin	Rats	251-800
	Rabbits	960
	Ducks	>10000
Fluvalinate	Rats	293
	Mice	105
Alphamethrin	Rats	700-5000*

* Depending on the excipient.

In rats, 52-54 % and 45-47 % *cis*-permethrin is excreted in urine and faeces, respectively whereas, 79-82% and 16- 18% of *trans*-isomer is eliminated in urine and faeces, respectively within 12 days after oral administration.

Mechanism of toxicity : The synthetic pyrethroids are neuropoisons. The main target of action is the nerve membrane sodium channels.

The non-*alpha*-cyano containing or type-I pyrethroids interfere with the axonal sodium gate causing delayed repolarization resulting in repetitive discharge of nerve impulses. The *alpha*-cyano containing or type-II pyrethroids extend the time constant for inactivation of the sodium gate by hundredth of a millisecond to seconds, producing a persistent depolarization and frequency-dependent conduction block in sensory and motor axons and prolonged repetitive firing of sensory nerves and organs and muscle fibers. The depolarizing action would have a dramatic effect on the sensory nervous system because such neurones tend to discharge when depolarized even slightly, resulting in an increase in the manner of discharges. A slight depolarization at presynaptic nerve terminals would result in increased release of neurotransmitter(s), serious disturbances of synaptic transmission and generation of the symptoms associated with type-II pyrethroids.

In addition to above mechanisms, permethrin, cypermethrin and deltamethrin inhibit Ca^{2+}-Mg^{2+}-ATPase, the effect of which would result in increased release of intracellular Ca^{2+} levels accompanied by increased neurotransmitter release. Deltamethrin has an inhibitory effect on the GABA-receptor-chloride channel complex; however, this mechanism has little importance compared to sodium channels.

Clinical signs : The clinical signs of acute type-I pyrethroid poisoning in laboratory animals are restlessness, incoordination, hyperactivity, tremors, prostration and paralysis. The type-I pyrethroids induced-tremors are referred to as 'T-syndrome'. Rats exhibiting 'T-syndrome' show aggressive behaviour and hypersensitivity to external stimuli. The body temperature is apparently increased during poisoning which is probably due to increased muscular activity associated with tremors.

The type-II pyrethroids cause burrowing behaviour, clonic seizures, writhing and profuse salivation which is also referred to as choreoathetosis-salivation or 'CS-syndrome' in laboratory animals.

In large animals, fenvalerate caused restlessness, frothing of mouth, dyspnoea, erection of ear and tail, mydriasis, regurgitation of ruminal contents, incoordination, tremors, clonic convulsions and recumbency. Deltamethrin spray causes hypersalivation, lacrimation, mucoid nasal discharge, excitement, incoordination, extension of limbs, anorexia and alopecia in buffalo calves. Pyrethroids are also known to cause contact dermatitis. Some pyrethroides are known to induce hepatic microsomal enzymes and suppress the immune responses in laboratory animals.

Diagnosis and differential diagnosis: Diagnosis may be based on:

i) History of exposure to synthetic pyrethroids.
ii) Clinical signs exhibited by the poisoned animals. Almost identical symptoms are also seen in organochlorine poisoning.

Treatment: There is no specific treatment. Following general and supportive line of treatment may be followed:

i) Removal of source: wash the animal if the route of exposure is through skin.
ii) Emetic: apomorphine at the rate of 0.1 mg/kg IM in dogs.
iii) Adsorbents: activated charcoal at the rate of 2g/kg.
iv) Purgative: magnesium sulfate or sodium sulfate at the rate of 5g/kg as 10% solution.
v) CNS depressants: diazepam at the rate of 0.2-2.0 mg/kg IV.
vi) Methocarbomol at the rate of 55-200 mg/kg by slow IV administration (speed should not exceed 200 mg/min.

6.6 HERBICIDES

M.A. Ayub Shah

Herbicides may be defined as any compound(s) that have the potential of either killing or damaging unwanted plants or weeds and are employed for the elimination of such plants or part(s) of a plant. The biochemical differences in plants make it possible to design herbicides that have selective toxicity potential against various species of plants. In early old days, simple chemicals like sulfuric acid, sodium chlorate, sodium arsenite, arsenic trioxide, petroleum oils etc. have been used to destroy plants. However, due to their non-selective effects, attention has been directed towards the development of newer herbicides with selective toxicity. These efforts resulted in the development of a series of chemical agents (Fig. 6.11) which destroy the plants/weeds selectively with no deleterious effects on the crops.

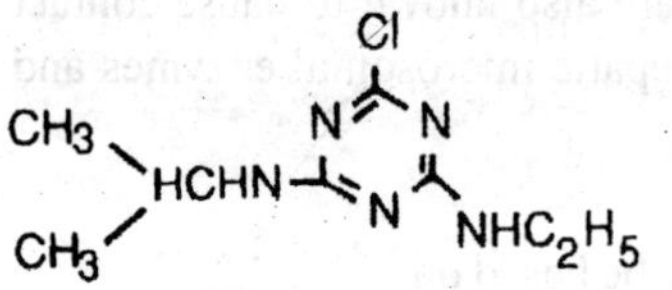

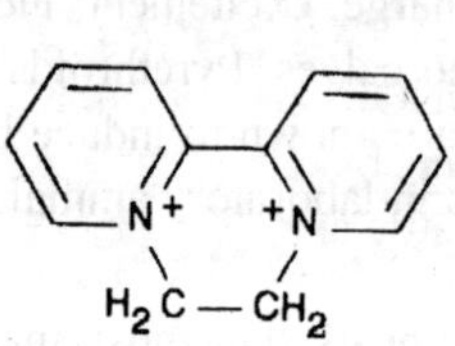

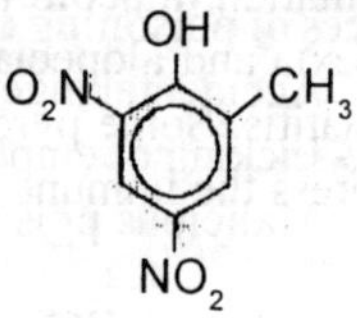

Fig. 6.11 : Chemical structures of some of the commonly employed herbicides

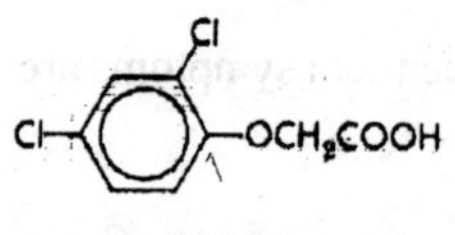

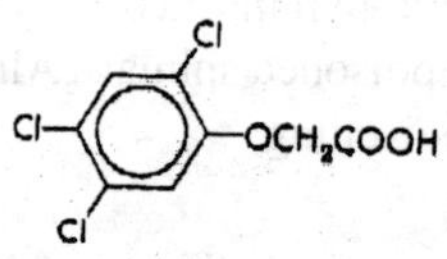

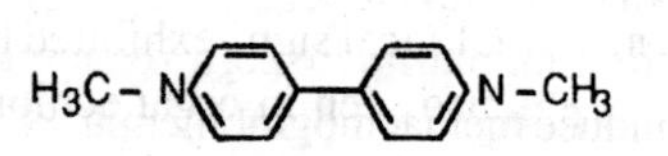

Herbicides may be categorized as below:

A. On the basis of their use:
  i)   Preplanting herbicides: mixed with the soil before seeding.
  ii)   Pre-emergent herbicides: they are applied before the emergence/ appearance of unwanted weeds.
  iii) Post-emergent herbicides: they are applied to the soil or foliage after the germination of crops and/weeds.

B. On the basis of their chemical nature:
  i)   Dinitro compounds e.g. dinitro ortho cresol (DNOC), dinitrophenol etc.
  ii)   Phenoxyacetic acids e.g. 2,4-D, 2,4,5-T etc.
  iii) Bipyridinium compounds e.g. diquat, paraquat etc.
  iv)   Heterocyclic compounds or triazenes e.g. atrazine, propazine, simazine etc.
  v)   Chloroaliphatic acids e.g. dalapon, sodium chloroacetate, sodium trichloroacetate etc.

vi)   Substituted ureas e.g. monouron, diuron, isoproturon etc.

vii)   Substituted dinitroanilines e.g. pendimethalin etc.

C. On the basis of their mechanism of action :

i)   Selective herbicides.

ii)   Contact herbicides.

iii)   Translocating herbicides.

Some of the commonly employed herbicides of different chemical groups are described here under:

(i) Dinitro compounds : Many dinitro compounds are used as herbicides. The commonly employed ones are dinitro ortho cresol (DNOC) and dinitrophenol.

Sources of poisoning :

a)   Accidental ingestion of DNOC-sprayed foliage by animals.

b)   Licking of empty containers by curious animals.

c)   Malicious poisoning in cattle and dogs.

Mechanism of toxicity : Dinitro compounds act by interfering with electron transport chain of energy metabolism. They uncouple the oxidative phosphorylation and convert all the cellular energy in the form of heat producing severe hyperthermia. In ruminants, the ruminal microflora reduce the dinitro compounds to diamine metabolites which induce methaemoglobinemia.

Post mortem lesions: There is rapid onset of rigor mortis. The dinitrophenol imparts a yellowish green colour to tissues and urine (mild jaundice). Degenerative changes in parenchymatous organs may be observed. Dark blood, gastroenteritis, hyperkeratosis of skin and hyperplasia of urinary bladder mucosa may also be recorded.

Diagnosis :

i)   History of exposure to dinitro compounds.

ii)   Clinical symptoms.

iii)   Post mortem lesions.

Differential diagnosis : Dinitro compounds poisoning should be differentiated from:

i)   Heat stroke;

ii)   Nitrate/nitrite poisoning -chocolate coloured blood (methaemoglobinemia) but absence of hyperthermia; and

iii)   Carbon monoxide poisoning-bright red blood but no pyrexia.

Treatment : There is no specific treatment or antidote available.

i)   Wash with soap and water if the source of poisoning is through skin contact.

ii)   Keep the affected animal in cool and calm place. To control hyperthermia, use cold bath or ice-water sponging. Use of antipyretics is contraindicated.

iii)   Saline purgatives or gastric lavage.

iv)   Intense oxygen therapy in carnivores.

v)   Dextrose saline to check dehydration.

vi)   Non-barbiturate tranquilizers and sedatives.

vii)   In ruminants, methaemoglobinemia may be treated with methylene blue solution (2-4%) -10 mg/kg, IV q8h for first 24-48 hours or ascorbic acid @ 5-10 mg/kg IV q8h for first 24-48 hours.

(ii) Phenoxyacetic acids : Chlorophenoxyacetic acids were the first commercially available herbicides in 1946. These are widely employed selective herbicides with specific translocation mechanism and are useful for the control of weeds in cereal crops, grasslands and meadows.

The most commonly used chlorophenoxyacetic acids are 2,4-dichloro-phenoxyacetic acid (2,4-D) and 2,4,5-trichlorophenoxyacetic acid (2,4,5-T). Both these agents are relatively harmless to mammals. These compounds are plant hormones which act as plant growth regulators selectively against dicotyledons. The use of these hormone herbicides to certain weeds increases their nitrate content, thus these may be harmful indirectly by increasing the risk of nitrate poisoning.

Sources of poisoning:

Direct poisoning: Due to accidental ingestion of herbicide itself. However, ingestion of recently treated plants may not be harmful if the agent is applied in accordance with the instructions of the manufacturer.

Indirect poisoning: Due to the ingestion of poisonous plants which have been treated with phenoxyacetic acids because these herbicides render the toxic plants more palatable which in normal course would have been avoided by the grazing animals and also increase the potential of the toxic principles present in certain plants e.g. increase in the content of nitrate/nitrite levels in many otherwise non-toxic plants.

Mechanism of toxicity: The exact mechanism of toxicosis of phenoxyacetic acids are not precisely known. However, they are known to produce reproductive toxicity in cattle and hepato-carcinoma in laboratory animals.

Clinical signs: Anorexia, ruminal atony, weight loss, occasional diarrhoea, depression, unthriftiness and muscular weakness of hind limbs may be observed. Abortion, irregular oestrus, anoestrus and ovarian atrophy may be recorded in cattle. In laboratory animals, they are reported to cause proliferation of hepatic peroxisomes and development of hepatocarcinoma.

Treatment: No specific antidote(s) are available. Symptomatic and supportive therapies include administration of diuretics and liver protectants.

(iii) Bipyridinium compounds : The bipyridinium or dipyrilidinium herbicides (paraquat and diquat) are rapidly acting broad spectrum desiccant contact herbicides which are extensively used in agriculture for preharvest of cotton and potatoes and weed control. Bipyridinium herbicides are highly toxic compounds. These are water soluble, non-volatile and stable at temperatures upto 300°C, but are rapidly inactivated by soil bacteria. The LD_{50} values of diquat and paraquat are given in Table 6.14.

Table 6.14 : LD_{50} values (mg/kg) of diquat and paraquat in different species of animals

Compound	Species	LD_{50} (mg/kg)	Lethal Dose(mg/kg)
Diquat	Rats	230-440	—
	Rabbits	190	—
	Dogs	100-200	—
	Cats	35-50	—
	Cattle	20-40	—
	Chickens	200-400	—
Paraquat	Rats	100-150	—
	Mice	120	—
	Rabbits	110	—
	Cats	35	—
	Cattle	35-50	—
	Sheep	—	8-10
	Pigs	—	75
	Ducks	—	100 (for 100 days)

Sources of poisoning:

(i)   Accidental poisoning due to direct ingestion of the compound or consumption of freshly treated vegetation.

(ii)   Malicious poisoning- principal targets are the dogs.

Mechanism of action: The bipyridinium compounds are caustic and irritant agents which cause ulceration and necrosis of the skin and mucous membranes in dogs, pigs and man. They also cause progressive irreversible pulmonary fibrosis. Paraquat is actively taken up by the alveolar cells via a diamine or polyamine transport system where it undergoes NADPH-dependent reduction, one electron reduction to form a free radical capable of reacting with molecular O_2 to reform the cation paraquat plus a reactive O_2 (superoxide anion; O_2^-). This superoxide anion is converted into H_2O_2 by the enzyme superoxide dismutase. The H_2O_2 and superoxide anion (O_2^-) can attack polyunsaturated lipids present in cell membranes to produce lipid hydroperoxides which in turn, can react with other unsaturated lipids to form more-lipid free radicals, thereby perpetuating the system and damaging the cellular membrane leading to reduction of the functional integrity of the cell and affects the efficient gas transport/ exchange and induce respiratory impairment. The severity of the cellular effects can be modulated by the availability of O_2, and animals kept in air with only 10 percent O_2 are far better than those kept in room air .

Clinical signs : Generally the clinical signs are seen after three days of exposure to paraquat. Symptoms of intoxication are emesis, anorexia, abdominal pain, dyspnoea, jaundice and CNS depression. If the affected animals survive for several days, animal may exhibit dehydration, pallor or cyanosis, tachycardia, uremia, moist rales, pulmonary oedema and emphysema.

Post-mortem lesions : Pulmonary congestion, oedema and haemorrhages, progressive intra-alveolar fibrosis, atelectasis, erosions and ulceration of buccal and pharyngeal tissues, haemorrhagic gastritis, congestion of liver, kidney and spleen with consistent histopathological changes. Brain damage characterized by spongy degeneration of cerebral white matter may be noted.

Diagnosis :
(i) History of exposure to bipyridinium herbicides.
(ii) Clinical symptoms.
(iii) Post-mortem lesions.
(iv) Analysis of the urine, suspected baits or the feeding material.

Differential diagnosis :
(i) Rule out pneumonia.
(ii) *Alpha*-naphthylthiourea poisoning- more acute and fatal in nature.
(iii) In cattle, it should not be confused with bovine pulmonary oedema.

Treatment : There is no specific treatment for diquat or paraquat poisoning. Symptomatic and supportive treatment includes:
(i) Check further absorption by giving emetics or gastric lavage or oral adsorbents (activated charcoal) followed by saline purgatives.
(ii) Demulcent or antacids to counter gastric irritancy.
(iii) IV fluids and diuretics.
(iv) Tranquilizers and sedatives.
(v) Vitamin A, C and E may have some beneficial effect.
(vi) Nonsteroidal antiinflammatory agents may be given to block the synthesis of inflammatory prostaglandins.
(vii) Corticosteroids
(viii) Biochemical antagonists like ascorbic acid, acetylcysteine, super oxide dismutase etc.

Note: Oxygen therapy is contraindicated because it will act as a ready source for the formation of more and more superoxides.

(iv) Heterocyclic compounds (Triazenes) : The heterocyclic compounds (triazenes) include atrazine, simazine, propazine, prometone and aminotriazole. This group of herbicides has a very low mammalian toxicity potential. Atrazine, the most widely used triazene herbicide is a selective herbicide for crops like maize, sorghum, sugarcane, pineapple etc.

The mechanism by which these agents produce toxicity is not well understood.

Clinical signs : Weakness, hypersalivation, ataxia and posterior paralysis appear three weeks after ingestion of the compound. Comparatively, prometone is more toxic than propazine. Aminotriazole poisoning is characterized by stimulation of GI and bronchial muscles and causes pulmonary oedema and severe gastric and intestinal haemorrhages.

Diagnosis : Based on the history of exposure and clinical signs.

Treatment : There is no specific antidote. Provide symptomatic and supportive treatment.

(v) Chloroaliphatic acids : The most commonly used chloroaliphatic acids are dalapon, sodium chloroacetate and sodium trichloroacetate. These agents are relatively harmless compounds.

Dalapon is a selective translocated herbicide for grass weed control. The most common source of poisoning is accidental ingestion. Cattle, sheep and goats are frequently affected animals. The LD_{50} values of dalapon are summarised in Table 6.15.

Table 6.15: LD_{50} and toxic dose (mg/kg) values of dalapon in different species of animals.

Species	LD_{50}	Toxic Dose
Rats	6000-8450	—
Guinea pigs	3400	—
Rabbits	3400	—
Chickens	5660	—
Cattle	—	>4000
Sheep	—	500 (for 7 days)

Clinical signs : Commonly observed clinical signs are anorexia, lassitude, diarrhoea and mild cyanosis.

Treatment : There is no specific antidote for dalapon poisoning. Symptomatic and supportive treatment may be provided. The animals may recover spontaneously even without any treatment .

(vi) Substituted ureas : Monouron, diuron and isoproturon are the most commonly used substituted ureas. The substituted ureas have low toxicity potential. Poisoning with these agents is rare.

Clinical signs: Anorexia, abnormal gait, excitability followed by depression and prostration, occasional respiratory difficulties, haematuria and hypersalivation may be noted.

Post-mortem lesions : Post-mortem lesions are non-specific. However, congestion of GIT, liver, kidneys and haemorrhages in the heart and lungs may be seen.

Treatment : No specific treatment is available. Provide symptomatic and supportive treatment.

(vii) Substituted dinitroaniline compounds : Pendimethalin and trifluralin are substituted dinitroaniline compounds having a broad spectrum weedicide activity and low mammalian toxicity. The acute oral LD_{50} values of pendimethalin are 733 and 1200 mg/kg while that of trifluralin are 5000 and 5000-10,000 mg/kg in mice and rats, respectively. Chronic oral administration at 24 to 48 mg/kg/day for 150 days did not cause mortality in rats.

Clinical signs : CNS depression, swelling of the face, incoordination, hypothermia, conjunctival haemorrhages, dyspnoea and death at very high dose have been recorded in laboratory animals. Cattle, sheep, goats and dogs are sometimes affected. General signs of toxicity are anorexia, tympanism, diarrhoea and prostration.

Treatment : Generally, prognosis is good; no specific treatment is available. Symptomatic and supportive treatment may be provided in addition to adsorbents, gastric demulcents and cardiopulmonary stimulants.

6.7 RODENTICIDES

M.A. Ayub Shah

Rodenticides are agents which destroy the rodent pests. Rodents such as black rats (*Rattus rattus*), mice (*Mus musculus*), squirrels etc. can consume substantial quantities of pre-harvest, post-harvest and stored grains, pulses, vegetables and fruits. They also render the foodstuffs unfit for consumption by soiling and contaminating with urine, faeces and pathogenic microorganisms which are capable of infecting other animals and man.

An ideal rodenticide should possess the following features :

(i) It should be potent and palatable to the target animals.
(ii) The bait should not induce bait shyness.
(iii) Death of the animals should occur in such a manner that the surviving rodents will not suspect bait as cause of death.
(iv) It should not make the intoxicated animal(s) to go out into open to die.
(v) It should be species specific with considerably lower toxicity to other animals that might inadvertently consume the bait or eat the poisoned rodent.

Some of the most commonly employed rodenticides are alpha-naphthyl thiourea, warfarins, zinc phosphide, fluoroacetate, red squill etc. (Fig. 6.12).

Alpha naphthyl thiourea (ANTU) : ANTU is relatively selective rodenticide. This agent is toxic to the rats but harmless to human beings.

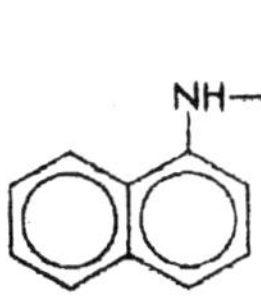

ANTU

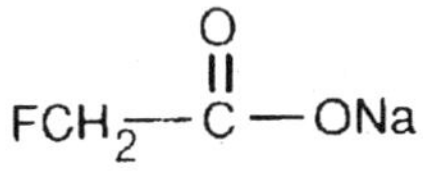

Fluoroacetate

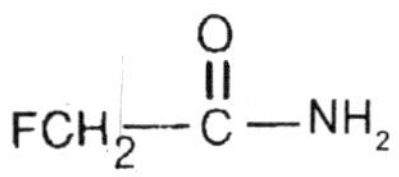

Fluoroacetamide

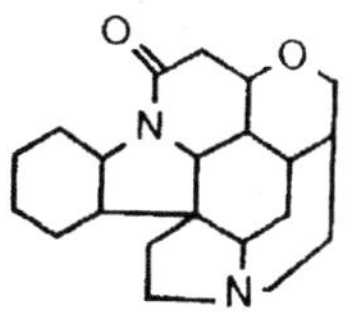

Strychnine

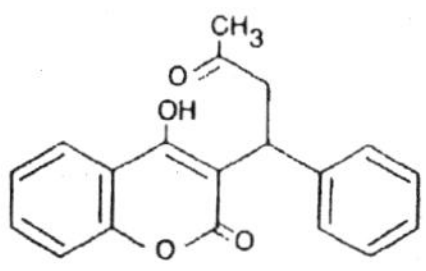

Warfarin

Zn_2P_3

Zinc Phosphide

Fig. 6.12 : Chemical structures of some of the common rodenticides.

Sources of poisoning: Accidental ingestion of baits intended for target rodents. Primarily affected animals are the dogs and cats.

Mechanism of toxicity: ANTU interferes with effective uptake of O_2 from pulmonary alveoli by producing extensive oedema of the lungs due to increased capillary permeability and seepage of fluid into the airways. This leads to formation of froth which further blocks the air passage and virtually the poisoned animal drowns in its own fluids.

ANTU induces vomiting in dog if the stomach is empty as it is highly irritant to gastric mucosa. However, ANTU will fatally poison an animal if the stomach is full.

Clinical signs: Clinical signs of ANTU poisoning include vomiting, dyspnoea, cyanosis, rales, tachycardia, anorexia, incoordination, prostration, cough, snort, asphyxiation, coma, clonic convulsions and death.

Post-mortem lesions: Cyanosis, dark coloured arterial blood, heavy and oedematous lungs, presence of blood-tinged fluid and froth in the bronchi, hydrothorax and hyperemic tracheal, bronchial and gastric mucosae, liver and kidneys etc.

Diagnosis:
(i)   History of employing ANTU as baits.
(ii)   Clinical symptoms.
(iii) Post-mortem lesions.

Treatment:
(i)   Emetics or gastric lavage.
(ii)   Sedatives like barbiturates.
(iii) Oxygen under positive pressure.
(iv)   Competitive ANTU antagonist, 1-ethyl-1-phenyl thiourea may reverse ANTU toxicosis.
(v)   *Alpha* adrenoceptor antagonists to dilate pulmonary vessels.
(vi)   Osmotic diuretics (50% glucose or mannitol) to reduce the pulmonary oedema.

Warfarins : Warfarin [3(*alpha*-acetonylbenzyl)-4-hydroxycoumarin], isolated from moldy sweet clover and its analogues are anticoagulant rodenticides. The warfarins are readily inactivated by cytochrome P450 enzymes of the liver. Therefore, repeated dosing for several days is recommended. Due to development of resistance to warfarins, second generation oral anticoagulant rodenticides like brodfacoum, difethialine, flocoumanafen and flupropadine have been developed. These newer agents are many fold more potent than warfarin and also have very long half-lives.

Sources of poisoning: The main source of poisoning is the ingestion of residues of the rodenticides or baits intended for target rodents. Pigs, dogs and cats often ingest warfarin poisoned dead rats or mice.

Mechanism of toxicity: Warfarin interferes with normal function of vitamin K and causes coagulation defects characterized by decreased blood concentrations of coagulation proteins - factor II (prothrombin), factor VII (proconvertin, autoprothrombin I), factor IX

(Christmas factor, autoprothrombin II) and factor X (Stuart factor, autoprothrombin III). The decreased coagulation factors cause massive internal haemorrhages. The onset time is 2-5 days and the affected animal dies due to tissue hypoxia.

Clinical signs: After ingestion of warfarins over a period of 2-5 days, the affected animals may exhibit signs of haemorrhages of gum, epistaxis, massive internal haemorrhages, blood discharge from body orifices, haematomas under the skin and at joints, dyspnoea, weakness, shock and death.

Post-mortem lesions : Massive internal haemorrhages, blood in the GIT, thorax, joints and pericardium may be recorded. Hepatic necrosis and jaundice may also be seen.

Diagnosis :
(i)   History of employing warfarin group of rodenticides.
(ii)   Clinical signs.
(iii)   Post-mortem lesions.

Treatment :
(i)   Sedatives or tranquilizers.
(ii)   Artificial respiration.
(iii)   Provide clotting factors by whole blood or plasma transfusion at the rate of 20 ml/kg or 9 ml/kg body weight, respectively.
(iv)   Injection vitamin K_1 at the rate of 5 mg/kg, slow IV in dogs and cats. Repeat for 2 days by IM route. In large animals, a dose of 0.5-1.0 mg/kg by slow IV followed by oral vitamin K administered daily for 4-6 days.

The rationale of giving vitamin K in warfarin poisoning is that vitamin K is biotransformed into vitamin K epoxide in the liver which is reduced by NADH to vitamin K hydroquinone. This vitamin K hydroquinone accelerates the synthesis of prothrombin.

Zinc phosphide: Zinc phosphide (Zn_3P_2) is commonly used rodenticide for destroying rats, mice, squirrels and dogs. This compound is widely employed in the developing countries because it is the cheapest rodenticide and is also quite effective.

Sources of poisoning:
(i)   Baits intended for target rodents may be eaten by other animals or birds.
(ii)   Malicious poisoning, particularly to kill dogs and cats.

Mechanism of toxicity : Zinc phosphide is directly irritant to gut and produces severe gastroenteritis. This action causes vomiting, a life saving reflex in animals that can vomit.

Acute zinc phosphide toxicosis is neither due to zinc nor phosphorus. The zinc phosphide liberates phosphine (PH_3) in the stomach following a hydrolytic reaction with water in the GIT as shown in figure 6.13. Any factor which stimulates the gastric hydrochloric acid concentrations, increases Zn_3P_2 toxicity. Absorbed phosphine is responsible for development of widespread cellular toxicity with necrosis of the GIT and other vital organs including liver and kidneys.

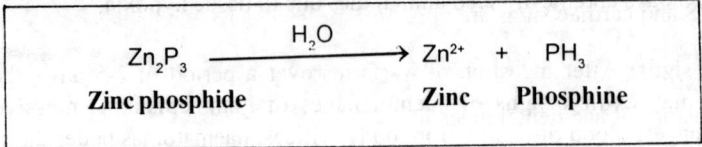

$$Zn_2P_3 \xrightarrow{\ H_2O\ } Zn^{2+} + PH_3$$

Zinc phosphide   **Zinc**   **Phosphine**

Fig. 6.13 : Production of phosphine from zinc phoshide.

Clinical signs : Anorexia, lethargy, increased rate and depth of respiration, stertorous respiration, abdominal pain, bloat (in ruminants), ataxia, weakness, prostration, dyspnoea, gasping, struggling, convulsions, coma and death in 4-48 hours.

Post-mortem lesions : Pulmonary congestion and oedema, pleural effusion, subpleural haemorrhages, congestion of liver and kidneys and gastroenteritis. Acetylene odour may be detected in stomach.

Diagnosis :
(i)   History.
(ii)   Clinical signs.
(iii)   Post-mortem lesions.

Treatment : There is no specific treatment. However, following symptomatic/supportive treatment may be given.
(i)   Gastric lavage with 5% sodium bicarbonate.
(ii)   Injection calcium borogluconate.
(iii)   Treatment for shock
(iv)   Anticonvulsants.

Fluoroacetate : Fluoroacetate as such is non-toxic but becomes highly toxic after its metabolic conversion in the body to fluorocitrate (a best example of lethal synthesis). Sodium fluoroacetate is commonly used as an effective rodenticide to destroy rats, mice and other rodent pests. This compound is extremely toxic and its LD_{50} value is 0.22 mg/kg in rats.

Mechanism of toxicity: Due to structural similarity, fluoroacetate competes with acetate in the tricarboxylic acid (TCA) or Kreb's cycle. In place of acetate, fluoroacetate condenses with coenzyme A (CoA) and oxaloacetate to form fluorocitrate. Fluorocitrate then competes with citrate for the active site of Kreb's cycle enzyme aconitase (the target receptor for fluoroacetate poisoning). This results in inhibition of aconitase leading to slowing of the Kreb's cycle and decreasing cellular respiration. The brain and heart functions are most severely affected.

Clinical signs: Clinical signs appear within 30 minutes to 2 hours after ingestion of fluoroacetate and include CNS excitation, restlessness, vomiting, diarrhoea, urination, hyperirritability, hypermotility, frenzied running, barking, frothing at mouth, clonic-tonic convulsions, terminal coma, gasping and death within 2-12 hours have been observed in dogs and guinea pigs. In horses, cattle, sheep and goats, colic, unrest, trembling, staggering, cardiac arrythmias, tachycardia, terminal ventricular fibrillation and death. Both CNS and cardiac signs are observed in swine, cats and hamsters.

Diagnosis:
(i)   History of accidental ingestion of fluoroacetate bait.
(ii)   Clinical signs.

Treatment : There is no specific antidote for fluoroacetate poisoning. Symptomatic treatment may be provided.
(i)   Barbiturates or benzodiazepines to control convulsions.
(ii)   Glycerol monoacetate at the rate of 0.1-0.5 mg/kg, IM.
(iii)   Treatment for cardiac arrythmia may be tried.

Red squill : Red squill is obtained from the plant *Urginea maritima* (sea onion). Red squill contains a cardiac glycoside-scilliroside. This compound is comparatively one of the safest rodenticide as it is non-toxic to poultry and causes emesis in animals that are capable of vomiting and very unpalatable to domestic animals. Red squill is more toxic to female rats than males.

Clinical signs: Small doses of red squill cause convulsions and higher doses may produce cardiac arrest before exhibiting convulsions. In livestock, CNS stimulation symptoms such as hyperaesthesia and incoordination may be observed.

Post-mortem lesions: Gastritis and enteritis (sometimes haemorrhagic), pronounced congestion of mesenteric vessels, swelling and congestion of kidneys, liver, lungs and myocardium may be seen. Histopathological examinations of these organs may reveal degenerative changes.

Diagnosis:
(i)   History of using red squill as rodenticide.
(ii)   Clinical signs.
(iii)   Post-mortem lesions.

Treatment: No specific treatment. Provide symptomatic and supportive treatment.

CHAPTER 7

TOXICITY CAUSED BY DRUGS

7.1 DRUG TOXICITY - I

Satish K. Garg and Rakesh K. Chaudhary

Carbon tetrachloride

Carbon tetrachloride (CCl_4) is used as a reference hepatotoxic agent besides its use as an anti-trematodal drug particularly against fascioliasis in ruminants. It is also used for fumigation of grains and as an insecticide. It is a very good fat solvent, and at one time, it was used for degreasing of the automobile engines. It is quite toxic to ruminants and all other animals and frequently causes fatalities. Swine is most susceptible of all the mammals, however, sheep appears to be quite tolerant.

Absorption following oral administration is minimal and slow. Following a single large dose, there is not much of toxicity, however, it is more toxic when administered in small and divided doses. Presence of large amount of fat in the ingesta promotes its absorption and toxicity. Carbon tetrachloride vapours ($> 1 : 1000$) inhaled through lungs also produce toxicity. Excretion of CCl_4 takes place through exhaled air, urine and milk. Toxicity of CCl_4 is influenced by several factors :

(i) **Dose :** Single large dose is less toxic compared to small repeated doses as toxicity depends on duration of exposure.

(ii) **Climate :** Toxicity is more pronounced during cold conditions.

(iii) **Diet :** High fat and protein intake always increases the chances of toxicity. Therefore, high fat and protein diets should be withdrawn 5-7 days prior to CCl_4 administration. Presence of certain other hepato-toxic agents in the diet also potentiate CCl_4- induced hepatotoxicity. Animals on calcium deficient diet are also more prone to CCl_4 toxicity.

(iv) **Condition of the animal :** Pregnant and lactating animals are more susceptible to toxicity than the nonpregnant non-lactating animals.

(v) **Hepatomicrosomal enzyme inducers :** Induction of hepatomicrosomal enzyme system as a result of pre-exposure/treatment of the animals with certain agents e.g. phenobarbital, DDT, chlordane, dieldrin etc. potentiate CCl_4- induced toxicity as these enhance the biotransformation process and so the generation of free radicals as discussed in the mechanism of action.

Mechanism of toxicity : Toxic actions of CCl_4 are mainly due to its fat solvent property. As a result, permeability of cell wall to certain essential substances is disturbed and these come out of the cell e.g. pyridine nucleotide coenzymes.

In carbon tetrachloride poisoning, reactive free radical - trichloromethyl (CCl_3^-) is generated as a result of hydrolytic cleavage of CCl_4. The CCl_3^- by reaction with oxygen results in the formation of trichloromethyl peroxy (CCl_3COO^-) free radicals as shown below:

$$CCl_4 \longrightarrow CCl_3^- \xrightarrow{O_2} CCl_3COO^-$$

These reactions are catalysed by cytochrome P450 dependent monooxygenase. Therefore, induction of this enzyme system by DDT, phenobarbitone etc. potentiate the CCl_4-induced toxicity. Free radicals (CCl_3^- or CCl_3COO^-) react with lipids and proteins and cause peroxidation of polyenoic lipids of endoplasmic reticulum and generation of secondary free radicals from these lipids. Thus, lipid peroxidation causes breakdown of membrane structure and function and cell death in more severe cases.

Toxicity and clinical signs : Transient exposure to toxic concentrations of CCl_4 vapours causes irritation of eyes, nose and throat, and thus nausea and vomition. However, continued exposure or following absorption of large quantities, acute toxicity may be produced.

Acute toxicity : Generally observed following a single large dose of CCl_4. It results in rapid accumulation of lipids in the liver (steatosis) before hepatic necrosis. Signs of poisoning are loss of appetite, gastrointestinal pain, diarrhoea and blood stained faeces after a few hours. Other signs are drowsiness, depression, stupor, incoordination, convulsions, cardiac collapse, coma and death within 12-24 hours. Hepatic steatosis i.e. fatty liver is defined as an increase in hepatic lipid content to greater than 50% by weight.

Delayed toxicity : It is because of extensive hepatic and renal damage. As the hepatic injury progresses, signs of renal damage are also observed. Hepatic insufficiency is a prominent feature but death generally occurs due to renal damage. Clinical signs mentioned above alongwith icterus are delayed by 2-3 days and the animals may collapse within 1-2 weeks, if not treated. There may be photosensitization too, on exposure to bright sunlight.

Post mortem lesions :
(i)   Gastroenteritis mainly in abomasum and small intestine.
(ii)   Congestion of liver and kidneys.
(iii)   Hydropic degeneration, fatty changes and centrilobular necrosis in liver.
(iv)   Fatty changes and necrosis of renal tubular epithelium.
(v)   Cirrhosis of liver in chronic conditions.

Treatment : No specific antidote is available.
(i)   If inhalation is the mode of exposure of animals or human beings, move the patient in open air.
(ii)   Gastric lavage and saline purgatives to prevent further absorption. Do not give oily purgatives as these enhance the absorption of CCl_4.

(iii)   Activated charcoal followed by glucose/dextrose infusion to combat the delayed hepato-toxic effects.

(iv)   Calcium borogluconate by slow intravenous infusion to take care of hypocalcemia.

(v)   Give cardiac and respiratory stimulants. Never use sympathomimetics to restore the cardio-pulmonary functions as CCl_4 sensitizes myocardium to the action of catecholamines.

Prevention : Risk of CCl_4 toxicity can be avoided by keeping the animals on low protein and low fat diet prior to CCl_4 administration and supplementing the diet with calcium for few days.

Phenothiazine

Phenothiazine is used as an anthelmintic. It was one of the early drugs with a fairly wide range of activity against gastrointestinal nematodes of different species of animals. However, its toxicity limits its use in swine and its use in dogs, cats and human beings has been almost completely stopped. Toxicity varies among different species of animals and even within a species depending on general condition of the animal. Horses are most susceptible, swine and dog come next, however, ruminants and birds are comparatively resistant. Amongst ruminants, cattle are more susceptible compared to sheep. Sheep are more resistant apparently due to greater efficiency of sheep liver to reduce phenothiazine sulfoxide to phenothiazine sulphate. Weak, emaciated and young animals are more susceptible to toxicity.

Phenothiazine is bluish dull coloured powder and almost insoluble in water. Following oral administration, 30-50% of the total drug is excreted unchanged in faeces. Some of the drug in the gut is oxidized to phenothiazine sulfoxide, a photodynamic agent/metabolite, by cellular enzymes of gut. It is a water soluble photosensitive pigment which reaches to liver through portal circulation. In the liver, sulfoxide is oxidized to leucophenothiazone and leucothionol. These colourless metabolites escape the liver and enter general circulation and get excreted in urine and milk. However, as soon as these metabolites come in contact with atmospheric air, get further oxidized to phenothiazone and thionol which impart red colour to urine and milk. Hair and wool of the animals may get stained. Tainted milk and wool are not acceptable.

Factors affecting toxicity :

(i)   **Particle size :** Toxicity of phenothiazine depends on the size of phenothiazine particles. If particle size is fine toxicity is more.

(ii)   **Size of the dose :** Single large dose is more toxic than the small repeated doses.

(iii)   **Digestive disturbances :** If peristalsis is low i.e. constipation, it augments toxicity

(iv)   **Diet :** If diet is deficient in proteins and glucose, toxicity is more and vice versa.

(v)   **Dehydration :** During drought or in dehydrated animals, toxicity is more.

(vi)   **Physical condition of the animals :** Debilitated animals are more prone to phenothiazine toxicity.

(vii)   **Age :** Young animals are more prone to toxicity compared to adults.

(viii)   **Pre-treatment with hepatotoxic substances :** Prior treatment of the animals with carbon tetrachloride aggravates phenothiazine toxicity.

(ix)   **Concurrent use of other agents :** Concurrent use of phenothiazine with organophosphate compounds potentiates toxicity.

Toxicity : Two type of toxicity is observed :

Haemolytic anaemia : Haemolytic anaemia is observed in all species of animals, except sheep and goats. Incidence is very high in horses. Exact pathogenesis of haemolytic anaemia is not known. Blood of horses contains an enzyme called lysolacithin. Phenothiazine activates this enzyme and causes haemolytic crisis. Secondly, blood of horses contains large amount of bilirubin which remains bound to plasma proteins. Phenothiazine also binds to proteins, thus it releases bilirubin and causes bilirubinemia which is responsible for icterus.

Clinical signs : Anorexia, dullness, depression, weakness of the hind limbs, staggering gait, jaundice, anaemia, haemoglobinemia, haemoglobinuria, weak and rapid pulse, dyspnoea, colic and prostration. Other signs are hepatitis, nephrosis and photosensitization in thin and light skinned animals on exposure to bright sunlight.

Post mortem lesions :

(i)   Enlargement of liver, kidneys and spleen.
(ii)   Yellow staining of subcutaneous tissues.
(iii)   Bladder full of dark red urine.
(iv)   Corneal opacity.

Photosensitization : Photosensitization is mostly seen in pigs and calves on exposure to bright sunlight 12 - 36 hrs. after a therapeutic dose of phenothiazine. Sheep are comparatively resistant. Photosensitization occurs due to production of phenothiazone sulfoxide - a photodynamic pigment in the intestine which on absorption reaches the liver. In the liver, two metabolites- leucophenothiazone and leucothianol are produced by enzymatic biotransformation which is a rate limiting process. Therefore, excess of phenothiazine sulfoxide escapes liver and enters the general peripheral circulation. Dermatitis develops on exposure to bright sunlight. The damaged or the liver of young animals is less efficient in eliminating sulfoxide, it enters the aqueous humour of eye, leads to keratitis which is manifested by ulceration of cornea and ultimately blindness, particularly in cattle within 36 hours. The condition is more worst when bright sunlight comes after reflection from snow covered areas. In non-pigmented animals, skin of the animals is also affected. Characteristic signs are similar to that of sun burns i.e. reddening of lightly pigmented areas (muzzle, nose, ears, eyes), inflammation and thickening of the area around eyelids, muzzle etc. followed by sloughing off of the skin leading to ulceration, shaking of head and rubbing of the ears against hard objects to get relieved of itching. In calves, it is in the form of keratitis, in sheep keratitis alongwith reddening and thickening of the muzzle and ears and pigs in the form of dermatitis.

Treatment : No specific antidote is available. Give symptomatic treatment.

(i)   Remove the source of poison/exposure.

(ii)   Keep the animals in dark and shady areas.

(iii)   Give gastric lavage and saline purgatives to check further absorption of the toxicant.

(iv)   Replenish lost red blood cells by blood transfusion in horses.

(vi)   Intravenous infusion of glucose - saline solution.

(vii)   Administer antihistaminics and topically also apply antiseptic/antihistaminic cream.

(viii) Broad spectrum antimicrobial drugs may be given to check the secondary bacterial infections.

Aminoglycosides

Aminoglycosides are broad spectrum bactericidal antimicrobial agents and are primarily used against aerobic gram-negative bacilli. These drugs interfere with protein synthesis in susceptible microorganisms. Though these have an excellent therapeutic potential, the margin of safety or therapeutic window is narrow as plasma concentrations exceeding 12-15 µg/ml have been reported to be highly toxic. Thus, the aim of therapeutic dosing is that the peak concentrations should not be allowed to exceed 12-15 µg/ml. Certain predisposing factors to aminoglycosides toxicosis are age, hypovolemia, renal insufficiency, prolonged drug administration, hepatic diseases, concurrent administration of loop diuretics, nephrotoxic drugs, cephalosporins, acidosis and sodium or potassium depletion.

Aminoglycosides mainly produce two types of toxicity : Ototoxicity and nephrotoxicity, however, at times neuromuscular blockade and hepato-toxicity may also be observed at higher doses.

Ototoxicity : It is largely irreversible and results from progressive destruction of vestibular or cochlear sensory cells due to progressive accumulation of these drugs in the perilymph and endolymph of the inner ear. When concentrations in plasma are too high, accumulation occurs in these fluids. Half life of aminoglycoside antibiotics in otic fluid is 5-6 times longer compared to plasma. Streptomycin and gentamicin largely cause vestibular toxicity while neomycin, kanamycin and amikacin produce cochlear toxicity and out of these neomycin is most toxic and thus precluded from systemic use, however, tobramycin affects both equally. Therefore, it is suggested that aminoglycosides should not be used for prolonged treatment and patients be carefully monitored.

Nephrotoxicity : Aminoglycosides have high affinity for accumulating disproportionately in high concentrations in the renal tissues particularly proximal convoluted tubules of renal cortex. Most of the drug is excreted unchanged in urine during the first 12 hours, however, aminoglycosides can be detected in urine for 10-20 days after discontinuation of treatment. Early manifestations of nephrotoxicity are enzymuria, hypokalemia, glucosuria, alkalosis and hypocalcemia. The nephrotoxic effects are dose- dependent and reversible with the cessation of treatment. Histopathologically, these produce severe degenerative changes in renal tubules and glomeruli, interstitial fibrosis and perivascular lymphocytic infiltration.

Cationic aminoglycosides bind to anionic membrane phospholipids, the so called aminoglycoside receptors. Biochemically, aminoglycosides inhibit various phospholipases, sphingomyelinases and ATPases and alter the functions of mitochondria and ribosomes. These also impair the generation of membrane derived autacoids and intracellular second messengers [prostaglandins, inositol-1,4,5 triphosphate (IP_3), diacylglycerol (DAG)] and these, particularly the decrease in PGs are responsible for tubular damage. Lysosomes are the prime target sites. Mitochondria are the second possible target of aminoglycosides and impair tubule cell bioenergetics profile. Third possible site is the proximal tubular cell membranes phospholipids and glomeruli of nephrons.

Neuromuscular blockade : Aminoglycosides inhibit the prejunctional release of acetylcholine (ACh) and also reduce post synaptic sensitivity to ACh and thus produce a competitive type of neuromuscular blockade but potency is less than that of d-tubocurarine. The order of decreasing potency is : neomycin, kanamycin, amikacin, gentamicin and tobramycin. Clinically, interaction of aminoglycosides with other neuromuscular blocking agents or anaesthetic agents is very important and therefore, these must be used cautiously.

Hepatotoxicity : Daily administration of aminoglycosides, particularly gentamicin even at therapeutic doses has been reported to produce hepatotoxicity. Liver exhibits severe fatty changes and hyperplasia of Kuffer cells.

Treatment: There is no specific antidote or the line of treatment, however, one of the most important point is that the drug should be immediately withdrawn. Excessive fluid therapy should be administered.

Sulfonamides

Sulfonamides are one of the oldest groups of antimicrobial agents and still most popular in antimicrobial armamentarium. These are bacteriostatic and have a broad spectrum of activity and are effective against gram-positive, gram-negative and many protozoan organisms. Though these are considered very safe, yet produce some toxic manifestations or adverse effects.

Factors affecting toxicity :
(i)   Solubility of the sulfonamides.
(ii)   Volume of urine secretion.
(iii)   pH of the urine.
(iv)   Amount of sulfonamide excreted.
(v)   Degree of acetylation (particularly non-existent in birds, dogs and foxes).

Acute sulfonamides toxicity is rare, except following overdosing or rapid intravenous administration. Dogs, cats, swine, horses and calves have been reported to be affected. Signs of toxicity are increased salivation, diarrhoea, vomition, excitement, muscular weakness, ataxia, incoordination of limbs and drop in milk yield. Goats are more susceptible compared to other ruminants. Other signs are listlessness and disinclination to feed in horses and sleepiness, anorexia and incoordination in dogs.

Chronic sulfonamides toxicity is characterised by renal obstruction due to crystalluria, haemolytic anaemia in calves, decrease in total leucocyte and total erythrocytes count and haemoglobin values and jaundice in cattle and cyanosis in dogs. There is degeneration of myelin sheath in spinal cord and peripheral nerves in cattle and chickens, reduced egg production and weight gain and egg shells are thin and rough in poultry. Hypo-prothrombinemia in dogs and chickens particularly with sulfaquinoxaline is observed in dogs and chickens as it is a potent inhibitor of vitamin K epoxide reductase. Signs of toxicity are loss of appetite, depression, haematuria, albuminuria, peripheral neuritis etc.

Crystalluria : In dehydrated animals and acidic urine, sulfonamides get ionized, their solubility is decreased which results in crystal formation and blockade of renal tubules. Thus, there is crystalluria, obstruction of renal tubules and haematuria. The yellowish crystals can often be seen in the renal papillae and pelvis or even forming pale radial lines can be appreciated with the unaided eyes. Sometimes their amount is so high that these act as obstructive calculi in the ureter. Microscopically, crystals in the renal papillae are seen to lie within the ducts of Bellini and collecting tubules. Sometimes, these changes are accompanied by formation of albumin casts and causing more renal injury and mild to moderate degenerative changes in the proximal convoluted tubules. Symptoms of this condition are anorexia, renal colic, frequent attempts to urinate and elevation in blood urea nitrogen. Alkalinization of urine takes care of the problem as it increases their solubility and promote urinary excretion of sulfonamides.

Haemotoxicity : Prolonged administration of sulfonamides results in agranulocytosis and haemolytic anaemia in individuals deficient in glucose-6-phosphate dehydrogenase, aplastic anaemia (though rare) and hypo-prothrombinemia. Sulfonamides are contraindicated in neonates as these drugs displace bilirubin from protein binding sites and cause jaundice, kernicterus, coma and some times even death.

Postmortem lesions : Crystals of sulfonamides in renal tubules and pelvis of kidneys.

Prevention and treatment :
(i)   No specific antidote is available.
(ii)   Withdraw the drug immediately.
(iii)   Provide adequate water alongwith sodium bicarbonate or any other urinary alkalizer(s) to increase pH of the urine, particularly in carnivores.

Fluoroquinolones

Fluoroquinolones form a promising family of new bactericidal antimicrobials and have emerged as the antibacterials of the decade. These have an excellent broad spectrum antimicrobial activity and are highly active against gram-positive, gram-negative aerobes, mycoplasma, rickettsia and even some are also active against mycobacteria. These have good activity against intracellular pathogens, the minimum inhibitory concentration (MIC) values range from 0.01 - 2.0 µg/ml against *Salmonella*, *Staphylococcus*, *Brucella*, *Listeria*, *Mycoplasma* and *Chlamydia* etc.

Fluoroquinolones are generally well tolerated drugs with good safety margins. With few exceptions, the adverse effects of fluoroquinolones are not of severe consequence when compared to the beneficial features they exhibit. Toxicity of fluoroquinolones is mild at therapeutic doses and generally consists of gastrointestinal disturbances such as nausea, vomition, abdominal pain and diarrhoea. With slightly higher doses, dizziness, restlessness, headache, depression, somnolence or insomnia may be seen. High serum concentrations may produce immediate toxic reactions, possibly due to excessive release of histamine, such as convulsions, defaecation, urination and emesis. Epileptogenic activity of fluoroquinolones has been suggested possibly due to their *gamma* amino butyric acid (GABA) receptor antagonist property.

Fluoroquinolones are principally eliminated by glomerular filtration and renal tubular secretion and very high urinary concentrations (upto 300 times of that in serum) have been found. These have poor water solubility at acidic pH, thus high concentrations in acidic urine may not be eliminated and result in the formation of crystals in the urinary tract i.e. crystalluria particularly in human beings and dogs and other species voiding acidic urine.

One of the most adverse effects of the antimicrobials of this class is their effect on joints. These agents may cause primary degenerative lesions in the cartilage of weight bearing joints in growing animals. Clinically, the condition is manifested by lameness and pain. The articular cartilage forms vesicles after a single very large dose or after several moderately large doses which progressively rupture and produce cartilagenous erosions. Following withdrawl of fluoroquinolones, some repair is probable, although complete recovery may not take place. Thus fluoroquinolones are contraindicated in children, equines and young and growing animals. Other notable toxic / adverse effects of fluoroquinolones are photosensitization, maternal toxicity, with some embryonic deaths and cataract in human beings following their prolonged topical application. Certain drug interactions of fluoroquinolones are important e.g. concurrent administration of theophylline or caffeine with fluoroquinolones is not desirable as these decrease the rate of elimination of drugs. Concurrent use of non-steroidal antiinflammatory drugs (NSAIDs) with fluoroquinolones is also undesirable as it has been reported to produce convulsions in human beings.

Treatment : There is no specific antidote or the line of treatment, however, one of the most important point is immediate withdrawl of the drug. Excessive fluid therapy should be administered to decrease the load of drug in the body.

Non-steroidal antiinflammatory drugs (NSAIDs)

Non-steroidal antiinflammatory drugs - aspirin, phenylbutazone, ibuprofen, naproxen etc. are potent antiinflammatory, analgesic and antipyretic agents. These interfere with the cyclooxygenase pathway of prostaglandins biosynthesis. PGI_2 and PGE_2 stimulate gastric mucous synthesis and have a vasodilatory effect on renal capillary beds. Prostaglandins are essential to maintain cellular integrity of the gastrointestinal tract.

All NSAIDs produce potentially life threatening side effects. These mostly cause irritation, bleeding and ulceration of the gastric mucosa. Most common clinical signs of NSAIDs toxicosis are nausea, vomiting, CNS depression, circulatory collapse, hypo-prothrombinemia etc. Toxic effects of most of the NSAIDs are related to their effects on the biosynthesis of prostaglandins. Toxic doses produce respiratory depression, respiratory acidosis, convulsions, coma and death. Toxicity is more severe in cats due to cumulative toxicity. Long term use results in fatty infiltration in liver and kidneys and finally irreversible nephropathy terminating into death in certain cases. Therefore, long term administration of NSAIDs is not recommended. NSAIDs mainly affect the gastrointestinal and haematopoietic systems.

Gastrointestinal erosion and ulceration, injury to mucosal cells and submucosal capillaries and mucosal bleeding are the main toxic lesions produced due to inhibition of PGE_2-mediated HCO_3^- and mucus secretion, epithelialization and blood flow. NSAIDs also inhibit thromboxane synthesis in platelets and thus impair blood platelet activity. Platelet aggregation defects caused by aspirin last for upto one week. Some also produce bone marrow dyscrasias.

Nephropathic effects of NSAIDs are rather infrequent. However, the condition is aggravated in the presence of other nephrotoxic agents e.g. aminoglycosides, amphotericin B, or diuretics etc. Acute renal failure may occur as a primary problem due to vasoconstriction of renal arterioles. Hepatic and renal damage produced by phenylbutazone is more compared to other NSAIDs.

Toxicity may be diagnosed based on history, metabolic acidosis, Heinz body anaemia, elevated BUN and creatinine.

Treatment: Withdraw the drug immediately. Give emetics, activated charcoal and saline purgatives to prevent further absorption from the gastrointestinal tract. Give enough intravenous fluids, bicarbonate or sucralfate and maintain respiration etc. Misoprostol- PGE replacement @ 2-5 µg/kg every 8 hrs may be used as a specific treatment. N-acetylcysteine is the specific antidote for acetaminophen and 1-methylcysteine for salicylates.

Barbiturates

Barbiturates, particularly the ultrashort acting and short acting barbiturates are used as anaesthetic agents, and intermediate and long acting barbiturates are used as hypnotics and anticonvulsants. However, barbiturates are no longer considered appropriate for the treatment of anxiety or insomnia as barbiturates have a narrow margin of safety.

In very low doses, barbiturates may produce hyperactivity of the cortex instead of sedation, however, large doses of barbiturates produce persistent depression; depression extends to the hypothalamus and medullary centres of cardiovascular and respiratory controls and it is, most of the times, very fatal and the subjects die of cardio-pulmonary failure, particularly depression of respiratory centre. Signs of barbiturates poisoning are slow and shallow respiration, dilatation of pupil as hypoxia develops, pulse becomes weak and rapid, reflexes disappear and skin becomes cold and cyanotic, coma and death.

Treatment :

(i)   Remove the drug by gastric lavage, if given orally.

(ii)   Give diuretic and urinary alkalizers to promote urinary excretion.

(iii)   Support the cardiovascular and respiratory functions, the latter by artificial respiration using oxygen containing 5% carbondioxide to prevent hypoxia.

(iv)   Do not give CNS stimulants.

7.2 DRUG TOXICITY-II (MISCELLANEOUS DRUGS)

H.S. PANWAR

Therapeutic use of drugs is alarmingly increasing not only in human medicine but also in veterinary medicine. Though the drugs are and should be used as per the directions of the physicians in proper dosages, yet sometimes toxicity/adverse effects may be observed in human beings and animals, may be due to their injudicious use, overdosage or idiosyncrasy. It is not possible to give detailed account of the toxicity of all the therapeutic drugs. Nonetheless, an attempt has been made to briefly describe the symptoms and line of treatment of toxicity due to some of the commonly used drugs in tabular form (Table 7.1).

Table 7.1 : Toxicity of some of the commonly used drugs, their symptoms and line of treatment

Drug/Compound	Symptoms	Treatment
Adrenaline	Tachycardia, palpitation, high B.P., dyspnoea dilatation of pupil, dizziness and collapse. Death due to acute cardiac dilatation, pulmonary oedema and ventricular fibrillation.	Piperoxan Phentolamine
Apomorphine	Dogs-violent vomiting, accelerated respiration, stimulation of motor areas of the brain causes the animal to run round (circling movement), tetanic convulsions, death may result from asphyxia. Horse-delirium, sweating and difficulty in breathing.	Supportive treatment, opioid antagonists are contraindicated.
Atropine sulphate	Dry mouth, thirst, dysphagia, constipation, mydriasis, tachycardia, hyperpnoea, restlessness, delirium, ataxia, muscle trembling, convulsions, respiratory depression, and respiratory failure leading to death.	Neostigmine Pentobarbitone to control convulsions, saline purgatives to increase peristalsis.
Carbondisulphide	Excitement followed by muscular weakness and possibly collapse, coma and death.	Symptomatic
Cardiac glycosides	Depression, vomiting, diarrhoea and bradycardia. In more pronounced cases, tachycardia, pronounced bradycardia, alternating weak and strong cardiac systole, beat becomes rapid and irregular, heart stops with auricle dilated and full of blood and the ventricle contracted and empty.	Gastric lavage, emetics, purgatives, demulcents, atropine to check excessive vagal stimulation.

(*Continued*

Chloralhydrate	Relaxation of the voluntary muscles, staggering, mydriasis, lowering, of body temperature, deep stupor, death results from respiratory failure.	No specific treatment, supportive treatment includes artificial respiration, analeptics and fluid therapy.
Chloramphenicol	Neonates and felines seem to be most sensitive to intoxication. In cats-depression, dehydration, reduced fluid intake, weight loss, emesis, diarrhoea, bone marrow hypoplasia and pancytopenia while in dogs there are milder signs of toxicity, mainly gastrointestinal.	Discontinue the drug, haematinics
Chloralose	It has both stimulant and depressant action on CNS, weak heart beats, coma, hyperexcitability and aggression accompanied by incoordination, profuse salivation and convulsions.	Analeptics such as bemegride or methyl-amphetamine may be used.
Diamfenetide	Temporary impairment of vision and loss of wool.	Symptomatic
Dichlorophen	Ataxia, mydriasis, vomiting, depression, tremors and hypersalivation in dogs and cats.	Symptomatic
Dimethylsulfoxide (DMSO)	Sedation, diuresis, intravascular haemolysis, haematuria, death is preceded by hypotension, prostration, convulsions and respiratory distress characterized by dyspnoea, pulmonary oedema, ocular toxicity and teratogenicity.	Symptomatic
Dinitro-compounds	Listlessness, loss of appetite and activity, rapid respiration, sweating (in some animals only), thirst, oliguria, muscular weakness, prostration, dyspnoea, death with terminal hyperpyrexia.	Symptomatic, keep animal in cool environment, barbiturate, glucose saline, I/V sodium methyl thiouracil to reduce BMR.
Disophenol	Tachycardia, polypnoea, hyperthermia, lenticular opacity and vomiting.	Symptomatic
d-tubocurarine, gallamine	Salivation, fall of blood pressure, paralysis of neck, limb and finally diaphragm.	Symptomatic, neostigmine @ 0.022 mg/kg I/V, 4-aminopyridine.

(Continued)

Ephedrine Amphetamine Methyl-amphetamine	Anxiety, restlessness, tremors, convulsions.	In amphetamine type drug poisoning, chlorpromazine I/V at high thera-peutic dosage.
Ethylene glycol	Depression, ataxia, weakness, flaccid paralysis, tachypnoea, progressive hypothermia, terminal clonic convulsions, coma and death.	Symptomatic Antiemetics, gas-tric lavage with 5 per cent sodium bicarbonate, I/V sodium bicarbo-nate, barbiturate to control convul-sions, peritoneal dialysis.
Ketoconazole	Anorexia, nausea, vomiting, pruritis, alopecia, lightening and drying of the hair coat and weight loss, fatal idiosyncratic hepatitis and teratogenicity. Cats are more sensitive than dogs, dose-related inhibition of testosterone resulting in gynecomastia, sexual impotence and azoospermia.	Symptomatic
Milbemycin oxime	Depression, salivation, coughing, tachypnoea and emesis.	Symptomatic
Monensin	The acute oral LD_{50} of monensin (mg/kg) in vari-ous species is : horse 2-3, cattle 21.9 - 80, sheep 11.9, goat 24-26.4, fasted swine 16.7, unfasted swine 50, rabbits 41.7; dog 10 to more than 20 (females are more susceptible). Diarrhoea, colic, anorexia, nephrotic signs-polyurea, (later oliguria or anuria) possible haematuria and polydypsia, tachypnoea, hyperpnoea or dyspnoea, reluc-tance to move, ataxia, stumbling, poor foot plac-ing, exaggerated stepping, knuckling of hind feet (later the front foot also), incoordination, poste-rior paralysis, clonic convulsions and death.	Withheld the source, mineral oil, fluid therapy, selenium, vitamin E injec-tion.
Naturally occur-ring cholino-mimetic alkaloids -pilocarpine, muscarine and arecoline.	Severe colic and diarrhoea, evoke exocrine glands secretions, marked miosis, dyspnoea, hypotension, extreme cardiac slowing.	Atropine is the specific antidote.

(Continued)

Morphine	In dog-vomiting, delirium, clonic spasms, stertorous breathing. In cat, there is mainly motor excitement.	Nalorphine, lavellorphan, emetics (but not apomorphine), stomach wash with 0.2% potassium permanganate solution, respiratory stimulant, artificial respiration.
Nicotine sulphate	Marked excitement, rapid respiration, salivation, irritation of buccal and pharyngeal mucosa, diarrhoea and vomiting. The transient phase of stimulation is followed by depression with incoordination, rapid pulse, shallow and slow respiration, coma, flaccid paralysis and death due to paralysis of thoracic respiratory muscles.	Artificial respiration, stimulants, gastric lavage with potassium permanganate, administration of strong tea.
Nitrofurazone	In chicks - depression, rough feathers, hyperexcitability with whirling in circles and ending with convulsive movement like those of a decapitated bird.	Symptomatic
Oil of Chenopodium	Gastroenteritis, muscular spasms, convulsions or paralysis, coma and death.	Symptomatic, saline purgatives
Oxyclozanide	Diarrhoea, depression, anorexia and loss of weight.	Symptomatic
Parbendazole	Anorexia, loss of weight, teratogenicity in sheep.	Symptomatic
Physostigmine	First stimulates then depresses the CNS, marked skeletal muscle weakness, nausea, vomiting, colic and diarrhoea, constriction of pupil, dyspnoea, bradycardia, lowered blood pressure, respiratory paralysis.	Atropine is the pharmacological antagonist.
Penicillins	Acute anaphylaxis, collapse, hypersalivation, shaking, vomiting, urticaria, fever, eosinophilia, neutropenia, agranulocytosis, thrombocytopenia, leukopenia, anaemia and lymphadenopathy.	Symptomatic

(Continued)

Piperazine	Although toxicity is low, but occasionally produces transient neurotoxic symptoms in small animals. In calves , tympany, diarrhoea and anorexia.	Symptomatic
Praziquantel	Transitory vomiting and depression in dog and cat.	Symptomatic
Rafoxanide	Inappetence, diarrhoea, occular pathological changes (cataract and optic nerve degeneration), blindness in sheep.	Symptomatic
Tetracyclines	Gastrointestinal upset- irritation of stomach and upper small intestine, hepatotoxicity, tooth mottling and phototoxicity. Oxytetracyline also causes nephrotoxicosis, vomiting and diarrhoea.	Symptomatic
Turpentine oil	Colic, nausea, vomiting, diarrhoea, excitement, delirium, locomotor disturbances and later coma. There may be albuminuria and haematuria.	Symptomatic
Vitamin A	Xerophthalmia, hyperaesthesia, shivering, incontinence of urine, extreme tenderness of all epiphyseal junctions. In rats - foetal resorption and embryo deformity.	Symptomatic

CHAPTER 8

MYCOTOXINS

Satish K. Garg and S.P. Verma

8.1 INTRODUCTION TO MYCOTOXINS

Mycotoxins are the secondary metabolites of toxigenic fungi on animal feed and food ingredients which cause adverse biological effects when consumed in sufficient quantities. Many ubiquitous genera and species of toxic fungi are known to grow which produce their toxins when temperature, moisture and aeration are favourable.

Mycotoxins - induced morbidity in animals was suggested in 1901, however, mycotoxicosis in man and animals was recognised during the years 1942-1944 in Russia when cereal grains could not be harvested from the fields due to preoccupation of people with the war. The left over crops, which remained under snow, when harvested in the following spring and used as feed, fatal outbreaks of "*toxic alimentary aleukia*" were observed. Dogs were found to suffer with "hepatitis X" after feeding of commercial dog feeds containing pea nut meal in 1955. In 1957, similar conditions were observed in cattle and swine. Later in 1960, in an outbreak of obscure etiology, 100,000 turkeys died in England, the condition was termed as "*turkey X - disease*" at that time and later it was found to be due to aflatoxins present in the Brazilian ground nut meal heavily infested with *Aspergillus flavus*. These outbreaks and mortalities marked the era of significance of mycotoxins in the health of man and animals. Aflatoxin was first chemically isolated in 1962. Molds grow on a variety of stored feeds and standing plants especially seed heads e.g. ergots on rye, phymopsis on lupins. A number of poisonings initially thought to be plant poisoning were rather due to mycotoxins. Today it is estimated that around 25% of the worlds cereals are contaminated with known mycotoxins.

The toxins are mainly produced by the fungi when these grow on crops in the field, at harvest, in storage or during processing of feed when favourable conditions, particularly sufficient moisture are available. Minimum 15% moisture (10-33 %) and 90-95% relative humidity (>70%) are necessary for mold growth alongwith an ambient temperature of 24-25° C (4 - 35 °C).

Moulds infested feed alongwith nutritionally poor and inadequate diet makes the animals more susceptible to mycotoxins-induced deleterious effects. More than 100 toxigenic fungi and mycotoxins have been identified world over during the last three decades. Some of the important genera of moulds producing toxins are : *Claviceps, Fusarium, Aspergillus, Penicillium, Pithomyces, Stachybotrys, Rhizopus, Rhizoctonia, Acremonium, Phomopsis* etc.

Formation of mycotoxins is a global problem. Some of the commonly encountered mycotoxins in various naturally contaminated food and feeds are aflatoxins, ocharotoxin-A, patulin, zearalenone, trichothecenes, citrinin etc.

Aflatoxins are not a major problem in colder, more temperate regions (Canada and Northern parts of USA) and most of the European countries. Most important mycotoxins in these region are zearalenone, ocharotoxin, T-2 toxin, HT-2 toxin. However, in warm and humid climates like those in Asia, Latin America and African countries, aflatoxins are most widespread of all the mycotoxins. During winter season coupled with high moisture conditions, other mycotoxins such as zearalenone, T-2 toxin, ocharotoxin etc. may also be noticed. In southern parts of India, hot and humid climate during most parts of the year also favours the presence of aflatoxins, ocharotoxin and T-2 toxins.

Mycotoxins adversely affect the health and productivity of almost all species of domestic animals and birds, however, growing, pregnant and lactating ones are worst affected. Aflatoxins affect all domestic animals, zearalenone and vomitoxin mainly affect swine and dairy animals, fumosins mainly affect swines and equines while T-2 toxin, ocharotoxin, diacetoxyscirpenol (DAS) affect mainly swine and poultry. Diagnosis of mycotoxicosis is generally difficult and multiple infections or combinations make it still difficult. However, progressive poor condition, performance and production of the animals, in the absence of specific well defined disease/signs gives an indication of mycotoxicoses and the same should be confirmed in the laboratory.

Some of the mycotoxins have well defined organ specificities e.g. aflatoxins for liver, ocharotoxin and citrinin for kidneys, ergot and zearalenone for uterus etc. and have accordingly been termed as hepatotoxins, nephrotoxins, oestrogenic toxins, neurotoxins, cytotoxins etc.

Mycotoxins in general affect the functioning and performance of almost all the body systems- gastrointestinal, reproductive, nervous and immune systems etc. The physical or apparent effects of mycotoxins range from reduced feed intake, poor food conversion efficiency and a general inability of animals to thrive. Symptoms of toxicity vary from toxin to toxin.

8.2 AFLATOXINS

Aflatoxins are the toxic metabolites of toxigenic fungi *Aspergillus flavus* and *A. parasiticus*. These moulds invade and grow on all sorts of stored feed ingredients and organic matters and produce mycotoxins under favourable aerobic conditions when sufficient moisture (>15%) and relative humidity (90 - 95%) and ambient temperature (24 - 25°C) are available, though toxin production to some extent can take place at lower or higher temperatures as well.

Aflatoxins are produced in varying quantities in a variety of grains and nuts, pea nut is the most important one. Cotton seed meal, cotton seed cake, maize corns, corn meals, silages, wheat, barley, oats, rice and other grains or cereal products are invaded by these fungal toxins.

Several aflatoxin fractions have been isolated and differentiated from one another by their fluorescence under ultraviolet, RF values on thin layer chromatography and bstructural identification and synthesis. Aflatoxins which fluoresce blue under ultraviolet light are termed as B_1 and B_2 while others which are dihydroderivatives of B_1 and B_2 and fluoresce green are designated as G_1 and G_2. Hydroxylated metabolites of B_1 and B_2 aflatoxins are excreted in milk and are termed as M_1 and M_2. Human exposure occurs on consumption of aflatoxins from the mycotoxins- infested feed stuffs as well as meat and milk of animals consuming toxigenic fungi contaminated feeds.

Out of the four main aflatoxins, aflatoxin B_1 (AFB$_1$) is most important because of its toxicity and high concentration in the contaminated feeds. B_2 and G_2 are present in very less concentrations. The order of toxicity is : $B_1 > G_1 > B_2 > G_2$. These toxins are relatively heat resistant and not soluble in water. The melting points for B_1, B_2, G_1 and G_2 aflatoxins are 269°, 238°, 245° and 239°C, respectively. Even the toxigenic fungi or toxins are not destroyed by milling of the grains.

Aflatoxins are a group of polycyclic, unsaturated, substituted coumarin derivatives with highly reactive bifuran nucleus on one side and a pentenone (B toxin) or a six membered lactone (G - toxin) ring on the other side as shown in Fig. 8.1.

Presence of additional oxygen in the AFG compounds results in reduction of their activity by a factor of 2 while unsaturated compounds are 4-5 times more potent compared to dihydro derivatives. The structure of B toxin resembles closely with the structure of pyrrozolidone which has got basic carcinogenic activity. Aflatoxins are potent mutagens, carcinogens, teratogens and liver damaging agents.

Mechanism of toxicity : Majority of the aflatoxins ingested in feed are physically bound to ruminal contents and as little as 2-5% reach the intestines. Aflatoxin B_1 in excess of 100 µg/kg of feed is considered to be toxic to cattle. Aflatoxins are rapidly absorbed from the intestine and bound to blood proteins. Liver removes most of the toxins from the blood stream. AFB$_1$ is primarily metabolized by a microsomal cytochrome P-450 dependent mixed function oxidase system in the liver and other organs to AFB$_1$-2, 3 epoxide mainly, which is a highly reactive intermediate. Other metabolites of AFB$_1$ are Q_1, P_1, B_2 and aflatoxicol. Epoxides are chiefly detoxified by glutathione

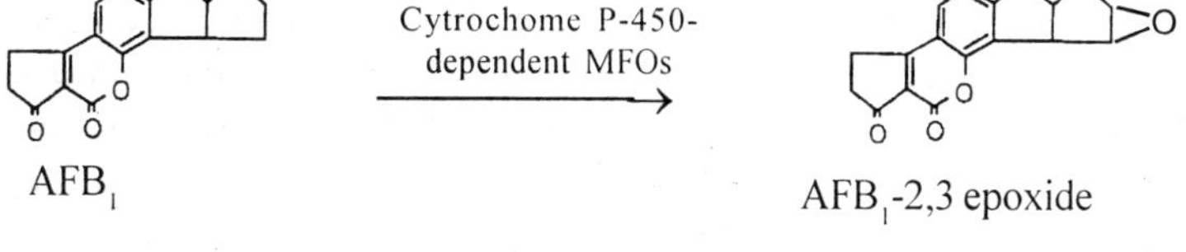

Cytrochome P-450-
dependent MFOs

AFB$_1$ AFB$_1$-2,3 epoxide

Fig. 8.1 : Chemical structure of Aflatoxin B$_1$ and Aflatoxin B$_1$-2,3 epoxide.

(GSH)-S-transferase and epoxide hydratase to aflatoxicol to some extent which subsequently are conjugated and excreted. However, detoxifying mechanisms are limited; when GSH-S-transferase gets exhausted due to excess of epoxides, aflatoxicosis is produced.

Excess of reactive products of aflatoxin B_1 interact with [7]N-guanyl residue of nuclear DNA of hepatocytes to inhibit synthesis of DNA, DNA-dependent RNA polymerase activity, messenger RNA synthesis and protein synthesis by interfering with transcription. AFB_1 also binds to endoplasmic steroidal ribosome-binding site causing disaggregation of ribosomes. Number of ribosomes decrease, there is reduced proliferation of smooth endoplasmic reticulum, loss of glycogen and degeneration of mitochondria. Impaired protein synthesis interferes with the formation of certain enzymes required for energy metabolism and fat mobilization. As a result, there is hepatic steatosis.

Cytotoxicity and carcinogenicity are due to macromolecular binding, not only of AFB_1 but also other metabolites. Haemorrhages, fatty liver and immunosuppression are due to inhibition of clotting proteins and protein synthesis.

Toxicity : The LD_{50} of AFB_1 is 0.3 - 9 mg/kg for all species of animals, birds and fishes tested. All species of animals including human beings are affected. Cattle, sheep and other ruminants appear less susceptible than the monogastric animals and poultry. Sheep and mice are comparatively most resistant whereas cats, dogs and rabbits are most sensitive. The susceptibility of various species of animals in decreasing order is : duckling > rabbit > turkey, chicken> neonatal rat > cat, pig, trout, guinea pig, rhesus monkey, adult rat, cattle and sheep. Besides species variations, age, sex, breed, strain of the animal, nutritional status of the animals also significantly affect the toxicity of aflatoxins e.g. diet rich in riboflavin, low in proteins, cholines, vitamin B_{12} and exposure to light enhance while high dietary proteins, lipids and carotenes lower the susceptibility to aflatoxins-induced toxicity.

Clinical signs : Aflatoxicosis may be acute, subacute or chronic but latter is the most commonly observed syndrome and is particularly characterised by high incidence of hepatic tumours.

Acute toxicity : Acute primary aflatoxicosis mostly occurs sporadically as a farm disaster when high to moderate amount of the toxin is consumed. Sudden deaths within 72 hours without much symptoms of toxicity like anorexia, depression, ataxia, dyspnoea, anaemia, haemorrhages, bloody faeces, tremors, convulsions and death. Most sensitive species to acute toxicity are rabbits, cats, ducklings, mink, trout, dogs, turkey and piglets. Most resistant species are monkeys, chicken, rats, mice and hamsters. However, horses, cattle, sheep, goats, adult pigs and guinea pig are moderately susceptible. A dose rate of 4 mg/kg of aflatoxin causes death of sheep, calves and pigs with in 15 -18 hours due to acute hepatic insufficiency while with 2 mg/kg, there is anorexia, increased respiration rate, increase in body temperature and diarrhoea with blood and mucus.

The LD_{50} values (mg/kg) of AFB_1 for some species of animals and birds are presented in Table 8.1.

Table 8.1: LD$_{50}$ values (mg/kg) of AFB$_1$ in some species of animals and birds.

Species	LD$_{50}$ value (mg/kg)
Rabbits	0.3 - 0.5
Ducklings	0.5
Cats	0.3 - 0.6
Dogs	0.5 - 1.0
Cattle	0.5 - 2.0
Horses	>2.0
Chicken	>2.0

The clinical signs in dogs appear in 2-14 days (average 5 days) and are characterized by anorexia, icterus, bile stained urine, prostration, occasionally blood in faeces, vomition (sometimes bloody), epistaxis and rarely convulsions.

Subacute toxicity: Consumption of sublethal concentrations of aflatoxins for several days or weeks causes this condition which is characterised by symptoms of icterus, hypoprothrombinemia, haemorrhages and haematomas.

Chronic toxicity : Chronic aflatoxicosis is the most commonly occurring syndrome in domestic animals, birds and human beings due to continuous intake of low levels of aflatoxins in feeds/foods for weeks and months. There are no overt signs of toxicity in the beginning, however, 1-2 months or more later, gradual decrease in feed efficiency, weight gain, productivity, icterus, ascites, oedema of lungs and abortion in pregnant animals may be observed. In cattle, there is blindness, twitching of ear, grinding of teeth, circling movement, frothing in the mouth, photosensitive dermatitis, keratoconjunctivitis, haematomas, ataxia, diarrhoea, anal prolapse, recumbency and convulsions, followed by death within 48 hours. Placental transfer of aflatoxins has been reported to cause cirrhosis of liver in calves. Tremors, lameness, recumbency and heavy mortality continuing for several weeks is observed in lambs. Pigs also exhibit similar signs; there are intermittent haemorrhages and diarrhoea. Animals exhibit seizures before death.

Incidence of aflatoxicosis in horses is comparatively less because these animals are not likely to be fed with damaged feeds. However, horses fed on a feed containing AFB$_1$, as low as 250 ppb, develop typical signs of hepatic toxicity and gastrointestinal upset. Death has been reported in 37-39 days in ponies following AFB$_1$ feeding @ 0.075 mg/kg/day, 26-32 days after 0.15 mg/kg/day and 12-16 days after 0.3 mg/kg/day. Sheep, however, are refractory to these effects as the ewes fed on highly toxic groundnut meal for 5 years failed to exhibit any clinical signs of aflatoxicosis except for reduced fertility. Probably, most of the aflatoxins appear to be destroyed in the body.

In poultry, aflatoxins interfere with absorption of lipids and transportation of yolk. Thus, the egg production and egg size are reduced. Aflatoxicosis increases susceptibility of turkeys to coccidiosis, pasteurellosis and salmonellosis and of chicken to coccidiosis and Marek's disease.

Laboratory investigations : AFBs inhibit transportation of fats from the liver, so there are fatty degenerative changes in the liver. Blood examination of aflatoxicosis affected animals would reveal elevation in blood serum levels of AST, alkaline phosphatase, isocitrate dehydrogenase, lactate dehydrogenase, *gamma*-glutamyl transpeptidase and glutamate dehydrogenase and bilirubin.

Blood clotting defects and extensive haemorrhages are observed in almost all species of animals due to impairment of blood clotting factors - II, VII, IX and X. Haemolytic syndrome is very common in poultry. Decrease in serum total proteins and elevation in blood urea nitrogen are also observed. Both the cell-mediated and humoral immune response are suppressed.

Post mortem lesions :
(i)   Liver is pale, firm and fibrosed.
(ii)   Microscopically, centrilobular necrosis, bile duct proliferation and veno-occlusion are the main changes. Hepatocytes are swollen, there are multiple foci of necrosis and fibrosis and hepatic carcinoma.
(iii)   Kidneys are yellow and surrounded by wet fat in young calves.
(iv)   Serous exudate in the body cavities.
(v)   Oedema and ascites of mesentery.
(vi)   Catarrhal enteritis.
(vii)   Eversion of rectum.
(viii)   Haemorrhages in thoracic and peritoneal cavities. Liver is yellow and mottled in dogs, however, it is off white to bright orange alongwith proliferation of bile duct, subcutaneous haemorrhages, ascites and oedema of mesentery in pigs.
(ix)   Diarrhoea and dysentery.

Diagnosis :
(i)   History.
(ii)   Clinical signs.
(iii)   Detection of AFTs in feed, blood and milk etc.
(iv)   Laboratory investigations.
(v)   Post mortem changes - both gross and microscopic lesions.

Differential diagnosis:
(i)   Warfarin or dicoumarol (oral anticoagulant rodenticides) poisoning.
(ii)   Copper poisoning.
(iii)   Poisoning by other hepatotoxins e.g. carbontetrachloride, pyrrolizidine alkaloids.
(iv)   Infectious hepatitis.
(v)   Coal tar poisoning.
(vi)   Other moulds toxins.

Prevention and treatment :

(i)   Contaminated feed must be withdrawn immediately.

(ii)   Provide easily digestible low fat and high protein diet/feed.

(iii)   Supportive therapy with multivitamins.

(iv)   Use 0.5% hydrated sodium calcium aluminosilicate as feed additive in the feed of pigs and lambs. It adsorbs aflatoxins thus reduces their absorption from the GIT.

(v)   Anabolic steroid stanozolol (2 mg/kg) by IM injection at 4-5 days interval to decrease hepatic damage.

(vi)   Activated charcoal @ 6.7 mg/kg intraruminally as 30% W/V slurry in M/15 phosphate buffer of pH 7.

(vii)   Oxytetracycline (10 mg/kg, IM once daily)

(viii)   Intravenously administer sufficient quantities of 5% dextrose.

(ix)   Supplement the diet with vitamin E and selenium to ameliorate the effects of aflatoxins.

(x)   Supplement the diet of affected animals with hepatotonics.

(xi)   Prevent undue stress to the affected animals.

Note: FDA considers aflatoxins in milk and meat as a serious threat to human health and a serious regulatory violation.

8.3 ERGOT ALKALOIDS

Ergotism, now rare, was first associated with the consumption of ergot-infested grains in the mid-sixteenth century. A more vivid account including human misery, the legal pat-a-cake, the dodging of responsibility and failure to remove the suspected food have been mentioned in the book entitled "*The Day of the St. Thomas Fire*" by Fuller.

Sources of poisoning : Ergot is a parasitic fungus (*Claviceps purpurea*) which invades the flowers and spikelets of cereals, particularly rye, oats, barley, wheat and grasses. Sclerotium, the toxic element of the fungus, is a hard, black elongated body which destroys and replaces the grains or seeds of the mature plants. If these are not harvested with the host plant, the sclerotia drop to the ground, overwinter and produce millions of spores which spread infection in the following spring. These sclerotia constitute the substance known as ergot which spreads on the grass. In case of livestock, losses are due to ingestion of fungus infested grasses while in humans due to contamination of wheat flour with the rye. Heaviest infestations occur in winter season.

Claviceps purpurea is the main fungus, other ergot moulds are *C. paspali* which infects seed heads of *Paspalum dilatatum* and *P. notatum* while *C. cinerea* infects *Hilaria mutica* and *H. jamesii*. Warm, moist and humid seasons are conducive for the growth of fungus and infection of cereals and grasses.

Ergotism commonly occurs in cattle, sheep and other animals and usually in stall-fed animals feeding on heavily contaminated grains over a considerable period of time.

Ergot contains a number of pharmacologically active alkaloids, namely - ergotamine (Fig. 8.2), ergometrine, ergotoxin (a mixture of ergocornine, ergocristine and ergocryptine) and others alongwith other amines such as acetylcholine, histamine etc. which vary with the maturity of the ergot. These ergot alkaloids are derivatives of lysergic acid (LSD), which contains the indole structure, many of which are CNS stimulants.

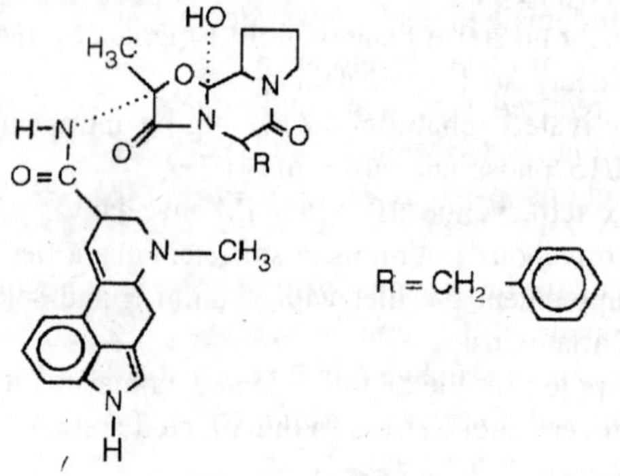

Fig. 8.2 : Chemical structure of ergotamine.

Ergot alkaloids are CNS - and smooth muscle stimulants and their basic actions are : vasoconstriction, uterine contraction, adrenergic blockade, serotonin antagonism, medullary effects (vomiting, bradycardia, inhibition of vasomotor centre and baroreceptors) and CNS stimulation.

Toxicity (Ergotism) : Ergot infested pastures may cause the disease. Animals may show early signs of lameness, irregular gait and evidence of pain in the feet, the posterior extremities being chiefly affected as early as within 10 days, but most of the animals do not become affected until 2-4 weeks after exposure.

There are two forms of ergotism :
    (i)   Gangrenous ergotism or chronic ergotism.
    (ii)   Nervous or convulsive ergotism or acute ergotism.

(i) Gangrenous ergotism and its pathogenesis : Chronic ergotism is mainly because of *C. purpurea* due to disturbances in the vasomotor system. The ergot alkaloids are potent smooth muscle stimulants, cause intense vasoconstriction, elevate blood pressure and induce strong uterine contractions (oxytocic effect). Excessive vasoconstriction damages capillary endothelium and causes vascular stasis and blockade of capillaries flow which results in dry gangrene formation due to thrombosis and subsequently sloughing off of hooves, ears and tail after several weeks of ingestion of small amounts of ergot alkaloids.

Clinical signs : Early stages of ergotism are initially characterised by reddening, swelling, coldness, loss of hair or wool and lack of sensation of the affected parts followed by lameness, irregular gait and evidence of pain in the feet. Palpation of the affected parts gives indication of cold and loss of sensation. Later, the affected parts look swollen, necrotic and eventually slough off if they consume a ration containing

0.3-0.5 % ergot sclerotia for several weeks. There is usually a clear line of demarcation and an inflammatory zone just proximal to the affected part. In animals, severe diarrhoea, stiffness of leg joints and cold feet and extremities is an indication of gangrenous ergotism. In pregnant animals, abortion is quite often. In birds, gangrene of the comb, wattles, tongue and beak are observed in addition to vesicular dermatitis.

In female piglets (sows), feeding of a ration containing ergot @ 0.5 - 1.0 %, there is lack of udder development and complete agalactia. Other associated general signs are reduced feed intake and weight gain, polydipsia and polyuria. Vitality of the new born piglets and weight gain is decreased. Subsequently, there is gangrene of the ear and tail tips.

(ii) Convulsive or nervous form of ergotism : Acute ergot poisoning is comparatively rare in animals, still commonly occurs in carnivores, horses and sheep and only rarely in cattle when animals have a free access to parasitised seed heads of grass (*Paspalum dilatatum*) with *Claviceps paspali* for few or several days. This form of ergotism not only depends on the rate of ingestion but also varies with the climate, location and species. Indoles and lysergic acid derivatives probably result in stimulation of CNS by interfering with the functions of brain neurotransmitters. Ergot alkaloids mimic the action of dopamine in CNS.

Clinical signs : Symptoms of acute ergotism are transient and characterised by hyperirritability, excitability, muscular incoordination, ataxia, aggressiveness, kicking, weakness, recumbency, tremors, fatal convulsions and death following respiratory failure. Increased heart rate, intermittent blindness and deafness are also observed. Appetite appears normal unless pharyngeal paralysis or the muscles involved interfere with ingestion. Skin may also indicate alternating periods of increased and decreased sensitivity. Gangrene of the extremities may also be observed in this form of poisoning. Affected cows show hyperthermia (105 - 107 °F), dyspnoea and hypersalivation. Milk production and growth rate are depressed. Hyperthermic syndrome is more severe in hot weather. Morbidity is about 100 per cent.

Diagnosis :

(i)   History.
(ii)   Clinical signs.
(iii)   Examination of the hay, straw or grains for the toxigenic fungi or mycotoxins.
(iv)   Detection of the ergot alkaloids in body fluids and tissues.
(v)   Sclerotia may be found on the grass heads, grains or hay.
(vi)   Abortion in pregnant animals alongwith other signs.
(vii)   Sloughing off of the hooves, feet or tail in severe cases.

Post mortem lesions :
(i)   Necrotic lesions - gangrene of the extremities is almost diagnostic of the condition.
(ii)   Grossly, mammary glands of females in late pregnancy are small and flaccid without any evidence of lacteal secretion.

(iii)   Evidence of congestion, arteriolar spasms and capillary endothelial degenerative changes.

(iv)   Ulceration and necrosis of the oral, pharyngeal, ruminal and intestinal mucosae are characteristic in sheep.

Differential diagnosis :

(i)   Rule out the infectious diseases, trauma, abscess, neoplasms, haemorrhages as the cause of CNS stimulation for acute ergotism.

(ii)   Deficiency or excess of selenium from chronic ergotism.

(iii)   Other mycotoxicoses due to *Fusarium* sp. or *Aspergillus* sp. etc.

Treatment : No specific antidote is there.

(i)   Offending feed, forage etc. should be immediately withdrawn.

(ii)   Provide a warm, clean and stress free environment.

(ii)   Give symptomatic treatment.

(iii)   Oral purgatives (e.g. magnesium sulphate) may eliminate some unabsorbed toxin from the gastrointestinal tract.

(iv)   Broad spectrum antibacterials for treating necrotic lesions and to prevent secondary bacterial infections.

8.4 HEPATOTOXIC MYCOTOXINS

Rubratoxins: Rubratoxins are the toxins elaborated by the fungi *Penicillium rubrum* and *P. purpurogenum* which infest maize, legumes, cereals, peanut kernels and pods, sunflower seeds and bran. Liver is the main target site for rubratoxins.

Rubratoxin A is the minor metabolite while rubratoxin B is the major metabolite and most toxic too of the *Penicillium rubrum* mycotoxins. These generally co-exist with aflatoxins and contaminate the feed stuffs. On equal weight basis, rubratoxin is less toxic compared to aflatoxins and both these mycotoxins probably potentiate the toxicity of each other i.e have the synergistic effect.

Rubratoxins are polycyclic compounds having *alpha* and *beta* unsaturated lactone rings. These are stable at room temperature but heating at 85 - 100° C for two hours may destroy or alter the *beta* toxins in feed stuffs.

Swine, dogs, cats, goats and horses are generally affected, swine, however, are most susceptible. Laboratory animals are comparatively resistant. The acute oral LD_{50} of rubratoxin B in rats is 400-500 mg/kg in dimethyl sulfoxide while in two days old rat pups is 6.4 - 6.9 mg/kg in corn oil.

Though rubratoxins are potent hepatotoxins similar to aflatoxins but unlike the latter, these have no carcinogenic activity. Rubratoxins produce extensive haemorrhages throughout the body but liver is the main site of injury. In mice, rubratoxins have been found to be teratogenic, embryocidal and growth- and immuno-suppressant. Detoxification mechanisms of rubratoxins are similar to that of aflatoxins.

Mechanism of toxicity : Mechanism of action of rubratoxins is similar to that of aflatoxins. Rubratoxin B or its toxic metabolites due to the presence of lactone moiety bind with cellular macromolecules DNA, RNA and others and alter the DNA, RNA polymerase or protein functions and also cause disaggregation of ribosomes and inhibit synthesis of certain proteins and enzymes including ATPases. Unlike aflatoxins, mitochondria are not the site of action of rubratoxin B. These block the electron transport chain between cytochrome C_1 or C and its terminus while aflatoxins block between cytochrome b and C_1 or C. Rubratoxin B-induced cell lethality and hepatic necrosis is because of its effects on ATPases and electron transport system.

Clinical signs : Anorexia, depression, dehydration, loss in weight, colic etc. are observed in affected animals. Pigs press their head against hard objects. Acute poisoning in horses is characterised by anorexia, depression, profuse bloody diarrhoea, foul smelling faeces, incoordination, recumbency on day 4 or 5 and terminal convulsions.

Post mortem lesions :
(i)   Icterus.
(ii)   Liver damage, necrotic foci and extensive haemorrhages in various body organs.
(iii)   Haemorrhagic enteritis.
(iv)   Microscopically, massive haemorrhagic necrotic changes.
(v)   Renal damage is mild or absent except in dogs where there are degenerative changes in the renal tubular epithelium.
(vi)   Massive ascites and oedema of the wall of gall bladder in cats.
(vii)   Brain haemorrhages in horses.

Diagnosis :
(i)   History.
(ii)   Clinical signs.
(iii)   Post mortem findings.
(iv)   Detection of rubratoxins in urine and feed.

Differential diagnosis : Rubratoxicosis may be differentiated from other mycotoxicoses, particularly aflatoxins.

Treatment : No specific antidote is available, give symptomatic and supportive treatment.

Sporidesmin : Sporidesmin is the toxic metabolite produced by the soil fungus *Pithomyces chartarum* (*Sporidesmius. bakeri*). The fungus grows and sporulates on perennial ryegrass and other pastures, infests dead plant materials in standing pastures when pasture is short and contains recently killed plant materials in abundance when climatic conditions are warm and humid. The mold lives saprophytically on the debris at the roots of growing grass. Rains help in percolation of the toxin from the spores and sunlight detoxifies the toxin.

Sporidesmin A to H are epipolythiadioxo-piperazine compounds and are potent hepato-toxic agents and known to cause facial eczema in sheep and cattle. Other species of affected animals are mice, rabbits, guinea pigs, goats and horses. Compared to sheep, goats are more resistant. Rats are much more resistant as liver damage is very little while pulmonary oedema is the most marked effect in rats. Morbidity rate in sheep is 70-80%, 5-50% of the affected animals may die and in the survivors, weight gain is poor. Morbidity rate is lower in cattle (50 %). Concomitant ingestion of *Tribulus terrestris* enhances toxicity of *Pithomyces chartarum.*

Mechanism of toxicity : Exact mechanism of toxic action is not known. Sporidesmin is a potent hepatotoxin and causes extensive liver damage and also damages biliary epithelium leading to acute biliary occlusion. The biliary secretions are not allowed to pass to the intestine. Phylloerythrin having photosensitizing properties are not eliminated into the intestine. Thus, phylloerythrin, the metabolic product of chlorophyll reaches the skin via circulating blood and causes secondary photosensitization, referred to as *facial eczema* in sheep and cattle.

Due to severe hepatic insufficiency, there is loss of condition and obstructive jaundice.

Clinical signs: This syndrome is most commonly observed in sheep and cattle. Onset is sudden, there is dullness, anorexia, jaundice and photosensitive dermatitis. Other associated signs are hyperirritability, lacrimation and nasal discharge. Eyes become inflammed and ears, eyelids and face are swollen. Many animals die during acute stage and in the survivors, condition of the animals become poor.

Post mortem lesions :
(i)   Liver is enlarged and mottled.
(ii)   Bile duct walls is thickened.
(iii)   In chronic cases, liver is tough and gets fibrosed.
(iv)   Microscopically, there is perilobular fibrosis, obliteration of the bile duct and atrophy of hepatocytes and spongy vacuolation of brain.

Diagnosis :
(i)   History.
(ii)   Clinical signs.
(iii)   Laboratory investigations indicating extensive liver damage.
(iv)   Post mortem findings.
(v)   Detection of sporidesmin in the feed and urine of affected animals.

Differential diagnosis :
(i)   *Lantana* poisoning.
(ii)   Photosensitization.
(iii)   Hepatotoxicity due to other agents.

Treatment :
(i)   Supportive treatment for hepatitis and photosensitization.
(ii)   Antibiotics and antihistaminics.
(iii)   Zinc sulphate @ 6 g /100 L for 28 days will help in recovery of affected animals.

Control :
(i)   Change the pasture.
(ii)   Use fungicides on the pastures/fields (carbendazin @ 0.15-0.30 kg/hectare or benomyl or thiophanate methyl @ 0.30 kg/hectare).
(iii)   Spray of zinc oxide on the pastures.
(iv)   Add zinc oxide to the drinking water of affected animals.

Other hepatotoxic mycotoxins :

(a) Luteoskyrin and cyclochlorotine : These mycotoxins are produced by *Penicillium islandicum.* These produce fatty degeneration of liver, primary malignant hepatomas and hyperplasia of bile duct, focal necrosis and haemorrhages in liver. These toxins caused great loss due to heavy mortality in chicken (>20%) in Japan in 1963.

(b) Sterigmatocystin : It is produced by the mold *Aspergillus versicolor* and *A. nidulans* and *Bipolaris* sp. Mice are very resistant to it, however, in rat it produces hepatic and renal necrosis and hepatocarcinoma while severe hepatic and renal damage in monkeys.

(c) Citrinin : This mycotoxin is elaborated by the fungus *Penicillium citrinum, P. viridicatum* and *P. palitans*, the common substrates for which are barley, oats, rye, wheat and corn. Citrinin occurs as a co-contaminant with ocharotoxins. Toxic effects of citrinin are similar to those of ocharotoxins. It mainly produces acute hepatic and glomerulonephrosis. Pigs are highly sensitive. It damages kidneys and causes enlargement of collecting tubules.

8.5 NEPHROTOXIC MYCOTOXINS

Ocharotoxins : It is produced by the mould *Aspergillus ochraceus* and *Penicillium viridicatum* infesting barley, corn, wheat, rice, oats, rye, green coffee beans and pea nuts. The production of toxins continue during long-term storage of grains. The toxins are stable and are slowly decomposed during prolonged storage.

Ocharotoxins is a group of nine isocoumarin derivatives linked with phenylalanine by an amide bond. Ocharotoxin A (Fig. 8.3) is the most common of all the ocharotoxins and has the greatest toxicological significance. These are potent nephrotoxins and the affected animals / birds suffer with renal damage and the syndrome is termed as *mold nephrosis* or *mycotoxic nephropathy.*

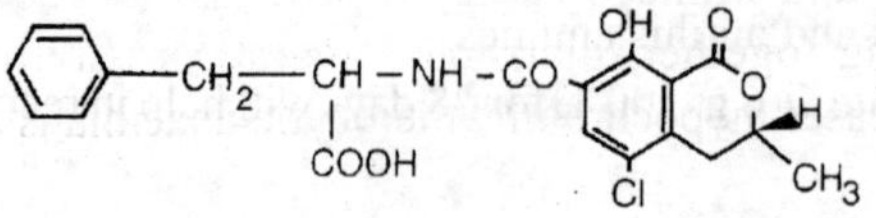

Fig. 8.3 : Chemical structure of ocharotoxin A.

Pigs, poultry, goats, horses, cattle and mice are affected. Though, pigs are most often involved, but birds are most susceptible to toxicity. The acute LD_{50} value ranges between 2 ppm for birds and 59 ppm in mice.

Ocharotoxins are immunosuppressant, reduce sperm quality in boars and cause foetal death and resorption and abortion in sows while liver and other organs are minimally affected. Its teratogenic effects in rodents include malformation of the head i.e. short head (hydrocephalus), jaws and tail, oligodactylia and heart. These are not mutagenic but appear to induce hepatomas and renal adenomas in 30% of the mice exposed to 40 ppm in the diet.

Ocharotoxin A is hydrolysed to nontoxic ocharotoxin *alpha* by carboxypeptidase A and *alpha* chymotrypsin. The absorbed ocharotoxin A mainly distributes to the kidneys and liver and is excreted in urine and faeces. It accumulates in the body tissues and fats of pigs and poultry. Ocharotoxin A disappears from the muscles and fat of affected pigs after two weeks and from liver and kidneys after 3-4 weeks, respectively. The half life of ocharotoxin A residues in swine tissue is 3-5 days and minute quantities are detected in kidneys for 30 days even after removal of the contaminated feed. Thus, it has been detected as a carryover in pigs and poultry meats which is important from public health point of view. It does not appear to cross the placental barrier and thus foetal pigs are not affected.

Mechanism of toxicity : Ocharotoxin A and ocharotoxin *alpha* inhibit mitochondrial respiration and oxidative phosphorylation and reduce ATP levels. It also interferes with the functioning of several mRNAs involved in the synthesis of various c-AMP mediated enzymes including phosphoenolpyruvate carboxykinase and other enzymes of gluconeogenesis pathway.

It being a potent nephrotoxin causes organellar damage, especially to tubular epithelial cells and impairs proximal tubular function. These decrease metabolic clearance and urine concentrating ability, inhibit anion transport and cause release of renal brush border enzyme e.g. leucine aminopeptidase. There is periglomerular and interstitial fibrosis, tubular atrophy, thickened basement membranes, glomerular sclerosis and fibrosis. Kidneys of affected pigs become enlarged and greyish in colour.

Clinical signs : Affected animals show signs of anorexia, depression, fatigue, lassitude, abdominal pain, diarrhoea, polydipsia, polyuria, dehydration, loss of weight and severe anaemia followed by renal damage. Death usually results from **uremia**.

In poultry, food consumption, growth rate, egg production and egg and body weight are reduced and kidneys are enlarged. However, in acute conditions, listlessness, crowding together, diarrhoea, ataxia, prostration and death may be observed. In severe cases, hypochromic microcytic anaemia is also noticed.

Post mortem lesions :
(i)   Dehydration.
(ii)   Enteritis, gastric ulcers are common in swine.
(iii)   Generalized oedema.
(iv)   Hepatic necrosis and depletion of lymphoid tissues
(v)   Kidneys are pale and enlarged and have a rough and irregular surface.
(vi)   Microscopically, renal tubular epithelial degeneration and fibrotic changes. Dilated and regenerating renal tubules are also seen.

Diagnosis :
(i)   History.
(ii)   Clinical signs.
(iii)   Post mortem findings.
(iv)   Detection of ocharotoxin in urine, faeces, feed and meat samples.

Differential diagnosis :
(i)   Other mycotoxins.
(ii)   Other agents/ drugs producing nephrotoxic lesions.

Treatment : No specific antidote is available
(i)   Remove the offending feed and change the feed and the pasture.
(ii)   Activated charcoal to reduce further absorption from the gastrointestinal tract.
(iii)   Reduce the intake of proteins in the feed/diet.
(iv)   Supportive therapy based on the symptomatic prescription as well.

Prevention: Most important for prevention of the condition is proper harvesting and drying of the possible substrates before storage.

Citrinin : Citrinin is produced by the toxigenic fungi *Penicillium citrinum, P. viridicatum, P. palitans, Aspergillus ochraceus etc.* which grow on a number of feed stuffs, mainly wheat, barley, rye and oats. It is found in combination with ocharotoxins and is a potent nephrotoxin. Citrinin produces a condition known as *mycotoxic nephropathy* or *mold nephrosis* and affects pigs, cattle and poultry. It probably also produces pyrexia-pruritis-haemorrhagic diathesis in cattle following feeding of moldy citrus pulp cubes containing citrinin @ 30-40 ppb. It is a substituted benzopyran carboxylic acid.

Citrinin-induced mycotoxic nephropathy is similar to that of ocharotoxin A. The mechanism of nephrotoxic action of citrinin is also similar to that of ocharotoxin A.

In subacute toxicity in chicks following dietary citrinin, signs of polydipsia, diarrhoea, haemorrhages in jejunum, mottled liver and enlarged kidneys are observed.

Feeding of *P. citrinum* contaminated maize in feed (62%), resulted in retarded growth, atrophy of glomeruli of kidneys, enlarged gall bladder, depletion of lymphoid tissues, cardiac and skeletal myopathies and centrilobular necrosis of hepatocytes.

In cattle, the clinical signs are pruritis, loss of hair, animals take roughages but refuse concentrates, pyrexia (104 - 107°F), petechial haemorrhages on conjunctiva and visible mucous membranes. Papular exudative dermatitic lesions are found on head, neck, perineum and udder. Itching is so severe that animals rub the affected parts against hard objects and induce lacerations and bleeding. The dermatitis subsides but fever persists. The animals loose their condition. Morbidity rate is 10 - 100 per cent and severely affected animals succumb to death.

Post mortem lesions :
(i)   Extensive and wide spread petechial haemorrhages in all body organs and tissues.
(ii)   Microscopically, interstitial nephritis is conspicuous.

Treatment : There is no specific treatment except for the symptomatic and supportive therapy and withdrawl of the offending feed.

Oosporein : It is produced by *Oospora colorans, Chaetomium* sp., *Acremonium* sp. and *Beauveria bassiana*. It is a substituted bisbenzoquinone and produces nephrotic gout in poultry by interfering with the excretion of uric acid. No specific treatment is available.

8.6 NEUROTOXIC MYCOTOXINS

Patulin : It is produced mainly by *Penicillium urticae* (or *P. patulum*) which is mostly associated with fruit products or malt feeds. Some other molds which produce patulin are *P. claviforme, P. expansum* and some of the *Aspergillus* sp. (*A. clavatus*). The oral LD_{50} of patulin in chicks is 170 mg/kg.

It has been reported to affect cattle, piglets, chicks and mice. In pigs, patulin causes vomiting, salivation, anorexia, polypnoea, weight loss, leucocytosis and erythropenia. Post mortem examination of the dead animals reveals haemorrhagic enteritis. In mice and bulls, there are nervous signs, cerebral haemorrhages and mortality. In acute poisoning, there are lung haemorrhages and congestion. It is known to be carcinogenic too.

Penitrem A : It is the toxic metabolite of the fungus *Penicillium cyclopium, P. palitans, P. viridicatum, P. puberulum* and *P. crustosum* growing on peanuts, peacons, corns, cream, cheese and walnuts etc.

Penitrem A affects sheep, horses, calves, cows, rats and mice. Chickens, rabbits, guinea pigs and hamsters are comparatively resistant; hamster being most resistant. It probably somehow alters the functions of central nervous system neurotransmitters which control muscle activity. It may act in the brain and increase the spontaneous release of glutamate and aspartate.

It is a neurotoxin and in sheep produces a condition similar to that of *rye grass staggers,* and characterised by ataxia and convulsions in cattle. Affected calves stand with legs wide apart and rhythmic swaying of the body, stiff gait, ataxia and falling. Later the calves become recumbent, there is pedalling of the limbs, tetanic convulsions, opisthotonus, nystagmus and profuse salivation. Nephrotoxicity is also observed.

Characteristic signs in chicken are dyspnoea, ataxia, impairment of righting reflex, listlessness and coarse tremors.

Post mortem lesions :
(i)   Nephrotoxicity.
(ii)   Fatty liver.

Treatment : No specific treatment is there. Offending feed should be removed. Symptomatic treatment may be given for controlling the convulsions, tremors and behavioural alterations. Magnesium sulphate should be given orally to remove the unabsorbed toxin from the gastointestinal tract.

8.7   CITREOVIRIDIN (YELLOW RICE TOXIN)

Yellow rice toxicosis was observed for the first time in Japan in the late 1800s and early 1900s in human beings. This condition was believed to be due to vitamin B_1 (thiamine) deficiency as the symptoms were similar to that of acute *Beri-beri* disease. Later it was found that it was not the avitaminosis but caused by consumption of moldy polished rice having a number of toxic compounds such as anthraquinones, islandicin, catenarin, luteoskyrin, rubroskyrin, iridoskyrin, skyrin, cyclochlorotine, islanditoxin and erythroskyrin from *Penicillium islandicum* and dark yellow toxic metabolite citreoviridin from *Penicillium citreoviride.*

These toxins have different target sites- e.g. luteoskyrin and islanditoxin are potent hepato-toxins while citreoviridin is a central neurotoxin.

Islanditoxin rapidly causes extensive haemorrhages, severe liver damage and death while luteoskyrin is a slowly acting lipophyllic toxin which causes centrilobular necrosis and fatty degenerative changes in liver.

Citreoviridin is a neurotoxin, affects medulla oblongata and spinal cord and results in respiratory paralysis and cardiac failure.

Exact mechanism of action of these toxins is not known. Clinical signs observed in the affected individuals are palpitation, nausea, vomiting, rapid and difficult breathing, cold and cyanotic extremities, abnormal heart sounds, rapid pulse, hypotension, restlessness, paralysis, convulsions and eventually death due to cardiac and respiratory failure.

8.8 TRICHOTHECENES

Human mycotoxicosis, a syndrome was observed in human beings during the second world war due to non-harvesting of wheat and linked to the consumption of over wintered cereal grains and wheat or bread made from the wheat contaminated with *Fusarium tricinctum, F. poae* and *F. sporotrichioides*. These molds produce several trichothecene toxins. In human beings the toxicosis, termed as alimentary toxic aleukia (ATA), is characterised by total atrophy of the thymus and bone marrow, agranulocytosis, necrotic angina, high fever, haemorrhagic diathesis, sepsis, skin rashes, vomition and diarrhoea, followed by death in 2-80% the affected individuals

There are several trichothecene toxins, namely T-2 toxin, deoxynivalenol (vomitoxin), diacetoxyscirpenol, neosolaniol, HT-2, T-2, verrucarin, roridin etc. These are very toxic and constitute the largest group of mycotoxins. Out of all these toxins, T-2 toxin is most cytotoxic. These toxins may occur in combination and cause a condition which is commonly referred to as T-2 *toxicosis* or *fusariotoxicosis*. Compared to human beings, farm animals are at higher risk of intoxication because of their greater chances of consuming trichothecenes in mouldy feeds. All species of animals including human beings are affected.

T-2 toxin : T-2 toxin (Fig. 8.4) is produced by the fungus *Fusarium tricinctium, F. nivale and F. sporotrichoides* which grow mainly on corn, and the toxicosis thus is generally referred as *moldy corn (maize) toxicosis*, but can also grow on other feed stuffs as well. This toxin is a sesquiterpene and very stable and persists in food/feed indefinitely. It can penetrate through intact skin and has got direct irritant action on the mucous membrane and skin and causes dermal necrosis. The acute oral LD_{50} values of T_2 toxin for rats, mice, guinea pigs, chicks, adult chicken and swines vary between 3.0 and 10.5 mg/kg. No effect level of T-2 toxin in ration of pigs is <1 ppm.

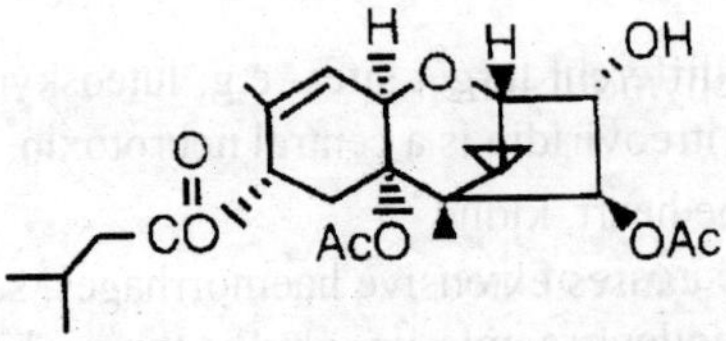

Fig. 8.4 : Chemical structure of T-2 toxin.

Trichothecenes are absorbed from skin and gut and distributed throughout the body but liver, kidneys and muscles have the highest concentrations. These are rapidly metabolized by hepatic microsomal enzymes by way of hydroxylation, acetylation and glucuronidation and the parent compound and metabolites are excreted in faeces and urine.

Mechanism of toxicity: T-2 toxin interacts with 60 S subunit of ribosomes and inhibits synthesis of DNA and protein by way of inhibiting initiation, elongation and termination of protein synthesis by inhibition of peptidyl transferase activity and also disaggregation of polysomes. In addition, T-2 toxin also acts on site I of electron transport chain and the cytotoxic effects may be mediated through free radicals.

Skin and gastrointestinal tract mucosal lesions are due to direct irritant effect of the toxin. The capillary permeability is increased and the clotting defects are probably due to inhibition of prothrombin synthesis in the liver.

Clinical signs : In domestic animals, dullness, loss of appetite, anorexia (feed refusal), poor weight gain, depression, general weakness, hyperemia of the buccal cavity, necrotic lesions in buccal cavity, splitting of lips and stomatitis leading to sloughing off of oral mucosa, excessive salivation and vomition and contact lesions on the snout, commissures of the mouth and prepuce in pigs are observed. There is mucus nasal discharge, colic, mucous and blood stained scouring, haematuria, epistaxis, decreased milk production, abortions, loss of hair, tremors, frequent urination, deep and laboured breathing, tachycardia, posterior weakness or paresis and paralysis followed by death in severely affected animals. These also cause leucopenia and lymphopenia.

In horses, the condition is characterised by retarded reflexes, bradycardia, disturbed respiration, cyclic movements, convulsions and death in 10-15 % of the affected horses.

In poultry, chicks are less sensitive than turkey poults. The characteristic lesions of trichothecenes toxicosis in poultry are anorexia, diarrhoea, panting, decreased feed conversion efficiency, weight gain and egg production, poor hatchability, thinning of egg shell, abnormal feathering, impaired righting reflex, abnormal wing posture, seizures, yellowish white necrotic oral lesions and deaths.

In laboratory animals, these cause foetal deaths and abortion, and tail and limb abnormalities.

Post mortem lesions :
(i)   Petechial and ecchymotic haemorrhages on the thoracic and abdominal viscera.
(ii)   Haemorrhages in the heart, kidneys, spleen and liver.
(iii)   Enlargement and necrosis of liver.
(iv)   Mesenteric lymph nodes are enlarged and hyperemic.
(v)   Focal yellow white caseous plaques at the margins of beak, on the hard palate and tongue and in the mouth in chickens.
(vi)   Atrophy of lymph nodes, thymus and decrease in size of spleen in chickens.

Diagnosis :
(i)   History.
(ii)   Clinical signs (mucosal and skin sloughing are characteristic of T-2 toxicoses.

(iii)   Post-mortem findings.

(iv)   Detection and estimation of T-2 toxin in the feed and faeces.

Treatment : Prognosis is poor in severely affected animals. There is no specific line of treatment. Only the symptomatic and supportive treatment need to be given. Activated charcoal, magnesium sulphate, dexamethasone , sodium bicarbonate and metoclopramide have been used in different species of animals.

Stachybotryotoxicosis: Horses , cattle, sheep and pigs are affected due to consumption of hay or use of hay contaminated with *Stachybotrys atra*. Acute condition is characterised by sudden onset of neurological signs such as loss of vision, uncontrolled movements and tremors, while chronic condition is characterised by fever, diarrhoea, dysentry, dermo-necrosis, leukopenia, gastrointestinal ulceration, epistaxis and haemorrhages. Post mortem lesions are wide spread haemorrhages and blood stained exudates.

8.9 ZEARALENONE (F_2 TOXIN)

Zearalenone or F_2 toxin (Fig. 8.5) is a phenolic macrolide nonsteroidal estrogenic mycotoxin produced by the mold *Fusarium roseum (F. graminearum)* or other species of *Fusarium* which infest maize, barley, wheat, sorghum and oats stored under warm and humid environment in small confinements. It is probably one of the most widespread and economically important mycotoxin which affects swine, cattle, sheep and other animals, however, pigs are most sensitive to its toxicity and the condition is termed as *porcine vulvovaginitis* or *hyperestrogenic syndrome in pigs*. The toxin is stable indefinitely and is not destroyed by roasting or other treatments. Though it has no structural similarity with estrogens but the uterotropic and anabolic effects are similar to that of excessive steroidal or synthetic estrogens.

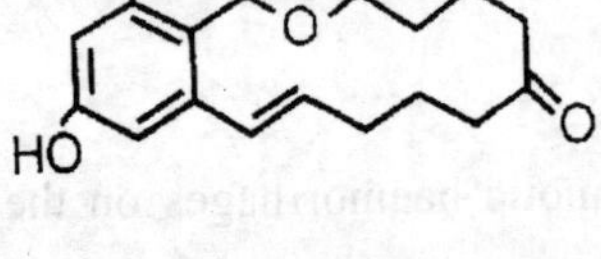

Fig. 8.5 : Chemical structure of F_2 toxin.

F_2 toxin @ 2.2 μg/g of mouldy wheat in sows and 25-100 ppm in poultry produce no adverse effects. The acute oral LD_{50} values of zearalenone are comparatively higher (5,000 - 20,000 mg/kg) in mice, rats and guinea pigs but 3.6 mg/kg of F_2 toxin produces toxic effects in sows and poultry. Feed should not contain > 10 ppb in ration of sows and cows.

After absorption from the gastrointestinal tract, zearalenone is distributed to different body tissues, rapidly metabolised to *alpha* and *beta* zearalenols which conjugate with sulphates or glucuronides and get excreted in faeces, urine or milk alongwith the parent compound.

Alpha zearalenol is the major metabolite in pigs and is 3-10 times more potent than the parent compound while the major metabolite in cattle is *beta* zearalenol and thus swine are comparatively more sensitive to zearalenone toxicosis. Due to rapid metabolism and excretion, consumption of meat or milk of affected animals seems to pose insignificant risk to humans within a week after cessation of ingestion of zearalenone. However, the parent compound or metabolite in red meat and poultry have been reported to cause premature development of breasts (thelarche) before 8 years of age, pubarche, gynecomastia and precocious pseudopuberty from Italy and Puerto Rico.

Mechanism of toxicity : Zearalenone and its active metabolite *alpha* zearalenol interact with the specific cytoplasmic estrogen receptors. The *alpha-* isomers of zearalenone metabolites have greater affinity for estrogen receptors than the *beta* isomers and thus the *alpha* isomers are more potent than the *beta* or parent compounds. The mycotoxin- receptor complex is translocated to the nucleus where it induces the synthesis of DNA polymerase I and II, nucleic acids and specific proteins and results in increased water and lowered lipid content in the muscles and increased permeability of uterus to glucose, RNA and protein precursors. It also enters the hypothalamus and pituitary and binds to estrogen receptors and results in prolonged estrous cycle due to persistent overstimulation of estrogen receptors.

Clinical signs: Zearalenone acts directly on the female reproductive tract, especially in pigs and induces *porcine vulvovaginitis syndrome* which is characterised by swelling (3-4 times the normal size), congestion and oedema of vulva and vagina, enlargement of vulva, uterus and mammary glands, atrophy of ovaries, swelling of uterine wall and hypertrophy of uterus, increased vaginal secretions, uterine bleeding, vaginal and rectal prolapse in severe cases, prolongation of reproductive cycle, pseudopregnancy, abortions and stillborn piglets, decreased litter size, small and malformed piglets, stunted growth of piglets and neonatal mortality. Piglets also show signs of vulvovaginitis, development of secondary feminization characters in males and atrophy of testes. Cornification of vulva and vagina, loss of mucosal epithelium of vagina and cervix is also observed. In cattle and sheep, however, the effect is less severe compared to sows. There is enlargement of the mammary glands and secretion of thin and watery milk. But the hyperestrogenism effect on the suckling calves and lambs is less.

Post-mortem lesions: There are no specific post-mortem findings except the enlargement of uterus and other changes in the reproductive tract.

Diagnosis:
(i)   History.
(ii)   Clinical signs.
(iii)   Detection of the specific fungus and estimation of F_2 toxins in the   feed, urine, faeces etc.

Differential diagnosis: Zearalenone toxicosis may be differentiated from other estrogenic toxicants.

Treatment: No specific treatment or the antidote is available. Give symptomatic and supportive therapy.

(i)   Withdraw the offending feed immediately.

(ii)   Dilute the offending feed with normal feed so that level of the mycotoxin may be reduced.

(iii)   Dehydrated alfalfa meal (15% in ration/feed) is protective as fiber reduces the absorption of mycotoxin from the gastrointestinal tract.

(iv)   Surgical repair of the prolapsed organ should be attempted, if possible.

CHAPTER 9

BACTERIAL TOXINS

N. K. Mahajan

The notion that pathogenic bacteria might produce their harmful effects by means of toxins/poisons is as old as the notion of pathogenic bacteria itself. Roux and Yersin towards the end of nineteenth century (1888) demonstrated that sterile (bacteria free) filtrates from cultures of the *Diphtheria* bacillus contained a toxin or poison which was capable of mimicking the symptoms on injection into guinea pigs, rabbits and pigeons as those produced by infection with the living organisms. This prompted the search for other toxins-producing bacteria without delay and tetanus toxin was discovered in 1890 independently by Knut Faber as well as by Briegel and Frankel. Proof of vital importance of bacterial toxins as disease producing agents was furnished by Von Behring and Kitasato, who in 1890, first prepared diphtheria and tetanus antitoxins and demonstrated their ability to specifically protect against lethal action of the homologous toxins and these are being used even today for the prevention of these diseases worldover. In the year 1897, botulinum toxin was discovered by Van Emergen. These all were the so called exotoxins derived from gram-positive bacteria.

In 1935, Boivin and Mesrobeanu reported a new type of toxin, the endotoxin, which they extracted from gram- negative organisms. Since then, many more bacterial toxins have been discovered and their number exceeds 160. Some toxins have been purified, characterized and studied in detail but their role in the pathology of many infectious diseases is still obscure.

Bernheimer (1976) described bacterial toxins as a collection of bacterial products whose principle common feature is their capacity to produce injury or to kill when administered in relatively small quantities in living entities. Bacterial toxins are very potent poisons e.g. 1.0 mg of tetanus or botulinum toxin is enough to kill more than one million guinea pigs. It is estimated that three kg of botulinum toxin could kill all inhabitants of the world. Because of their high potential to kill, bacterial toxins are considered very important as biological warfare agents.

Bacterial toxins are relatively high molecular weight substances in the form of proteins (simple or more complex 'conjugated' forms), peptides or lipopolysaccharides. The exact degree of characterization whether physical, chemical or biological of many known toxins vary greatly. Therefore, general statements about many bacterial toxins may change as the degree of understanding may increase with the availability of more and more data in the years to come. With the advancements in biotechnology and molecular biology, our understanding about the role of bacterial toxins in the pathogenesis of various diseases will also increase.

A wide range of bacterial toxins are responsible for food poisoning in human beings. The major factors contributing to bacterial contamination during food preparation for humans include inadequate cooking, improper cooling of cooked food, improper storage, cross contamination between cooked and raw food, inadequate reheating, inadequate cleaning of equipment and utensils etc. All these factors facilitate multiplication of bacteria and production of toxins. However, most of these factors generally do not apply to animal food.

Sources and classification : Sources and classification of important bacterial toxins according to cellular location alongwith their chemical nature and toxic concentrations (where ever known) have been shown in Table 9.1. Table 9.2 enlists the major toxin classes on the basis of their role in pathogenesis.

Mechanism of action : Different bacterial toxins have very diverse biological activities. Each exotoxin may have its own particular pharmacology or toxicology, while all the endotoxins act on a defined substance or an undefined component of tissue, cell or an organ. Many produce necrosis, oedema, haemorrhages, haemolysis, interfere with protein synthesis and some are neurotoxic. Many exotoxins exert their toxic effect because of catalytic (enzymatic) action e.g. *alpha*-toxin of *Clostridium welchii*, a phospholipase C catalyses the cleavage of phosphorylcholine from phosphatidylcholine.

(A) Protein toxins : General characteristic of any toxin is its toxicity and its range of activity, from doses which produce some physiological change to doses which are lethal. Differences in species susceptibilities are most pronounced e.g. guinea pig is most susceptible to diphtheria toxin followed by rabbit, dog, monkey, rat and mouse. Similarly, horse is most susceptible to tetanus toxin followed by guinea pig, monkey, sheep, mouse, goat. Cat and dog are least susceptible to tetanus toxin. Age of animals, physiological status and environmental temperature also influence the activity of toxins.

As it would be very difficult to discuss the mechanism of action of various toxins, some representative examples are discussed here under.

(i) Diphtheria toxin : The growth and proliferation of *Corynebacterium diphtheriae*, the causative agent of diphtheria, in the upper respiratory tract of humans results in the formation of a characteristic pseudomembrane. The organism does not disseminate but the toxin disseminates and damages a large variety of organs such as heart, liver, lungs and kidneys. When injected into experimental animals, the toxin produces symptoms similar to those in natural disease. Immunization with diphtheria toxoid results in complete protection against the disease.

Diphtheria toxin is highly lethal to most animals and man. It acts by inhibiting protein synthesis, which requires a cofactor NAD and is the result of inactivation of elongation factor-2 (EF-2), the protein that catalyzes the translocation step during peptide chain elongation. The whole toxin is cytotoxic as it arrests cellular protein synthesis. It is now thought that the NAD : EF2 -ADP ribose transferase activity of fraction A is the lethal activity of toxin for cells and animals.

Table 9.1 : Classification of bacterial toxins according to cellular location

Class	Group and sources of toxin	Toxin	Chemical nature	Toxic dose (µg)	Mode of action
A. Cell bound toxins	I. Intracytoplasmic toxins		Protein		
	Gram negative bacteria				
	Shigella dysenteriae	Neurotoxin, enterotoxin		2.3×10^{-3} (Lethal, rabbit)	ADP ribosylation
	Vibrio cholerae	Cholera toxin (enterotoxin)			
	Bordetella pertussis	Pertussal toxin		0.3 µg/mouse	
	Yersinia pestis	Plaguetoxin			
	Other sp.	Neurotoxins			
	Gram positive bacteria	Not known			
	II. Surface bound toxins (cell wall components)		Lipopolysaccharide, protein complex		
	Gram-negative bacteria (Most species)	Conventional endotoxins			
	Gram positive bacteria	Not known			
B. Entirely or partly extracellular toxins	III. True exotoxins	Not known	Protein		
	Gram-negative bacteria				
	Gram-positive bacteria (aerobic and anaerobic)				
	Corynebacterium diphtheriae	Diphtheria (Cytotoxic)		6×10^{-2} (lethal, mouse),	ADP ribosylation of EF2
	Staphylococcus aureus	Enterotoxins		2.5 (emesis, monkey)	
		α-toxin		5 (lethal, rabbit)	
		δ-hemolysin		2.5 per ml (50%,hemolysis, 1% RBCs)	
	Staphylococcus aureus	Leukocidin		4×10^{-4} per ml (kills macrophages)	
	Streptococcus pyogenes	Streptolysin O. Erythrogenic toxin		5×10^{-6} (skin reaction, man)	

(Continued)

Class	Group and sources of toxin	Toxin	Chemical nature	Toxic dose (ug)	Mode of action
	Listeria monocytogenes	Listeriolysin			
	Clostridium perfringens	α-toxin, τ-toxin,			
	Clostridium tetani	Tetanolysin, tetanospasmin, (lethal, hemolytic cardiotoxic)		4×10^{-5} (lethal, mouse)	
	Probably other species	Several toxins	Protein		
	IV. Toxins with both intracellular and extracellular location during logarithmic growth				
	Gram-negative bacteria Gram positive bacteria Aerobic- Bacillus anthracis	Not known			
		Complex oedema producing toxin			
	Anaerobic- Clostridium botulinum	Botulinum toxins (neurotoxins) Botulinum A,		2.5×10^{-5} (lethal mouse)	
		Botulinum B,		2.5×10^{-5} (lethal, mouse)	
		Botulinum D,		0.8×10^{-5} (lethal, mouse)	
		Others: C1, C2, D, F, G			

Table 9.2 : Major bacterial toxins and the bacteria responsible for their production

A. Enterotoxins :	*Vibrio cholerae* (diarrhogenic)
	Escherichia coli (diarrhoegenic)
	Bacillus cereus (diarrhoegenic, emetic)
	Clostridium perfringens (diarrhoegenic, emetic)
	Salmonella sp. (diarrhoegenic)
	Staphylococcus aureus (emetic)
	Shigella sp. (diarrhoegenic)
	Aeromonas hydrophila (diarrhoegenic)
B. Hemolytic :	*Streptococcus* sp.
	Staphylococcus sp.
	Clostridium perfringens
	Vibrio parahaemolyticus
	Bacillus cereus
	Aeromonas hydrophila
	Clostridium novyi
C. Neurotoxin :	*Clostridium botulinum*
	Clostridium tetani
	Shigella dysenteriae
	Staphylococcus sp.
D. Cytotoxic, Cytolytic :	
	Streptococcus sp.
	Staphylococcus sp.
	Shigella dysenteriae
	Aeromonas hydrophila
	Vibrio parahaemolyticus
	Clostridium difficile
	Legionella sp.
E. Necrotising :	*Clostridium novyi*
	Clostridium perfringens
	Clostridium septicum
	Clostridium welchii
F. Direct inhibitors of macromolecular synthesis :	
	Clostridium diphtheriae
	Bacillus thuringiensis
	Yersinia pestis
	Pseudomonas sp.
	Vibrio cholerae
	Escherichia coli
G. Endotoxins :	All gram-negative bacteria

(ii) Botulinum toxin : Botulinum toxin is a potent neurotoxin which is lethal to all vertebrates through paralysis of skeletal muscles, particularly of respiratory muscles. It is now known that the toxin interferes with transmission at the peripheral cholinergic motor nerve terminals of parasympathetic nervous system. Most studies on the action of botulinum toxin have been made with type A toxin. It is believed that the site of action of botulinum toxin is located on the presynapse. Action involves three steps (i) initial binding of the toxin to membrane, (ii) translocation of the toxin from binding site to lytic site, and (iii) lytic step, which is temperature-dependent and requires calcium ions and by which release of acetylcholine is inhibited.

(iii) Enterotoxins : Few enteric diseases have been as well characterized as gastroenteritis caused by *Vibrio cholerae* and enterotoxigenic *Escherichia coli* (ETEC). Cholera toxin is composed of two distinct subunits, called A and B, which are held together by strong noncovalent bonding. The A subunit of molecular weight 28000 is responsible for biological activity of the molecule. After adhesion or colonization of vibrios to gut epithelium, the cholera enterotoxin is assembled and secreted, which binds to specific receptors on epithelial cell membranes, internalized and finally exerts its biological activity. The net result is hypersecretion of chloride (Cl^-), bicarbonate (HCO_3^-) and water, which define the copious rice water diarrhoea characteristic of cholera. It causes increase in levels of cyclic AMP (cAMP) in intestinal epithelial cells via enzymatic transfer of ADP-ribose from NAD to a regulatory protein (RP) component of the cyclase complex. The events after elevation of the levels of cAMP lead to hypersecretion of Cl^- and HCO_3^- ions by intestinal crypt cells.

Other toxins have also been shown to act via mechanisms similar to that of cholera toxin. Of these, most closely related are *E. coli* heat labile enterotoxins (LTs) which are structurally, functionally and immunologically very similar to cholera toxin. Diphtheria toxin and *Pseudomonas aeruginosa* exotoxin A, while synthesized as a single peptide chain, are also activated by proteolytic nicking to form active A and binding B subunits and enzymatically transfer ADP-ribose from NAD to an acceptor protein, which is EF2 and their activity, unlike that of cholera toxin, results in cell death through inhibition of protein synthesis.

(B) Endotoxins of gram negative bacteria

The active principle (endotoxin) participating in the induction of toxic effects (fever, headache, changes in blood cell count and blood pressure, diarrhoea etc.) due to infection with gram-negative bacteria reside in their cell envelop (tightly associated with the cell wall). Chemically, the endotoxins are lipopolysaccharides with diverse composition and structure, but all express typical endotoxic activities. It consists of a lipid region, lipid A, and a covalently bound hydrophilic heteropolysaccharide chain, often being branched, which is subdivided into the core and O-specific chain.

Endotoxins act through the activation of a considerable number of mediators of host origin, each exhibiting a selective specific activity. Endotoxins exert their actions in most instances because the higher animals react with the production of such mediators as endogenous pyrogen, tumour necrotising factor, superoxide anion, colony stimulating factor, interferon, interleukin-1, glucocorticoid antagonising factor, prostaglandins, histamine, serotonin, activated hormonal and enzymatic systems and others. Humoral systems and target cells of the host thus make endotoxins an active biological principle. Macrophages could be identified as important cell types which on incubation with lipopolysaccharide are stimulated for synthesis and excretion of prostaglandins of E_2 and F_{2alpha} type. There are great species differences with regard to sensitivity or resistance to one and the same lipopolysaccharide preparation. Rabbits, dogs, horses or humans are highly sensitive, mice, rats and guinea pigs are medium in sensitivity, while primates like baboons or vervets are highly insensitive.

Pathogenesis : In infections caused by bacterial pathogens, the complex series of interactions that compose pathogenesis often involve toxin-mediated damage to the host tissues. In some diseases like cholera, diptheria, tetanus, botulism, clostridial and staphylococcal food poisoning etc., the role of protein toxin(s) in pathogenesis is well established. In others, where several toxins and enzymes are produced, it is not easy to pin point the role of individual toxic factor, indeed synergism between various toxins occurs.

Some of the representative toxins are discussed here :

Organisms of *Clostridium* sp., the anaerobes, are widely present in nature and cause a number of disease conditions in animals and human beings. The pathogenicity of clostridia appears to depend almost entirely on their toxin production. *Clostridium tetani* for example, multiplies locally and does not invade the body. Similarly *Clostridium botulinum* rarely parasitises animals or man but its preformed toxins in food are the cause of disease.

(i) Tetanus : Tetanus bacilli remain localized at their site of introduction and start to proliferate and produce neurotoxin only if local tissue oxygen tension is lowered. The toxin reaches the central nervous system by passing up the peripheral nerve trucks and not through blood-brain barrier. The exact mechanism of action of the toxin is not known. No structural lesions are produced but there is central potentiation of normal sensory stimuli so that a state of constant muscular spasticity is produced. Death occurs by asphyxiation due to fixation of the muscles of respiration.

Clinical signs : Incubation period varies between 1-3 weeks. A general increase in muscle stiffness, accompanied by muscle tremors, prolapse of third eyelid, stiffness of hind legs, exaggerated response to normal stimuli, tetany of masseter muscles, lock jaw and saliva may drool, constipation and bloat. As the disease progresses, muscular tetany increases, tetanic convulsions begin, opisthotonus, sweating and rise in temperature. Duration of fatal illness in horses and cattle is usually 5-10 days.

Post -mortem lesions : There are no gross or histological findings by which a diagnosis can be confirmed but search should be made for the site of infection and isolation of the organisms should be attempted.

Diagnosis : Fully developed tetanus is very distinctive. The muscular spasms, prolapse of third eye lid and recent history of accidental injury or surgery.

Treatment and prevention : Main principles in the treatment of tetanus are to eliminate the causative bacteria, neutralize residual toxin, relax the muscle tetany to avoid asphyxia and maintain relaxation until the toxin is eliminated or destroyed. Animal should be kept at a quite and dark place.

Long acting penicillin, tetanus antitoxin, chloral hydrate, magnesium sulphate or chlorpromazine injection are useful. For prevention, in enzootic areas, all susceptible animals should be immunized with 'toxoid'-an alum-precipitated, formalin-treated toxin. Animals who get some injury or wound or after surgery be given antitoxin or toxoid injections.

(ii) Enterotoxaemia : Enterotoxaemia is an acute toxaemia of ruminants caused by the proliferation of *Clostridium perfringens* type D in their intestines and liberation of toxins. Clinically, disease is characterised by diarrhoea, convulsions, paralysis and sudden death within 1-2 hours.

The *epsilon*-toxin of *Clostridium perfringens* type D increases the permeability of intestinal mucosa to this and other toxins. It causes a profuse, mucoid diarrhoea, produces CNS stimulation followed by depression. There is severe hyperglycemia. On post mortem, the kidneys appear pulpy , thus the syndrome is referred to as *'pulpy kidney disease.* Further, there is rapid autolysis of the renal tissues, however, rest of the carcass is usually in good condition.

Treatment : Hyperimmune serum may be injected or sulfadimidine be administered orally. Chelating agents are useful in neutralizing the toxin.

Prevention : Do not overfeed the animals particularly with grains and give vaccination.

(iii) *Escherichia coli* enterotoxins : Enterotoxigenic *E. coli* (ETEC) strains are responsible for diarrhoea in new born calves, piglets, lambs and foals. Marked heterogeneity with respect to physico-chemical, biological and immunological properties exists in enterotoxins produced by *E. coli*. On the basis of heat sensitivity, the toxins can be distinguished into heat labile (LT) and heat stable (ST) toxins. Besides ETEC, enteroaggressive *E. coli* strains produce a partially heat stable enterotoxin and a cytotoxin (verotoxin). Faeces of infected animals is the primary source of infection.

In order to produce diarrhoea, ETEC must bind to enterocytes and release enterotoxins, which enzymatically mediate the secretion of electrolytes and fluids. The mechanism of action of heat labile (LT) toxins is similar to that of cholera toxin discussed earlier i.e. by stimulating adenyl cyclase activity.

Pathogenesis : After colonization, ETEC produces enterotoxins, which results in net secretion of fluid and electrolytes from the systemic circulation into lumen of the gut, thus resulting in varying degrees of dehydration, electrolyte imbalances, acidosis, hyperkalemia, circulatory failure, shock and death.

Clinical signs : Most common form of colibacillosis is observed in new born calves, primarily 3-5 days of age. Presence of a single strain of ETEC may cause a state of collapse called enteric toxaemia. There is severe weakness, coma, subnormal temperature, pale mucosae and mild convulsions. Diarrhoea usually may not be evident. Such calves die within 2-6 hours after onset of signs.

In a more common form of the disease in affected calves, there is diarrhoea in which faeces are profuse, watery or pasty, usually pale yellow or white in colour and very foul smelling and the defecation is frequent and effortless . Affected calves may not suck or drink depending upon degree of acidosis, dehydration and weakness. In terminal stages, there may be bradycardia with arrythmia and hyperkalemia.

Post mortem lesions : There is dehydration, the intestinal tract is distended by yellow watery contents and gas, abomasum is distended and may or may not contain milk clot. Abomasal mucosae may show petechial haemorrhages.

Diagnosis : The definitive diagnosis may be difficult and often inconclusive. Detailed epidemiological investigation at the farm will be very useful, besides, microbiological and pathological findings.

Treatment and prevention : Consideration for treatment and prevention of acute neonatal diarrhoea include- change of diet, fluid and electrolyte replacement, antimicrobial therapy, parasympatholytics/antimuscarinic drugs and intestinal protectants. Colostral immunity besides hygiene plays important role in prevention.

(iv) Endotoxins : These are toxins of a very different kind and are released only after natural autolysis or artificial disruption of gram-negative bacterial cells.

Fever or pyrogenic effect is the toxic effect produced by the smallest doses of endotoxin. As little as 0.002 µg endotoxin per kg body weight injected intravenously, in highly susceptible species such as rabbit or man, causes an elevation of body temperature within 15 minutes that lasts for several hours. The action of endotoxin (exogenous pyrogen) is to cause polymorph leucocytes, macrophages and perhaps other tissue cells to release an endogenous pyrogen which passes via blood to act on the thermoregulatory centre in hypothalamus of brain.

Endotoxic shock is caused by intravenous injection of a large dose of endotoxin in experimental animals. It also occurs as a fatal, terminal condition in gram-negative bacteremic infections in man. The LD_{50} of endotoxin for rabbits is in the range of 1.0 to 100 µg/kg. The injected endotoxin causes vasomotor disturbances within 4 to 18 hours and brings about a drastic fall in blood pressure, collapse of the circulation and death.

In addition to pyrogenicity and shock, endotoxins may also result in necrosis of bone marrow, complement activation, platelet aggregation, induction of prostaglandin synthesis, depression of blood pressure etc.

Diagnosis : Limulus amoebocyte lysate test.

Treatment : A number of endotoxic effects such as fever, early hypotension, shock and abortion can be suppressed by non-steroidal antiinflammatory drugs (NSAIDs).

Prevention : The fever - inducing pyrogen sometimes found in fluid preparations used for therapeutic injections or intravenous therapy consists of endotoxins derived from saprophytic gram-negative bacilli that have grown in small or large numbers in water or other components of the preparation. This endotoxin is not inactivated by autoclaving process used to sterilize infusion fluids. Care must be taken to ensure the absence of bacterial contamination in such preparations. Use freshly distilled water and pure chemical ingredients and clean the glassware thoroughly. Care must also be taken to sterilize the fluid and protect it from subsequent contamination as gram-negative contaminants might readily multiply in the fluid during storage.

It is certain that bacterial toxins play a major role in the pathogenesis of nearly every disease attributable to bacteria. When one includes endotoxins, peptidoglycans and exoenzymes which facilitate bacterial penetration of the host defences, in the definition of toxin, then it can be said that all bacterial diseases develop either directly or indirectly through these substances.

CHAPTER 10

ZOOTOXINS

Satish K. Garg

Zootoxins are the toxins produced by lower animals e.g. snakes, fish, toads, scorpions, bees, wasps, spider, ticks etc.

Venom : It is the poison or toxin secreted by specialized glands of an animal.

Venomous animals : These are the animals or creatures that are capable of producing a poison in a highly developed secretory gland or group of cells and deliver the toxin during a stinging or biting act.

Venom may be composed of proteins (polypeptides and enzymes) of both high and low molecular weight. These may be amines, lipids, steroids, aminopolysaccharides, quinones, 5-hydroxytryptamine (5 HT), glycosides or other substances. Action and toxicity of venom depends on :

i)   Species of the snake or other venomous animal.
ii)   Route of entry into the body.
iii)   Location / site of bite or stinging.
iv)   Quantity of the venom injected.
v)   Absorption from the site of entry into the body.
vi)   Distribution.
vii)   Accumulation and action at the receptor site.
viii)   Biotransformation.
ix)   Excretion.
x)   Species of animal affected.

10.1 SNAKES

Snake bite in animals generally occurs while grazing or hunting. Most of the cases of snake bite have been reported in dogs and horses, however, other species of animals are also affected.

Snake venoms are complex mixture of toxins- consisting of amino acids, polypeptides, glycopeptides and biogenic amines in addition to certain cations such as K^+, Na^+, Ca^{2+}, Mg^{2+}, Ni^{2+} etc. Most active component of the venom is a peptide or polypeptide which mostly acts enzymatically and at times nonenzymatically too. Generally, most of the snake venoms produce two type of toxicity i.e. neurotoxicity and cardiotoxiciy or haemotoxicity. Neurotoxicity is mostly produced by enzymatic portion of the venom.

There are more than 3500 different species of snakes out which more than 400 have been found to be poisonous and dangerous. Most of the poisonous snakes have been found to belong to mainly six families :

(i)   **Elapidae** : Coral snakes, cobras, kraits, mombas.
(ii)   **Crotalidae:** Rattle snakes, water moccasins, copper heads, bush master, pitvipers.
(iii)   **Viperidae** : The old world vipers and adders.
(iv)   **Hydrophidae:** The true sea snakes.
(v)   **Laticaudidae:** Sea kraits.
(vi)   **Colubridae:** Boomslang, bird snake, rednecked, keelback snake.

Toxicity due to snake bite generally depends on :

(a)   Toxicity of the venom and the quantity of venom injected.
(b)   Ratio of animal i.e. size of the animal and venom injected.
(c)   Species of snake.
(d)   Location of bite.
(e)   Species of animal involved. On the basis of body weight, horses are most susceptible, followed by sheep > cattle > goat > dogs > pig > cats.
(f)   Prompt availability of the appropriate therapy.

Venoms of snakes contain different fractions - necrotizing, anticoagulant, coagulant, neurotoxic, cardiotoxic and haemolytic fractions. The venoms of cobra and krait are mainly neurotoxic while that of vipers and rattle snakes are haemotoxic in nature.

Some of the important enzymes present in the venoms of snakes belonging to different families are listed in table 10.1.

Haemotoxins of pitvipers induce local tissue damage and injury to blood vessels, change in blood cells, blood coagulation defects and shock. Cardiopulmonary functions are also adversely affected. There is hypotensive crisis due to acute blood loss due to haemolysis. Rattle snake bite in calves causes restlessness, teeth grinding, hypersalivation, dyspnoea, ataxia and convulsions. Local swelling at the site of bite is generally observed. Animals generally die without any previous history of illness.

Neurotoxins of cobra and krait induce paralysis of cranial nerve of eyelids resulting into ptosis, paralysis of muscles of eyeballs and respiration. Neurotoxins bind specifically to cholinergic receptors at neuromuscular junction and produce curare like effect and respiratory paralysis.

Clinical signs : Clinical signs of neurotoxins containing snake bite are salivation, hyperexcitability, mydriasis, asphyxia, gasping, recumbency, convulsions and death within 2-4 hours. Regurgitation of ruminal contents, paralysis of tongue, oesophagus and larynx is also observed in calves.

Table 10.1: Important enzymes found in the venoms of snakes of different families

Enzyme	Families of poisonous snakes				
	Crotalidae	Elapidae	Hydrophidae	Laticaudidae	Viperidae
Proteolytic enzyme	++++	– / +	– / +	?	++
Arginine ester hydrolase	+	–	+	?	+
Thrombine like enzymes	++++	+ / –	+ / –	?	++++
Collagenase	+	?	?	?	+
Hyaluronidase	–	+	+	–	–
Phospholipase A$_2$	+	+	+	+	+
Phosphodiesterase	+	+	+	+	+
Acetylcholinesterase	–	+	+	?	–
5' Nucleotidase	++		?	?	++

Diagnosis : Snake bite can be diagnosed by the history of sudden death, observing the fang marks, local swelling, oozing of blood at the site of bite, cyanosis etc. Identification of the snake in the vicinity also helps in making diagnosis.

Differential diagnosis :
(i)   Black leg.
(ii)   Anthrax.
(iii)   Non-specific phlegmonous infections.
(iv)   Botulinum.
(v)   Myasthenia gravis.
(vi)   Polyradiculoneuritis.
(vii)   Tick paralysis.

General management of snake bite :

(i)   Keep the animal undisturbed.
(ii)   To restrict the further absorption and distribution of venom, apply a tight torniquet above the site of bite.
(iii)   Incise the local area of snake bite in the direction of blood vessel and go for sucktion and infiltrate the area with 5% soap solution.
(iv)   Inject antivenin, antibiotics and antitoxins.
(vi)   If snake has not been identified, give polyvalent antivenin intravenously and also infiltrate the antivenin locally around the site of bite.

(v)   Polyvalent antisnake venom should be administered intravenously @ 10-20 ml per animal depending on the weight of the animal.

(vi)   Give supportive treatment for managing shock and cardio-pulmonary disturbances by administering corticosteroids.

Precautions/contraindications :

(i)   Do not use potassium permanganate.

(ii)   Do not give extreme hot or cold treatment at the site of bite or incision.

10.2 TOADS

Dogs and cats may play with toads and get exposed orally to the toxins of toad secreted by the glands in their skin located above and posterior to the eyes. Different toad toxins are bufogenins (bufodienolides) - bufotalin, bufotenidin, bufotenin, bufoviridin, serotonin and calecholamines. Bufogenins are cardiotoxic glycosides and have effect on the heart and other smooth muscles. The toxins differ amongst different species of toads. Some of the poisonous/toxic species of toads are *Bufo vulgaris* (common toad), *Bufo marinus*, *Bufo regularis* and *Bufo alvarius*; out of these *B. alvarius* (river toad) and *Bufo marinus* (marine toad) are the most toxic ones.

Bufo vulgaris is least toxic and recovery of affected animals is very quick. Clinical signs of toad poisoning are hypersalivation (some times foamy), vomiting and pawing at the mouth, cardiac irregularities such as cyanosis, weakness, pulmonary oedema, convulsions, prostration and collapse. In case of *B. marinus*, death occurs within 15 minutes while in *B. regularis* poisoning there is paralysis as well and the animals generally die within 2-6 days.

Diagnosis :

i)   History of pet playing with a toad.

ii)   Clinical symptoms.

Treatment : No specific treatment.

i)   Wash the mouth with plenty of water.

ii)   Give activated charcoal and osmotic purgatives.

iii)   Administer atropine sulfate intravenously to check excessive salivation and bronchoconstriction.

iv)   Give antihistaminics, sedatives or tranquilizers.

v)   Administer corticosteroids.

vi)   Administer cardioprotective agents. Give large doses of propranolol (0.2 mg/kg) to control cardiac arrythmias and myocardial fibrillation.

10.3 BEES, SCORPIONS AND WASPS

Venom of bees, scorpion and wasps is a complex mixture of peptides, nonenzymatic proteins such as apamin, melittin or kinins, enzymes such as phospholipase A and B, hyaluronidase, formic acid and biologically active amines, such as histamine and 5-HT etc. Melittin is a protein mainly found in honey bees and is antigenic in nature and produces allergic reactions mainly in human beings and horses.

Severity of toxicity varies from individual to individual in different species of animals. Anaphylaxis and death from a single sting occurs in hypersensitive animals. However, in others, only mild to moderate inflammation and painful swelling is observed. Following single insect bite, there is extreme serous exudation. This accumulated fluid exerts protective effect by diluting the poison, exerts local pressure on the circulation thus reducing the dissemination of poison. However, in multiple bee or wasp stings, there is severe local inflammation and oedema at the site of sting, intense pain and pronounced excitement. In severe cases in horses, there may be diarrhoea, methaemoglobinemia, bilirubinemia, jaundice, haemoglobinuria, tachycardia, dyspnoea and followed by death, though in rare cases.

Treatment : No specific antidote is available. Only symptomatic treatment need to be given.
i)   Local application of weak solution of ammonia and sodium bicarbonate.
ii)   Nervine tonics and stimulants, if there is prostration.
iii)   Tracheotomy, if asphyxia is severe.
iv)   Emergency supportive therapy for restoration of cardiopulmonary functions and management of anaphylaxis.

10.4 SPIDERS

Numerous species of spiders have been implicated in bites of human beings. These are- black widow spider, brown widow spider, red legged spiders, brown recluse spider, desert violin spider, Arizona violin spider, cobweb spiders, jumping spiders, running spiders etc. However, in animals black widow spider (*Latrodectus mactans*) has been mainly found to bite dogs and cats. Venom of this spider, *alpha* latrotoxin, is a potent and labile neurotoxin which affects neuromuscular junctions and also binds to calcium channels and increases the membranes permeability to calcium ions and enhances depolarization. It is also known to have a high content of isoleucine and leucine and low of tyrosine in addition to lipoproteins and hyaluronidase. Spider venom is 10-15 times more potent than rattle snakes.

Clinical signs of spider sting are intense pain at the sting site. On pressing, jelly like oedematous swelling around the bite, whole integument is painful and hypersensitive to pressure, cramping spasms of abdominal muscles, nervous excitement due to reflex contraction of muscles, rigidity of abdominal muscles, emesis, loss of appetite, weakness, dyspnoea and hypertension. Paralysis occurs in acute cases and death in 4-6 hours due to paralysis of respiratory muscles, but in mild cases it may take days. The venom of brown recluse spider (*Loxosceles reclusa*) is rich in hyaluronidase, proteases and other spreading factors and haemolysins. The toxin probably damages endothelial cell membranes. The toxin is known to cause local swelling, vascular thrombosis and necrosis.

Post mortem lesions : Post-mortem lesions are not specific except for venous congestion.

Treatment :

i)   No specific treatment is available, however, if available, specific antivenin may be injected.

ii)   Symptomatic treatment needs to be given.

iii) Give calcium gluconate, sodium pentobarbitone and muscle relaxants intravenously.

iv)   Administer antihistaminics, intravenous fluids, respiratory stimulants, corticosteroids and atropine sulfate.

10.5 TICKS (TICK TOXINS)

Ticks are important as vectors of many diseases and generally do not pose any immediate or serious effects upon the host except when present in large numbers on a particular site. In addition, ticks are involved in venom poisonings.

Tick paralysis is a paralytic disease in man, cattle, sheep and bison. When one or more *Dermacentor andersonii* (wood ticks) are attached to skin, produce paralytic symptoms which disappear when offending ticks are removed. However, may result in death of animals if ticks are not removed. Symptoms of toxicity are due to a neuroparalytic toxin of ticks.

Ixodes holocyclus is another tick which causes tick paralysis in sheep and cattle. Cats are comparatively more resistant. Similarly, *Amblyomma* sp. and *Ornithodorus* sp. also cause tick paralysis. Symptoms of tick paralysis are violent retching, anorexia, lethargy, drooling of saliva, muscle weakness, incoordination, extensive dehydration and complete ascending flaccid paralysis. The sequence of paralysis is : hind limbs, fore limbs, chest muscles; respiration is affected due to paralysis of respiratory muscles including diaphragm. All limb and eye touch reflexes are absent, pupils dilate widely and death occurs due to respiratory paralysis in calves and yearlings. Vomiting, loss of voice and secondary aspiratory pneumonia are observed in dogs.

The toxins are secreted by salivary glands of female ticks. Dogs are most commonly affected but losses also occur in lambs, calves, goats, foals and children. Severity of paralysis is independent of the number of ticks involved, susceptible animals may be affected seriously by a few ticks. The toxin causes paralysis of skeletal muscles by interfering with the release or synthesis of acetylcholine at the motor end plates of muscle fibers. Four to ten adult female *Ixodes holocyclus* ticks produce paralysis within 6-13 days of infestation. However, 35-150 ticks of *Dermacentor andersonii* attached for 5-8 days produce paralysis and complete recovery takes place within 24 hours after removal of ticks.

Treatment : No specific treatment or the antidote is available.

i)   Administer hyperimmune serum in dogs.
ii)   Remove the ticks either manually or by making use of acaricides.

Control : Eradicate the ticks from the animals and the premises by periodic use of effective acaricides.

10.6 FISH (ICHTHYOTOXINS)

Fish is the principle source of proteins for inhabitants of the coastal areas and also the animals and birds in the form of fish meal in their feed. Fish poisoning is often observed in human beings. Worldover about 125 people are estimated to die of puffer fish poisoning per annum, however, fish poisoning is not common in animals, though dogs and certain animals and birds being fed fish or fish meal in their feed, respectively, may at times be poisoned.

Toxins may be produced by fish itself or by the marine plankton or algae consumed by the fish. Fishes may contain a toxin in their musculature, viscera or skin or fish produce toxin that is related to their gonads or fishes have a toxin in their blood. About 700 different species of marine fishes are known to be toxic. However, some are extremely potent and highly toxic. Some of the poisonous fishes are shell fish, puffer fish, sting fish, scombroid fish, ciguatera and Moray eel etc. Certain fish toxins are very potent e.g. the intraperitoneal LD_{50} value of tetrodotoxin in mice is 8 μg/kg. The lethal dose of saxitoxin in human beings is approximately 4mg. Some of the commonly observed fish poisonings alongwith their mechanism of action and clinical signs are summarised in Table 10.2.

Table 10.2 : Some of the common fish poisonings and their clinical symptoms.

Name of fish	Toxin	Mechanism	Clinical signs
Shell fish	Saxitoxin	Inhibits inward current of Na^+ across the axonal membrane.	Burning sensation in lips and gums, tongue, face and numbness of these parts, pain in joints, difficulty in movement, thirst, progressive generalized paralysis and death due to respiratory failure.
Puffer fish (fugu fish)	Tetrodotoxin	Alters neuronal membrane permeability to Na^+ and K^+.	Rapid onset of weakness, dizziness, paresthesia around lips, tongue and throat, increased salivation, diaphoresis, hypotension, bradycardia, intense cyanosis, dyspnoea, epigastric pain, ataxia and flaccid paralysis.
Moray eel	Ciguatoxin	Increases neuronal membrane permeability to Na^+ and causing depolarization of nerves. It also has anticholinesterase activity.	Tingling of lips, tongue and throat, numbness of these parts, nausea, vomition, abdominal pain, diarrhoea, pruritis, bradycardia, pain in joints and muscles, paresis and temporary blindness.

CHAPTER 11

ENVIRONMENTAL TOXICOLOGY

Satish K. Garg

Man has been conscious of his environment since prehistoric times as a law for disposal of the waste at a designated place outside the city walls was passed in Athens as early as 500 BC. It is believed that approximately 80% of the newly reported cases of cancer are environmentally-induced.

Despite the growing concern and realization for human and animal health, environmental pollution remains a major public health issue. To consider air, water and land pollution separately does not seem logical as the pollutants and their effects are interchangeable. Air pollutants combine with rain or snow and come to the ground and become water or soil pollutants and similarly, water pollutants on evaporation become air pollutants and so the cycle is repeated. The pesticides sprayed on crops are land pollutants as these may be carried by wind to become transient air pollutants which eventually settle on the land or water. The pollution of environment, which includes air, water and land is attaining alarming proportions with the scientific advancements and increasing industrialization and ever increasing use of chemicals.

Chemical fertilizers, insecticides, herbicides and high yielding varieties of wheat and rice have revolutionised the world agriculture production. With the extensive use of chemicals, general health of the individuals has improved as some of the vector-borne diseases have been controlled, however, air and water quality deteriorated. The great air pollution disasters of Meuse Valley of Belgium in 1930, Donora Pennsylvania in 1948, London smog of 1952 and Bhopal gas tragedy of December 3-4, 1984 drew attention of the world to the dangers of emission of toxic substances from combustion of fossil fuel, industrial wastes, effluents etc.

Air, water and land act as sink for disposal of toxicants. Atmospheric carbondioxide is increasing by combustion of fossil fuel and disappearance of forests thus resulting into climatic changes - depletion of stratospheric ozone, warming of the earth's surface etc. The resultant increase in ultra-violet radiations reaching earth's surface may increase the incidence of skin tumours, retard growth of crops and affect food chain of marine species.

Pollution of air, water and land is more in urban and suburban areas. Important sources of air pollution are emissions from automobiles, industrial units, power generation units, residential heating, waste incineration and other modern development programmes in society which pollute the air, water and soil around us. Bhopal gas tragedy of 1984 was due to industrial release of 30 tons of methyl isocyanate vapours into air resulting into death of over 3000 people and about 200,000 people were injured and/or permanently impaired.

Human beings and animals are exposed to chemicals in the air they breath, the food/fodder they eat and the water these drink. The toxic effects of chemicals are dependent on the dose.

11.1 AIR POLLUTANTS

Air pollution is mainly an urban problem. Air pollutants originate from geophysical, biological and man-made sources. There are numerous air pollutants- carbon monoxide (CO), carbon dioxide (CO_2), sulfur dioxide (SO_2), a mixture of nitrogen oxides, mixture of hydrocarbons - volatile organic compounds, metals, non metals and other suspended particulate matters. The concentrations and proportion of air pollutants varies greatly depending on sources, season, ambient temperature, relative humidity etc. e.g. in the vicinity of a smelter, the main pollutants are sulfur oxides, metals and particulate matters while in urban and suburban areas automobile emissions having CO_2, volatile organic compounds and nitrogen oxides are the predominant pollutants.

Air pollutants enter the body predominantly through inhalation i.e. lungs. Some of these chemicals are absorbed into the blood and some that are not absorbed are retained by the lungs. The hazards and toxicity of air pollutants may be acute due to exposure to high concentrations of pollutants (e.g. Bhopal gas tragedy due to methylisocyanate) or may be chronic due to long term exposure to low levels of pollutants. Respiratory ailments are more common in people living in urban areas than those in villages as air pollution is the precipitating factor.

Five pollutants, namely CO (52%), sulfur oxides (18%), hydrocarbons (12%), particulate matter (10%) and nitrogen oxides (6%) account for almost 98% of air pollution.

Depending on the release, there are two types of air pollutants :

(i) Primary air pollutants : The pollutants which are directly relased in atmosphere in enough concentrations without modification e.g. CO, hydrocarbons, particulates, SO_2 and nitrogenous compounds (NO, NO_2).

(ii) Secondary air pollutants : The pollutants which interact with each other in the presence of certain compounds, particularly energy sources, e.g. nitrogenous compounds, ozone, peroxyacetyl nitrate (PAN) etc.

Depending on the chemical nature, two types of pollution is recognised :

(i) Reducing type of pollution : Pollution due to SO_2 and smoke is due to incomplete combustion of coal, fog and cool temperatures.

(ii) Oxidizing type of pollution or photochemical air pollution : Pollution due to hydrocarbons, oxides of nitrogen and automobile exhaust where intense sunlight causes photochemical reactions in polluted air masses.

The major pollutants are discussed briefly in this chapter.

Carbonmonoxide : Carbonmonoxide (CO) is the most notorious and abundant pollutant present in the lower atmosphere ambient air. Of the total global emission of CO, 60-90 per cent is from the natural sources particularly combustion of fossil fuel, atmospheric oxidation of methane, forest fires, terpene oxidation and the oceans where microorganisms produce CO. Underground garages and tunnels also result in CO production. Rest 10% of CO is from human activity including the most important source of automobile emissions.

Carbonmonoxide is highly toxic as it has high affinity for haemoglobin (Hb), thus displaces Hb-bound oxygen and increases carboxy-haemoglobin (CO-Hb) concentration with the concurrent decrease in oxygen-carrying capacity of blood and thus causing dyspnoea. Affinity of Hb for CO is 220 times greater than for oxygen, therefore, CO is dangerous even at very low concentrations.

Degree of intoxication depends on the concentration of CO, duration of exposure, minute volume of respiration and amount of CO-Hb. Dissociation of carboxy- Hb is a slow process. Carboxy-haemoglobin also exerts inhibitory effect on the dissociation of oxy-Hb (O_2–Hb) still available, thus further reducing the availability of oxygen to tissues. Toxicity of CO is not only due to interference with the delivery of O_2 by blood to tissues but also its direct toxic effect by binding to cellular cytochromes, such as those contained in respiratory enzymes and myoglobin.

On acute exposure to CO, deaths have been reported in human beings and animals, however, effect on health following prolonged chronic exposure to low levels of CO is not well documented. The symptoms of CO poisoning are dyspnoea, diarrhoea, sweating, headache, dizziness, thirst, loss of weight, irritability, insomnia and blurred vision. In addition, long term exposure may increase Hb and haematocrit values. As the gas readily crosses the placental barrier, foetus are more prone to toxic effects as these are extremely susceptible.

Sulfur dioxide : Global emission of SO_2 is more or less equal from natural and anthropogenic sources. Natural sources of SO_2 are volcanos and decaying organic matter while anthropogenic emissions are by combustion of sulfur containing coal and smelting of nonferrous ore. Sulphur dioxide is readily absorbed on tiny particles of coal. Sulfurdioxide when comes in contact with oxygen or water in air (in the presence of moisture) is oxidized to sulfur trioxide, sulfuric acid, ammonium sulfate, or other sulfates and may be transported to distant places. Sulfuric acid comes on the earth in the form of acid rain. In the atmosphere, this reaction is catalysed by manganese and vanadium. Iron and copper help in formation of stable sulfite complexes.

Sulfurdioxide is a mild respiratory irritant and predominantly acts on the upper respiratory tract and causes bronchoconstriction. It increases mucus secretion due to proliferation of goblet cells and finally produces bronchitis. It also impairs macrophage-dependent host defence mechanism. Air flow resistance is increased and thus, it is more poisonous or toxic in people who have some other respiratory problem.

In high concentrations, SO_2 is injurious to eyes, causes inflammation of conjunctiva and irritation in the nose and throat. Long term exposure to low levels of SO_2 is recognised as industrial or occupational hazard as there is alteration of mucus secreting cells and thickening of mucus layers in the trachea. In rats, daily exposure to 10 ppm of SO_2 for 1-2 months has been found to thicken the mucus layer in the trachea by about 5 fold.

Like SO_2, sulfuric acid (H_2SO_4) is also a respiratory system irritant, the latter being 4-20 times more irritant than the former. Sulfuric acid is strongest of the atmospheric acids. It induces bronchoconstriction by causing hyperplasia of airway mucosecretory cells. And, on chronic and prolonged exposure, sulfuric acid would deposit deeper along the respiratory tree and cause bronchitis. Asthmatics are more sensitive to both sulfuric acid and sulfurdioxide.

Hydrocarbons : Natural and anthropogenic sources of hydrocarbons (HCs) are vegetation, microbial decomposition, forest fires, natural gases, incomplete combustion of fossil fuel, evaporation of liquid fuels and solvents and motor vehicle exhausts.

Different hydrocarbons formed are low molecular weight aliphatic, olefinic and aromatic compounds. Olefins and dienes tend to polymerise via free radical formation to form polycyclic aromatic hydrocarbons (PAH). Biochemically, the aliphatic and alicyclic hydrocarbons are generally inert but not biologically. Most of the effects of hydrocarbons on health are not direct but due to derived compounds from atmospheric reactions of HCs with other substances in air e.g. formaldehyde, other aldehydes, ketones and ozone etc. The olefins and aromatic hydrocarbon vapours are more irritating to the mucous membrane and systemic injury occurs on inhalation of aromatic vapours. Ingestion of hydrocarbons is more hazardous as the liquids have a low surface tension and can be easily aspirated into the respiratory tract by vomiting or eructation. The hydrocarbon air pollutants promote the formation of photochemical smog. Some of the PAH(s) are potent carcinogens and mutagens.

Photochemical smog constituents produce eye irritation. The carcinogenic PAH particles having a diameter of 1 μm or less may penetrate the lungs, however, large particles (2-5 μm) do not reach alveoli. Large particles enter circulation via gastrointestinal tract through food chain as these settle on the ground and become water and land pollutants.

Gasoline contains approximately 20% benzene. Benzene is a bone marrow poison and a carcinogen and produces myelocytic and acute nonlymphocytic leukemia.

Ethylene, a major constituent of automobile exhaust or formed by combustion, is a normal constiuent of plants and is a plant growth regulator and also controls falling of leaves and ripening of fruits, however, is toxic for the plants in higher concentrations.

Particulate matter : Both organic and inorganic particulate materials of different diameter are there. Virtually all metals are found in some concentration in the atmospheric particles, but most common metals released with oils and coal combustion are lead, mercury, copper, manganese, chromium, cadmium, beryllium, iron, magnesium, nickel etc. The general toxicity of most of the common metals is discussed in details elsewhere in this book. Most of the lead comes from the combustion of leaded gasoline and waste incineration while mercury and beryllium are from coal combustion. Major toxic effect of beryllium is pneumonitis. It is also a carcinogen.

Fluoride is one of the important byproducts of coal combustion and is released in large quantities entirely in gas phase. It is a highly reactive non-metal air pollutant and combines with other metals or elements and is a stronger respiratory irritant and phytotoxin and causes leaf damage and eventually defoliation.

Carbonaceous matter constitutes about 50-60% of the total mass of fine particulate matter. The sources are forest fires, fly ash and diesel exhaust. At high concentrations, normal mucocilliary clearance mechanisms gradually get exhausted resulting into progressive build up of particles in the lungs, followed by cessation of clearance, epithelial hyperplasia, adenosarcomas and squamous cell carcinoma. Particles of the size of > 30 µm are not hazardous as these rapidly settle on the ground, however, particles of < 5 µm enter the tracheo-bronchial tree and irritate respiratory system and aggravate respiratory problems.

Nitrogen oxides : Nitrogen oxide is formed by lightening and microbial digestion of organic matter and by high temperature combustion of cellulose nitrate films. On microbial digestion, first nitrous oxide (N_2O) is produced which is oxidized to nitric oxide (NO) and further to nitrogen dioxide (NO_2) in the atmosphere. The mixture of N_2O, NO and NO_2 is termed as nitrogen oxides.

Nitrogen dioxide is a deep lung irritant and thought to penetrate alveolar capillary membranes where it is converted to nitric acid and produces lung oedema. It increases the resistance of airways by damaging type I cells of the alveoli. This pollutant is of particular risk to farmers because very high amounts of NO_2 are liberated from ensilage. *Silo filler's disease* due to NO_2 and CO poisoning is well documented.

Photochemical oxidants : Photochemical oxidant pollutants arise from a series of complex atmospheric reactions between hydrocarbons and nitrogen oxides. Ozone is one of the most important constituents of this group. The mixture of ozone, peroxyacetyl nitrate (PAN), aldehydes and ketones form a haze which is termed as photochemical smog - a reddish brown haze in the atmosphere.

Ozone : The oxidant found in highest concentrations in polluted atmosphere is ozone (O_3). Nitrogen dioxide is most efficient in absorbing short wave ultraviolet light in the atmosphere, several miles above the earth's surface, as a result a complex series of reactions take place as shown in Fig. 11.1.

The liberated oxygen atoms react with hydrocarbons in the atmosphere resulting into production of oxidised compounds and free radicals that react with NO to produce NO_2. This way cyclicity of the production of atmospheric pollutants NO_2 and O_3 is maintained which result in accumulation of NO_2 and O_3 and depletion of NO in the atmosphere.

$$NO_2 \xrightarrow{\text{UV light}} NO+O$$
$$O+O_2 \longrightarrow O_3$$
$$O_3+NO \longrightarrow NO_2+O_2$$

Fig. 11.1 : Recycling of nitrogen dioxide and ozone in the atmosphere.

Ozone is a respiratory toxin. It penetrates deep into the bronchioles and alveoli of lungs and deposited in the acinar region of the lungs, damages respiratory system and produces pulmonary oedema and emphysema. Ozone produces morphological, functional, immunological and biochemical alterations. It causes desquamation of the epithelium throughout the ciliated airways or rupture of capillary endothelium in the alveoli, type I cells are replaced by type II cells and proliferation of type II cells is an indication of ozone toxicity. Long-term exposure to ozone may cause thickening of the terminal respiratory bronchioles. Chronic bronchitis, pulmonary fibrosis and emphysema are the main clinical findings. Ozone also produces irritant effect on the eyes and skin. Pulmonary injury by ozone may be due to formation of reactive free radical intermediates by its interaction with sulfhydryl groups. Sulfhydryl compounds and antioxidants protect against ozone toxicity. Ozone, sulfur dioxide and nitrogen oxides act synergistically and not only produce harmful effect on human beings and animals but also plants.

Ozone depletion : Ozone layer forms a shield and filters part of the solar ultraviolet light thus reducing the amount of radiation reaching earth's surface. Thinning of ozone layer will result in increased radiation of life on the earth. Environment protection agency (EPA) has predicted that a 10% depletion of ozone layer by the middle of the next century will result in two million additional skin cancer cases every year. Chlorofluorocarbons (CFCs) cause extensive depletion of ozone layer. Ozone depletion also causes global warming i.e. there is rise in the temperature of gases present in atmosphere. If production/generation of green house gases is not reduced, global warming will reach 2-5 °C over the next century and thus put stress on natural and social systems.

Green house effect : Carbondioxide and water vapours are responsible for maintaining earth's temperature. About 50 per cent of the solar energy is absorbed by earth and rest by the atmosphere after reflection. Carbondioxide, water vapours and other gases in atmosphere absorb large part of energy and emit it towards the ground where it warms the surface of earth and referred as natural green house effect. More is the production of carbondioxide, more is its concentration in the atmosphere, more is the earth temperature. Carbon dioxide is the main green house gas contributing about 49% towards green house effect. Other gases are methane (18%), N_2O ((6%), chlorofluorocarbons (14%) and others (13%).

11.2 WATER POLLUTANTS

Water is the most basic need second to air and indispensable in sustaining plant and animal life. The concept of relationship between water and health is not new. Foul air and foul food have been recognised as the cause of disease for centuries. The Aldogate pump in London was closed in 1876 after. John Snow's investigation that contaminated water of the pump was responsible for the epidemic of cholera. Thereafter it was realized that clean drinking water is required for healthy living.

Any molecules present in water which are not water and are detrimental to health are termed as water pollutants. The pollutants released into or those present in the environment may be partitioned amongst different environmental compartments e.g. air, water or soil. Contaminants or pollutants enter the hydrosphere by direct application, wet or dry deposition, spills or through acid rains.

It has been recognised over the years that presence of toxic chemicals in water adversely affects both the plants and animals. Another important facet of water pollution and human health is chemical contamination of food chain of humans and animals. e.g. Minimata disease episode due to high levels of mercury in sea water.

Sources of water pollution : Sources of water pollution may be natural or anthropogenic. Discharges of municipal sewage, industrial units, cooling systems, electricity generating plants etc are added to rivers, lakes, pipes or channels; these constitute the point sources of water pollution. Non point sources of water pollution are agricultural land run-offs containing pesticides, fertilizers, nutrients, phosphorus, nitrates, salinity, acidity etc. and urban run-offs entering into ditches and streams. Main urban water pollutants are municipal sewage, industrial effluents, run-offs of city streets and landfills.

Type of water pollutants : Depending on the nature, water pollutant(s) have been divided into three types :
(i)   Physical pollutants.
(ii)   Chemical pollutants.
(iii)   Biological pollutants.

(i) Physical pollutants : Water may be polluted with mud, grit, stone, lime, ash, mining wastes, aquatic fauna or heat/thermal pollution due to absorption of solar energy, release of industrial cooling or power plants water discharges. These result into increased ambient temperature of water and decreased solubility of oxygen in water and hence oxygen-deficit in water which will have adverse effects on aquatic flora and fauna. Similarly, soil erosion, animal wastes, street refuse, water run-offs from mines, soil tops, construction sites, brick kilns, industrial effluent detergents, municipal wastes etc. contaminate water and make it unfit for human and animal consumption.

(ii) Chemical pollutants : A large number of mineral salts and metallic ions find their way into drinking water by natural or other means e.g. carbonates, bicarbonates, sulfates, and chlorides of calcium, magnesium and sodium. Lead, iron, arsenic, zinc, manganese, cadmium, fluorides etc. present in water produce severe toxicity in consumers, both human beings and animals. Other common sources of chemical pollutants are petrochemicals due to large spills from the tankers or near shores, aromatic hydrocarbons, phenols, sodium chloride (brine) washed into streams, hydrogen sulfide, inorganic compounds- sulphur in ores, fuels, chromium, cadmium, nickel, lead, mercury, zinc, copper, fluoride, phosphorus, chlorine, nitrate, nitrities, pesticides, soil nutrients, air-borne pollutants/acid rains, metals, chemicals due to leaching of soil nutrients. All these contaminate standing surface and ground waters. As a result of acid rains pH of water of rivers, lakes and soil is reduced. Acid rain also leaches calcium, magnesium, potassium and sodium out of the soil into the ground water and thus results into deficiency of these metals/ions into the soil and making the soil infertile. Similarly, excessive levels of salts in water result in salinization of soil and inhibit plant growth. Air-borne toxins e.g. metals, pesticides, organic chemicals etc. are carried by air to distant places come down to earth with rains or snow and pollute soil and water.

(iii) Biological pollutants : Aquatic fauna contribute significantly towards water pollution. Certain plants and animals found as aquatic bottom dwelling organisms in natural waters are known as planktons. These include a wide variety of saprophytic, holozoic and chlorophyllous forms of life, sponges, worms, algae, creepers (snails, insects) etc. These zooplanktons may prevent propagation of beneficial aquatic fauna, facilitate insect breeding and also reduce oxygen content of water due to decomposition of dead plankton. Planktons protect microorganisms in water from disinfection. Some algae planktons by producing potent toxins cause toxicity. In addition, certain bacteria (*Salmonella typhi, Shigella* sp., *E. coli*), fungi, parasites and viruses also contaminate water and these may not actually produce overt disease on first exposure, however, become potential threat to human and animal health on prolonged consumption of such polluted water.

The adverse effects/toxicity due to most of the common pollutants have been described elsewhere in the book. However, toxico-epidemiological data and impact of environmental pollutants, including air and water pollutants, on health of livestock is not very well documented.

11.3 FOOD TOXICANTS

Food of animals and humans contains several naturally occurring substances which are toxic at high concentrations. These can be classified according to their origin - fungal, bacterial, environmental contaminants or natural toxins present in plants. Sometimes, some exogenous agents are used as food additives and preservatives to improve their flavour, keeping quality or to give it a better look in the food prepared for human beings. Similarly, some of the feed additives are added in feeds of animals and poultry as growth promotor and for prevention of diseases; the details of the toxic manifestations of some of the food additives and preservatives are discussed in chapter 12.

Natural contaminants of food :

Food toxicants are unavoidable as lot of anutrient substances are present in plants which are vital for the growth and survival of plants but produce deleterious effects on the health of both humans and animals e.g. goiterogens in *Brassica* sp., *Arachis* sp., *Linum* sp., trypsin and chymotrypsin inhibitors in soybeans, antithiamine in fish and ferns (*Pteridium aquilinum*). Similarly, heavy metal contaminants (lead, arsenic, cadmium), chlorinated organic compounds (DDT, PCBs), food born molds and mycotoxins (*Aspergillus* sp., *Penicillium* sp., *Fusarium* sp., *Claviceps* sp.), bacteria and bacterial toxins (*Clostridium* sp., *Salmonella* sp., *E.coli*, *Vibrio* sp.) and natural constituents of plants like alkaloids, glycosides, tannins, saponins etc. are toxic. In addition, plants also have toxic chemicals termed as phytoalexins, which are a major mechanism of plant defence.

Phytoalexins are low molecular weight, antimicrobial agents which are synthesized by and accumulated in plants after exposure to microorganisms. Some of the common chemical groups of phytoalexins are coumarins, glycoalkaloids, isocoumarins, isoflavonoids, terpenoids etc. Induction of phytoalexin synthesis takes place after exposure to bacterial or viral infection, exposure to cold, UV light, heavy metals, salts, antibiotics, fungicides, herbicides etc. Some of the phytoalexins of food plants are given in Table 11.1.

Table11.1 : Some of the food plants and their phytoalexins

Plant	Phytoalexin(s)
Pea	Pisatin
Soybean	Glyceollin
Bean	Phaseollin
Cotton	Gossypol, hemigossypol
Rice	Momilactones, oryzalexins
Castor bean	Casbene
Carrot	6-methoxymellein, falcarinol
Alfalfa	Vesitol, salivan, medicarpin

Depending on the origin, food toxicants may be divided into :

i) Toxicants of plant origin
ii) Bacteria and bacterial toxins
iii) Molds and mycotoxins
iv) Environmental contaminants
v) Toxicants of animal origin

(i) Plant toxicants : Most of the food proteins (> 70%) originate from plant sources all over the world. However, there are certain antinutritional or toxic factors/substances in addition to the nutrients present in food of vegetable origin. Such anutrient factors in plants have been/are excluded from human diet, however, some of the accidental toxic manifestations may still be observed in human beings. On the contrary, animal feeds may contain such factors/agents as animals mainly depend on the feeds of vegetable origin and have a indiscriminate type of grazing /feeding behaviour. Thus, animals are the worst victims of toxicants of plant origin. Some of the toxic principles of plants are discussed here in this chapter (Table 11.2), though details of their chemistry and some of the common and important toxic plants have been discussed in details in chapter 4 (Toxicity of plants).

Table 11.2 : Toxicants of plant origin.

Plant toxins	Name of plants	Active principle
Alkaloids	*Heliotropium europaeum*	Heliotrine, lasiocarpine
	Conium maculatum	Coniine
	Nicotiana tobacum	Nicotine
	Nicotiana indicum	Nicotine
	Atropa belladona	Atropine
	Datura stramonium	Atropine
	Papaver someniferum	Morphine
	Rauwolfia serpentina	Reserpine
	Chenodrodendron	Curare (d-tubocurarine)
	Strychnos nuxvomica	Strychnine
	Solanum sp.	Solanine
Glycosides	*Digitalis purpurea*	Digitoxin, digitalin
	Prunus sp	Amygdalin
	Sorghum sp.	Hydrocyanide
	Nerium oleander	Oleandroside, nerioside, thevetin, convallarin
	Hypericum perforatum	Hypericin
	Thevetia peruviana	Thevetin
Phytoestrogens	*Trifolium subterraneum*	Genistein, genistin, diadzein, coumestrol
	Fusarium roseum	Zearalenone
Lectins	*Ricinus communis*	Ricin
Oxalates	*Beta vulgaris,* *Halogeton glomeratus,* *Sarcobatus* sp.	Oxalates
Phenolics	*Quercus incana* and other sp.	Tannins
	Melilotus officinalis, M. alba	Coumarin
	Hypericum perforatum	Hypericin
	Gossypium sps.	Gossypol
Saponins	*Aleurites fordii, Lucerne* sp.	Saponin
Proteins	*Abrus precatorius*	Abrin
	Ricinus communis	Ricin

(Continued)

Plant toxins	Name of plants	Active principle
Aminoacids	*Leucaena leucocephala*	Mimosine
	Mimosa pudica	Mimosine
	Pithocolobium lobatum	Djenkolic acid
	Indigofera endecapylla	Indospicine
Antivitamins	*Pteridium aquilinum*	Thiaminase and cyanogenetic glycoside
Cyanogens	*Linum catharticum,*	Dhurrin
	L. usitatissimum,	Linamarin
	Trifolium repens,	Lotaustralin
	Acacia leucophloea	HCN
Lipids	*Brassica napus*	Erucic acid
Mineral toxins	*Chenopodium album*	Nitrate
	Astragalus sp.	Selenium
Miscellaneous toxins	*Eupatorium rugosum*	Trematol
	Cannabis sativa	Tetra-hydrocannabinol
	Aethusa cynapium	Aethusin
	Carrots	Carotatoxin
	Perrila frustescens	Volatile oils

(a) **Alkaloids :** Different types of alkaloids present in various plants - *Conium maculatum, Nicotiana tobacum, Astragalus lentiginosus, Senecio, Crotolaria, Halotropicum, Echium, Lupinus, Phalaris tuberosa, Datura stramonium* (Table 11.2) are consumed by grazing animals and produce toxicity. Some of the alkaloids are even secreted in milk of animals and human beings too are poisoned on consumption of such milk.

(b) **Glycosides :** Cyanogenic glycosides in *Sorghum vulgare*, Sudan grass, Johnson grass, almond, apricots, peaches, white clover and linseed, cardiac glycosides in *Digitalis purpurea, D. lanata, Strophanthus kombe and S. gratus*, sapogenic glycosides in soybeans, alfalfa and common beans, coumarin glycosides in sweet clover (*Melilotus officinalis, M. alba*) and mustard oil glycosides produce toxicity in different species of animals. Some of the common glycosides with their plant sources are listed in Table 11.2.

(c) **Phytoestrogens :** These are non-steroidal estrogenic substances of vegetable origin which cause infertility in animals on grazing. Some of the pasture species which cause problems in livestock because of phytoestrogens are sub-terranean clover (*Trifolium subterraneum*), red clover and alfalfa (*Lucerne* sp.). Isoflavoids (genistein, genistin, diadzein), coumestans (coumestrol, 4-o-methyl coumestrol), zearalenone and zeralenol on mold infested corn (Table 11.2) produce effects similar to that produced by estrogens, however, their affinity for the natural estrogen receptors is less compared to that of estradiol. Sheep are the worst affected after several years of grazing on such pastures, fertility of the flock goes down.

(d) Lectins or phytohemagglutinins : Lectins are high molecular weight proteins or glycoproteins of plant origin, and have the ability to produce agglutination of RBCs and stimulate lymphocytes to undergo mitosis, e.g. ricin from castor beans (Table 11.2). Depending on the dose, source and route of exposure, the toxic effects vary. Lectins have been found in 800 edible plants species of which 600 belong to the *Leguminosae* family. These were first isolated from castor beans and are also found in soybeans. These are also present in sponges, crustaceans, mollusks, fish blood, amphibian eggs and mammalian tissue. These suppress growth, induce necrosis of intestinal epithelial cells, diarrhoea, immuno- and hepato-toxicity.

(e) Oxalates and phytates : Certain plants are very rich in oxalates and phytates (Table 11.2), consumption of which results in hypocalcemia, tetany and deposition of insoluble calcium oxalate crystals in the kidneys and blood vessels. Similarly, phytates also bind with di- and trivalent ions (Cu^{2+}, Zn^{2+}, Co^{2+}, Mn^{2+}, Fe^{2+}, Ca^{2+}) and result in mineral deficiencies. Some of the oxalate containing plants are listed in Table 11.2 and are discussed in details elsewhere in this book. *Halogeton* sp. and *Sarcobatus* poisoning are common in cattle and sheep.

(f) Plant phenolics : Some of the plant phenols are very toxic. Common plant phenols are tannin, lignin, nitrogenous phenols (tyrosine, mescaline), gallic acid, mycotoxins, coumarins, hypericin, gossypol etc. (Table 11.2).

(g) Saponins : These are the glycosides which are bitter in taste, have foaming properties and cause gastric irritation. Saponins are widely distributed among plants such as soybeans, alfalfa and common beans (Table 11.2). These are surface active agents and produce stable foam and cause bloat in ruminants. Alfalfa is very rich in saponins, the use of which in the ration of chicks and swine results in lower growth rate and anaemia.

(h) Toxic amino acids and proteins: Some of the plants contain a number of amino acids (Table 11.2) which are vital for plant defense against predators and have a potentially high nutritive value for animals and human beings. However, widespread use of these in the diet/ration may be toxic e.g. *Leucaena leucocephala* is a tropical legume which is drought tolerant, rapidly growing, palatable and high yielding and its leaves contain 25-30% crude protein. However, its use as a rich source of protein in animal feed is limited because of the presence of toxic amino acid mimosine in it. Mimosine exerts goiterogenic effect in ruminants in addition to destruction of hair follicle matrix and alopecia, poor growth, swollen and rough coronets above the hooves, lameness, mouth and oesophageal lesions, depressed serum thyroxin level and development of goiter in non-ruminants.

Certain plants also contain some proteins (Table 11.2) which produce toxic effects in animals, the toxicity of these is discussed elsewhere in this book.

(i) Vitamins and antivitamins : Certain plants contain very high concentrations of vitamins and antivitamins (Table 11.2) e.g. vitamin A or carotenoids in excess are teratogenic in many animal species (rats, mice, hamsters, rabbits, guinea pigs, dogs, pigs). Inclusion of raw soybeans (30%) lowers vitamin A levels in body, dicoumarol causes fatal haemorrhages in cattle and other species of animals, antithiamine factor in bracken fern causes vitamin B_1 deficiency in horses.

(j) Miscellaneous plant toxicants : Some of the toxic principles are unsaturated alcohol (trematol), acetylene derivatives (aethusin), cannabinols, carotatoxin, volatile oils, ketone-substituted furan etc. present in a variety of plants (Table 11.2), consumption of which produces toxicity in animals. Similarly, certain plants also contain mineral toxins (Table 11.2), the toxicity of which is discussed elsewhere in the book.

ii) Bacteria and bacterial toxins : Bacterial food-borne disease may result from the presence of bacteria in food. In human beings all over the world, bacterial contamination of food represents the greatest food-borne risk (74%) of the total food-borne illness. On the contrary, other contaminants are more important in animals except in pets where microbial contamination is of particular concern.

Some of the important agents associated with food poisoning and localized gastrointestinal discomfort, diarrhoea, dysentery, fever, etc. and other systemic effects and sometimes even death are *Salmonella typhi, E. coli, Clostridium botulinum, C. perfringens, Staphylococcus aureus, Bacillus cereus, Shigella sonnei* etc. The poisoning generally results as a result of improper preparation, handling, storage and cooking of food. Signs of poisoning are mainly diarrhoea, vomition, fever, shock and sometimes death. Details of toxicity of bacterial toxins are discussed elsewhere in the book (see Chapter 9).

(iii) Molds and mycotoxins : Details of various molds growing on different foods and feed stuffs and the conditions associated with their consumption have been dealt in details in the separate chapter on mycotoxins in this book (see Chapter 8).

(iv) Environmental contaminants : Some of the common heavy metals, namely-lead, arsenic, cadmium and mercury, organometallic compounds, organic compounds-pesticides (DDT, PCBs), nitrosamines, nitrosamides and N-nitroso substances, nitrites in soil, water, sewage etc. enter the food chain and pose threat to human and animal health. Some of the environmental pollutants persist in the environment and depending on their persistence, these are divided into three groups : Persistent which decompose by 75-100% within 2-5 years, moderately persistent which decompose within 1-18 months and nonpersistent which decompose in 1-12 weeks. The persistent contaminants having high lipid solubility have a greater bioaccumulation potential and pose serious health hazards. Toxicity of most of the environmental contaminants/pollutants has been discussed in details in different parts of this chapter of this book.

(v) Toxicants of animal origin : Drugs and chemicals are used in animals not only for prevention and treatment of diseases but also for promotion of growth. Residues of antibiotics, other drugs or pesticides in milk, meat or eggs of animals and birds have been detected; and many a times certainly much above the permissible limits as the proper withdrawl periods are not observed. Consumption of such animal products is important from public health point of view, both in human beings and animals and may result in chronic toxic effects. Similarly, contamination of food with microorganisms or chemicals during growth, processing, preparation and storage also pose threat to human and animal health. Environmental pollutants like heavy metals, other inorganic metals, agrochemicals have also been identified to be present in animal products.

11.4 RADIATION HAZARDS AND TOXICITY

Radiation is produced through decomposition or disintegration of unstable naturally occurring or synthetic elements. These not only always produce deleterious effects, but at times also the beneficial effects. Radiations have been used to improve the growth of seeds. Radiation therapy is being extensively used in diagnosis and treatment of diseases, particularly cancer chemotherapy both in veterinary and human medicine.

Disintegration of unstable nuclei results in a gradual loss of radioactivity which is characteristic of each nuclide. The radioactive nuclear species not only differ in the type of radiation emitted but also the energy with which radiation is emitted. The period during which the nuclide loses half of its activity is termed as radioactive or physical half life.

Radiation toxicology is the study of adverse effects of radiation on living organisms. Exposure of animals and humans to radioactive material results in radiation injury. Exposed animals are reservoirs of radioactivity which could be passed on to human beings through milk, meat or other animal products and pose a serious threat to human health.

Source of radiation exposure may be directly ionizing or indirectly ionizing. Radiation that on passage through matter produces ions by knocking electrons out of their orbits is called ionizing radiation. The directly ionizing radiations carry an electric charge. These charged particles include *alpha* particles, electrons, accelerator particles, charged mesons etc. that directly interact with atoms in the tissue or medium by electrostatic attractions or repulsion. Indirectly ionizing radiations are not electrically charged and include all photon radiations, neutrons and neutral mesons.

Penetration of ionizing radiation depends on the type of radiation i.e. its mass, charge and energy. When a particle or a ray travels through matter, it gradually loses energy by transferring it to the matter. Radiation injury depends on the distance from and time after the emission of radiation and differs due to depth of penetration and the degree of injury caused. Most of the damaging events are radiochemical in nature and occur after the initial ionization which occur in femta seconds (10^{-13} to 10^{-14} sec.) after physical interaction.

Sources of radiation : Radioactive materials are being increasingly used in medicine, industry, agriculture and power generating reactors. Sources of environmental radiation pollution may be natural or man made or anthropogenic. Both humans and animals are being continuously exposed to external and internal sources of radiation.

(i) Natural sources : Cosmic rays from the space and the external terrestrial radionuclides composed mainly of the emission from ^{238}uranium, ^{235}uranium and ^{232}thorium in the geochemical environment consisting of certain rocks, soil and phosphate deposits. Two decay products ^{222}radon and ^{220}thoron of ^{238}uranium and ^{232}thorium, respectively are responsible for 54% of the earth's background radiation and dangerously pollute the atmosphere. Solar radiation has assumed special significance because of depleting ozone layer around the earth. Other natural sources of radiation are ^{40}potassium, ^{87}rubidium, ^{3}hydrogen, ^{14}carbon etc. Some of these have a very long half life values- 13×10^9 years for ^{40}K and 4.89×10^{10} years for rubidium and thus exist in soil, water and muscles of body for very long time as these are distributed almost uniformly throughout the body while others in lungs and bones and serve as internal sources of radiation exposure for pretty long time.

(ii) Anthropogenic sources : Nuclear reactors (fall out contamination of pastures and fields), nuclear explosions (effluents and discharges into water or air), industrial units using radioactive materials and medical units using radioisotopes for diagnosis and cancer chemotherapy and research units handling radioisotopes. Radiation-induced cell changes may result in death of cells, death of the organisms, modulation of the physiological activity, mutagenesis, teratogenesis and carcinogenesis etc. Most of the damaging events are radiochemical in nature and occur after initial ionization within femta seconds (10^{-13} -10^{-14} sec) after physical interactions.

Mechanism and pathogenesis : Radiation toxicity represents a dynamic interaction with matter by direct or indirect processes to form ion pairs, some of which are free radicals. These ion paires rapidly interact with themselves and surrounding molecules to produce free radicals (OH and H) which react with cellular macromolecules or with each other to form hydrogen peroxide, a strong oxidising agent. It is through the agency of these free radicals that ionizing radiation produces the observed effects on biological systems. The interaction of these free radicals or H_2O_2 with cellular macromolecules such as nucleic acids, amino acids, proteins, lipids, thiols or carbohydrates leads to a variety of damages; DNA strands break, point mutation and chromosomal aberrations with the subsequent loss of those gene products coded for by that portion of DNA. If these gene products are required by the cell to maintain itself, it will eventually result in cell death. The rate of chromosomal aberrations is directly related to radiation dose. Chromosomal aberrations are still evident in the lymphocyte chromosomes of Japan's atomic bomb survivors or in uranium miners.

Whole body is exposed to radiation, however, some organs are comparatively more susceptible to radiation damage. Generally the rapidly dividing and undifferentiated cells are most sensitive. Skin, gastrointestinal tract and haematopoietic system are worst affected with the exception of human lymphocytes. On whole body exposure of human beings to 300 rads, there is rapid decline in nucleated formed blood elements and lymphocytes disappear completely within two days followed by quick drop in neutrophil and platelet count .

Biological systems irradiated in the presence of oxygen are more susceptible to injury than when irradiation occurs without oxygen as it results in the formation of hydoperoxy or hydrogen peroxide radicals which are more damaging because of their longer half lives and resistant to biochemical repair processes, the response is termed as oxygen effect. Ethanol, narcotics, leukotrienes and some thiols are radioprotective as these induce hypoxia, probably suppress the respiratory centre in CNS.

Radiation by free radicals (OH or superoxide) also result in peroxidation of lipids, as a result, the lipid radicals interact with another organic molecules and convert the organic molecules to free radical state and propagate the damage. However, the primary free radical scavengers (Vitamin A and E) and thiols in the biological systems influence cellular radiosensitivity. In addition, 8 enzyme systems-glutathione (GSH), reduced NADPH-dependent GSH, selenium-dependent glutathione peroxidase, selenium-independent glutathione peroxidase, ferric superoxide dismutase, manganese superoxide dismutase, zinc-superoxide dismutase and catalase are associated with detoxification and repair of free radical injury.

Radiation toxicity : The animals get exposed to radioactive materials and suffer from radiation injury and hazards. Sensitivity of different species of animals to radio-toxicity varies greatly e.g. donkeys, rabbits and poultry are less susceptible than man, dogs, pigs and goats, The LD_{50} value for a 30 day exposure to X-rays in rats, rabbits, goats and dogs are 796, 751, 237 and 244 rads, respectively. Developing organisms are more radiosensitive compared to adult ones e.g. LD_{50} for fish embryo is 50 rads while adult fish can tolerate 800-900 rads. In human population too, foetus and children are more sensitive and even small doses may cause mental retardation, stunted growth, deformation and cancer.

Depending on the level of exposure and the lesions inflicted, acute, subacute and chronic conditions are encountered.

Acute radiation toxicity : Direct and acute exposure to high doses of irradiation results in radiation sickness which is characterised by signs of acute irritation of the alimentary tract resulting in intense and refractory diarrhoea, dehydration and death in some cases, while redness of skin, lot of thirst, weakness, recumbency, rapid respiration, panting, profuse and blood stained nasal discharge in others. Almost after one week, there is severe depression of bone marrow as reflected by lymphopenia and depression of granulocytes and thrombocytes. Blood clotting and antibody production is impaired, followed by anaemia and necrosis of gastrointestinal tract wall, hair start falling (depilatation), ulceration of skin, secondary bacterial infections, degenerative changes in lens of eye (cataract), high rate of mutations, tumours mostly of haemopoietic system, particularly leukemia. Death in a few days due to dehydration and salt depletion. Most of the deaths in acute and subacute cases occur in 1-4 weeks of irradiation.

If the dose of radiation is 5000 rads or more, the animals suffer of central nervous system syndrome which is characterised by convulsions, loss of muscular coordination, coma and death within three days or less. There is excessive intracranial pressure due to radiation induced oedema.

Subacute radiation toxicity : Following exposure to median doses of radiation, there is initial phase of radiation sickness which is characterised by anorexia, vomition, depression and weakness lasting for several hours to several days. It is followed by fever, knuckling at the fetlock, swelling of legs, diarrhoea, dysentery, polydipsia, recumbency and hyperirritability and severe anaemia and septicemia in terminal stages and death in about 3 weeks time. Foetal death and resorption, teratogenesis, decreased survival of the young ones born alive and stunted growth are observed. Other signs in the survivors are alopecia, sterility, tumours, especially of the haemopoietic system and skin injury.

Chronic radiation toxicity : It is generally due to continuous contamination of pasture with the fall out of radioactive material. Effects depend on the level or concentration of the radioactive material consumed. The effects differ depending on the radioactive material e.g. radioactive iodine damages thyroid glands, strontium bone tissues etc. Both these are excreted in the milk of animals and sometimes, their levels are too high. Consumption of milk and meat of such animals in turn adversely affects human health.

Alopecia, sterility, mutational changes, tumour of haemopoietic system (leukaemia), thyroid, breast and lungs and teratogenesis and stunted growth are the main clinical signs of chronic exposure.

Local irradiation :

Skin : High dose exposure to irradiation causes severe skin burns, initiated by oedema of the dermis, followed by destruction of the cells in the basal layer of epidermis. Hair follicles and sebacious glands are destroyed. Compared to pig and rabbits, sheep are more susceptible. Cattle also develop severe skin lesions.

Gut : Local irradiation of gastrointestinal tract also causes ulceration of intestinal lining, acid producing cells get lost and villi become swollen.

Lungs : Radiation pneumonitis is characterised by respiratory depression, hypoxemia, decreased pulmonary blood flow, alveolar epithelial and endothelial damage followed by pulmonary fibrosis, decreased lung volume and breathing capacity.

Thyroid : Radiations destroy thyroid glands, there is hypothyroidism, lethargy, constipation and dryness of skin.

Bones : Bone seekers - ^{89}Strontium, ^{90}Strontium, ^{140}Barium, radium isotopes etc. substitute calcium in the bones, suppress bone marrow and result in lymphopenia, leucopaenia and anaemia.

Reproductive organs : The cell division of testicular germinal epithelium is stopped. An acute dose of 600 rads to testes produces permanent sterility. However, fully developed sperm cells and primary spermatocytes are relatively radio-resistant. Atrophy and degeneration of ova takes place in the ovary. However, radiation sensitivity of ova changes with maturity of follicles, the increasing order of sensitivity is - small, mature and intermediate follicles. Exposure of conceptus to ionizing radiation prior to implantation causes detectable abnormalities at later stages of development and result in death of the embryo.

Liver : In acute phases, there is hepatomegaly, ascites and jaundice. However, in later phase, may be after six months, liver is pale and shrunken, there is portal hypertension, right sided congestive heart failure, obstruction of hepatic out flow and periportal fibrosis.

Pituitary : Radiation induces hypopituitarism, there is loss of weight, body hair, skin is dry, pulse is slow, hypothermia , dwarfism, genital atrophy, failure of development of sexual organs in males and females.

Adrenal : Hypertrophy of adrenal cortex.

Pancreas : It is relatively radio-resistant. *Beta* cells of islets are more resistant compared to *alpha* cells.

Post mortem findings :
(i)   Haemorrhagic to ulcerative gastroenteritis.
(ii)   Ulceration of pharyngeal mucosa.
(iii)   Pulmonary oedema.
(iv)   Extravasation of fluids.
(v)   Fibrinous pneumonia, pericarditis.
(vi)   Degeneration of bone marrow and lymphoid tissues.

Diagnosis :
(i)   History of exposure to irradiation or radioactive materials.
(ii)   Clinical signs.
(iii)   Post mortem lesions.

Differential diagnosis :
(i)   Bracken fern in cattle.
(ii)   Trichloroethylene extracted soybean meal.

Prevention : Prevention is the best strategy.

Treatment : No specific treatment is available. Give only the symptomatic treatment alongwith supportive therapy.

CHAPTER 12

FEED ADDITIVES AND PRESERVATIVES

I. A. Siddiqui

Feed additives are defined as drugs, chemicals or biological substances added directly to animal feeds in small quantities, usually in concentrations of a few ppm for the purpose of modifying some aspects of performance or production. Flavouring agents, preservatives, emulsifiers, sweetners, vitamins and colouring agents etc. are added to human food to increase its palatability, acceptability, taste, keeping quality and nutritive value. However, use of such agents in animals or poultry feed is not there except for the pets feed. Rather, to increase the feed efficiency, growth rate and production of animals and birds, certain agents which are not nutrients and not considered as dietary essential are used as feed supplements. Such substances / agents are termed as feed additives. In veterinary practice, growth promoters - such as anabolic steroids, antibacterials, antiparasitic drugs, nonspecific chemicals and rumen fermentation modifiers etc. are added to animals and poultry feeds.

Certain feed additives are absorbed and persist in the body system and can be detected in sufficiently high concentrations in milk, meat or eggs of animals or birds. Consumption of such animal/bird products is not recommended from public health point of view. Indiscriminate use or inappropriate levels of these agents, instead of increasing growth rate and efficiency, may cause toxicity and economic losses and even be fatal at times.

For convenience, feed additives may be classified in various classes as shown in Table 12.1.

Table 12.1 : Different classes of feed additives.

Antibacterials	Ionophores	Hormones and anabolic steroids	Chemicals/ drugs/metals	Miscellaneous
Tetracyclines	Monensin	Thyroxine	Copper sulfate	Rumen
Nitrofurans	Lasalocid	Estradiol	Cobalt oxide	fermentation
Penicillins	Narasin	Diethylstilbestrol	Arsanilic acid	modifiers e.g.
Bacitracin	Salinomycin	Trenbolone	Tranquilizers	Monensin
Chloramphenicol		Zeranol	Sodium	
Neomycin		Hexoestrol	bentonite	
Tylosin		Progesterone	Selenium	
Erythromycin		Testosterone	Organic iodine	
Lincomycin		Iodinated casein		
Sulfonamides		Thyroprotein		
Tiamulin		Pituitary growth hormones		

Antibacterials : Antibacterial growth promoters play an important role in the effective use of animal feeds, both in poultry and pig production. Only those antibacterial agents should be used as feed additives which have little or no therapeutic application in animals and man and those that will not impair the efficacy of antibacterials through the development of resistant strains of microorganisms.

The growth promoting action of antibacterials is due to :
(i)   Reduced activity of pathogens.
(ii)   Eliminate bacteria which produce toxins that reduce the growth of animals.
(iii)   Stimulate the growth of those microorganisms which synthesize nutrients.
(iv)   Increase the absorptive capacity of intestine.

Toxic effects : Antibacterials are generally added in calf starter, poultry feed and pig diet. Feeding of antibacterials at higher than the recommended amount may cause toxicity such as scour in calves, and other side effects such as allergic reactions, eosinophilia, aplastic anaemia, hepato- and nephro-toxicity etc. in other species of animals. Continued and prolonged administration of antibacterials through feed in farm animals could pose a human health hazard because of the potential for development of resistant strains of enteric bacteria in human beings. Lincomycin produces severe gastrointestinal haemorrhages and necrosis. Sulfonamides result in anorexia, depression, haematuria, albuminuria and oliguria. Concurrent use of tiamulin with ionophores causes ionophore toxicity, myonecrosis and death.

Ionophores : Monensin, lasalocid, salinomycin, narasin and maduramicin are used in feeds as anticoccidial agents and for promotion of growth in ruminants, pigs and poultry. Supplementation should be gradual. These should not be given to horses and other equines.

Toxic effects : Feeding of monensin in feed in excess of the recommended doses or improper mixing of feed may produce toxicity. Signs of acute monensin toxicity are dyspnoea, tachycardia, congestive heart, diarrhoea, cardiac failure etc. Feed consumption in cattle is reduced, somtimes by 90% within three days of feeding.

Hormones and anabolic steroids : Hormones having growth promoting properties such as estrogens, androgens, progestrogens, thyroxine and pituitary growth hormones are used as growth stimulants. Anabolic steroids increase nitrogen retention and protein deposition in animals, thereby increasing feed conversion efficiency, lean content and growth rate. In U.K. about 25-30% of steers and heifers are implanted with anabolic agents. Iodinated casein is a commercial product which has activity several times greater than that of dried thyroid gland.

Toxic effects : Anabolic steroids delay the age of puberty, impair udder development and incidence of dystocia is increased. Other effects are restlessness and milk secretion from rudimentary teats. These side effects are more liable to occur with excessively high doses of hormones. Besides side effects in animals, human health is also adversely affected due to carcinogenic potential of the residues of certain synthetic estrogens in milk or meat.

Non specific chemicals : Certain chemicals which are used as feed additives are arsenicals in the form of arsanilic acid, copper sulfate, cobalt oxide, selenium, tranquilizers, organic iodine etc.

Arsenicals : Arsenic compounds have long been used in veterinary medicine as tonics to improve the general health and general appearance of animals. Care is required in the use of arsenicals in animals / birds feed because of the potential risk of building up of dangerous levels of arsenic in the body.

Detailed toxicity of arsenic is discussed elsewhere in the book (see Chapter 2.1).

Copper sulphate: Copper sulphate is used as a growth promotor mainly in the pigs ration. Copper content in majority of the forages and grains is below the dietary requirement of most of the animal species. Soil deficient in copper, reduced biovailability of copper inspite of adequate levels in the soil and certain elemental interactions with sulfur and molybdenum are responsible for further deficiency of copper in forages and grains. Low roughage and high concentrate diets also interfere with the availability of copper as these promote sulfide production. Therefore supplementation of copper in diet of animals is recommended.

Toxicity of copper is discussed in details elsewhere in the book (see Chapter 2.5).

Tranquilizers : Certain tranquilizers such as reserpine- the natural alkaloid of *Rauwolfia serpentina* and metoserpate have been tried to improve the growth rate in fattening beef cattle and poultry. Chlorpromazine alongwith reserpine has been used in poultry to reduce the stress as well. Chlorpromazine in high doses / concentrations may produce toxic manifestions characterised by variable degree of shivering, lethargy, relaxation of anal sphincters, diarrhoea, loss of righting reflex etc. Similarly, reserpine or related compounds due to their catecholamines depleting action may result in severe depression, diarrhoea, polyuria etc. in animals and birds.

Organic iodine: Iodine is a trace element and is added to the ration of animals particularly cattle to correct iodine deficiency. Potassium iodide is used for the treatment of actinobacillosis while ethylenediamine dihydroiodide is used for prevention of foot rot, lumpy jaw and as a mild expectorant. Acute toxicity is rare, but subacute and chronic toxicity are more likely on long term use. Iodine toxicity or adverse effects are characterised by focal dermal exfoliation, congestion of conjunctiva and tracheal mucosa, nasal discharge, bronchopneumonia and enlargement of mediastinal lymph nodes.

CHAPTER 13

CHEMICALS, DRUGS AND PLANTS INDUCED CARCINOGENICITY AND GENOTOXICITY

13.1 CARCINOGENICITY

Yogeshwer Shukla

Cancer is a general term that covers a wide range of nearly 200 diseases, caused by various biological, physical and chemical agents, alone or in combination and characterised by abnormal cell growth. It may be defined as "an abnormal mass of tissue, the growth of which is abnormal and is incoordinated with that of other normal tissues and persists even after the cessation of the stimulus that evoked a change." Induction of cancer is an irreversible and often life threatening process. The development of cancer following exposure to certain chemical(s) is termed as chemical carcinogenesis and the chemical(s) inducing cancer in animals including human beings are referred to as carcinogens. Majority of human cancers are caused by various environmental chemicals.

Carcinogen is an agent administration of which to previously untreated animals results in statistically significant increased incidence of neoplasms of one or more histogenetic types. Several epidemiological studies show that prolonged exposure to benzene causes the induction of acute myelogenous leukaemia in humans. The agents having the capacity or capability of inducing tumour are termed as carcinogens.

Since most of the carcinogens are present in the environment in procarcinogen form, to exert their carcinogenic effects their metabolic activation (also termed as biotransformation) is required. The basic purpose of biotransformation reactions is inactivation or detoxification of the toxicant to facilitate its elimination from the body of organisms. But there are some cases, where the product of biotansformation is more toxic than the parent compound. Details of the biotransformation processes are described in Chapter 1.5 of General Toxicology of this book.

Types of carcinogens :

Carcinogens are classified into several types. Those that interact with and alter the genetic material (DNA) are classified as "genotoxic". This category contains the classic organic carcinogens that are electrophilic reactants either in their parent form i.e. direct acting carcinogens or after metabolism i.e. procarcinogens. Permanent change in DNA is the first key event in the initiation of carcinogenesis by these compounds. The second broad category of the carcinogens is designated as epigenetic agents, comprise those chemicals for which no evidence exists of direct interaction with genetic material, but which produce another biological effect that could be the basis for their individual carcinogenicity. This category contains cytotoxic agents, solid state carcinogens, hormones, immunosuppressants and tumour promoters. The third category includes "unclassified carcinogens" which can not be classified directly under the above two categories. This class includes radioactive compounds (e.g. uranium, polonium and radium), metals (e.g. arsenic, nickel, chromium, cobalt, lead), peroxisome proliferators as certain phthalate plasticizers, hypolipidemic drugs etc.

A more simple classification of the types of carcinogens is also there ; according to which, there are three main types of carcinogens:

(i) Primary or directly acting carcinogens : The agents of this group by virtue of their specific structure(s) react directly and specifically with some particular biological constituents of the body mainly DNA and result in the formation of tumour e.g. organic compounds like nitrogen and sulfur mustards, epoxides, metallic ions such as arsenic, cadmium, lead, chromium, beryllium etc.

(ii) Secondary or procarcinogens : Biologically and chemically, these agents are inert and in the host by metabolic activation are converted to active carcinogens e.g. benzo(a) pyrene by catalytic action of cytochrome P450 is converted to benzopyrene 7, 8, diol-9,10 epoxide which is an ultimate carcinogen. Other procarcinogens are- plant alkaloids, mycotoxins, tannins, nitrosamines, organochlorine compounds etc.

(iii) Cocarcinogens or tumor promoting agents: These compounds are not carcinogenic or mutagenic in nature but can potentiate the primary or procarcinogens-induced carcinogenicity e.g. coal tars, phorbol, myristyl acetate in croton oil, phenol, anthralin etc.

Causes of cancer : There may be enormous causes and factors responsible for cancer which are difficult to enumerate, still these are classified mainly as :
a)   Genetic predisposition
b)   Environmental factors (tobacco, smoke, carbon tetrachloride)
c)   Chemical factors
d)   Unknown factors

These factor(s) solely or in combinations may be responsible for carcinogenesis. Here in this chapter, we are restricting our discussion mainly to chemicals, drugs and plants-induced carcinogenicity which encompass a part of environmental factors.

Chemicals-induced carcinogenicity: Carcinogenesis by chemical agents was first reported by Sir Percival Pott in chimney sweepers over two centuries ago. It is possible to induce tumours experimentally in laboratory animals by a wide variety of chemicals both synthetic and natural. It is often difficult to define precisely the causes of naturally occurring tumours in man and animals. Today, there is a strong evidence that exposure to atleast 20 different chemicals or chemical mixtures in certain industrial, medical or societal situations induce cancer in humans (Table 13.1, 13.2). Majority of known chemical carcinogens are organic compounds.

Table 13.1 : Chemicals associated with occupations and established to be inducing carcinogenicity

Agent	Industries and trades	Primary affected sites
Para-aminodiphenyl	Chemical manufacturing	Urinary bladder
Asbestos	Construction, asbestos mining and milling, production of friction products and cement	Pleura, peritoneum, bronchus
Arsenic	Copper mining and smelting	Skin, bronchus, liver
Alkylating agents (mechloroethamine hydrochloride and bis {chloromethyl} ether)	Chemical manufacturing	Bronchus
Benzene	Chemical and rubber manufacturing, petroleum refining	Bone marrow
Benzidine, *beta*-naphthylamine and derived dyes	Dye and textile production	Urinary bladder
Chromium and chromates	Tanning, pigment making	Nasal sinus, bronchus
Isopropyl alcohol manufacture	Chemical manufacturing	Cancer of paranasal sinuses
Nickel	Nickel refining	Nasal sinus, bronchus
Polynuclear aromatic hydrocarbons from coke, coal tar, mineral oils and creosote	Steel making, roofing, chimney cleaning	Skin, scrotum, bronchus
Vinyl chloride monomer	Chemical manufacturing	Liver
Wood dust	Cabinet making, carpentry	Nasal sinus

Table 13.2 : Chemicals associated with occupations and suspected to be inducing carcinogenicity

Agent	Industries and trades	Suspected sites
Acrylonitrile	Chemicals and plastics	Lung, colon, prostate
Beryllium	Beryllium processing, aircraft manufacturing, electronics, secondary smelting	Bronchus
Cadmium	Smelting, battery making, welding	Bronchus
Ethylene oxide	Hospitals, production of hospital supplies	Bone marrow
Formaldehyde	Plastic, textile and chemical production, health care	Nasal sinus, bronchus
Synthetic mineral fibres (e.g. fibrous glass)	Manufacturing, insulation	Bronchus
Phenoxyacetic acid	Farming, herbicide application	Soft tissue sarcoma
Polychlorinated biphenyls	Electrical equipment production and maintenance	Liver
Organochlorine pesticides (e.g. chlordane, dieldrin)	Pesticide manufacture and application, agriculture	Bone marrow
Silica	Casting, mining, refracting	Bronchus

It is known through many studies that chemical carcinogenesis is a multi step process mainly defined by initiation and promotion. The initial step, initiation is an early, rapid and an irreversible process involving a single subcarcinogenic application/ exposure of a chemical or physical agent, which causes certain molecular changes in the exposed cells. These initiating agents in themselves, however, are insufficient for tumour formation. The process of tumour promotion involves application of promoting agents. These agents which include phorbol esters, bile salts and various hormones lead to full blown carcinogenic effect. Thus it is clear that for the transformation of a normal cell into tumour cell, both initiation and promotion steps are essential, the detailed mechanism of which will be discussed later on in the chapter.

Drugs-induced carcinogenicity : The continuous use of certain drugs is associated with the increase in incidence of some types of cancer (Table 13.3). The most important among these are the anticancer drugs, which are cytotoxic, immunosuppressive and anti-inflammatory, used in the management of organ transplant recipients and patients with autoimmune and collagen diseases. Majority of the anticancer agents have the property of killing cells in cycle by specific interference with the metabolic machinery of the cell. This may lead to the unwanted side effect of increased incidences of certain cancers. Three major classes of anticancer drugs are known for their carcinogenic properties : alkylating agents (cyclophosphamide), antimetabolites (methotrexate) and naturally occurring agents (clofibrate) which may cause various types of acute leukaemia, followed by non-Hodgkin's lymphomas (NHL's), carcinomas of the urinary bladder and other malignancies.

Plants-induced carcinogenicity : Over the past few years, many plants or plant products have been identified to be carcinogenic (Table 13.4). The major sources of carcinogens in plants are natural biosynthetic processes. These processes include synthesis by the plant itself or by fungi that infest the plant, its fruit or its seed. Among the most potent carcinogens of plant origin includes aflatoxins and some pyrrolizidine alkaloids. Besides, other natural products as safrole (the active ingredient of oil of *Sassafras*); cycasin, (water soluble compound present in the nuts, leaves and roots of cycads) etc. have been claimed to possess a carcinogenic potential. Carcinogens of fungal origin (which infest the plants) are aflatoxin, penicillic acid, anthraquinones, crotosin etc. All of them are carcinogenic causing hepatoma, adenoma and renal tumours in animals.

Epidemiology : Epidemiological studies have emerged gradually as a result of series of observations on populations with particular cancer problems. One early observation by Percival Pott in 1775 showed the occurrence of scrotal cancer among chimney sweepers. Since then various types of cancers have been recognized. The epidemiological approach is now widely used in teratological and carcinogenic studies e.g. high levels of catechin in plants have been found to be associated with high incidence of oesophageal cancer in man, cattle and sheep, however, to depend solely on epidemiological data may sometimes be misleading. The International Agency for Research on Cancer (IARC) has published a detailed evaluation of chemicals strongly associated with carcinogenic risk. The risk associated with these chemicals are evaluated on the basis of all types of study but majority of the substances examined, however, lack sufficient data for evaluation.

Table 13.3 : Carcinogenic risk of some of the drugs used in medical therapy

Drug	Associated neoplasms	Evidence for carcinogenicity
Alkylating agents (cyclophosphamide, melphalan)	Bladder, leukaemia	Sufficient
Inorganic arsenicals	Skin, liver	Sufficient
Azothioprine (an immunosuppressive drug)	Lymphoma, reticulum cell sarcoma, skin, Kaposi's sarcoma (?)	Sufficient
Chlornaphazine	Bladder	Sufficient
Chloramphenicol	Leukaemia	Limited
Diethylstilbestrol	Vagina	Sufficient
Estrogens (premenopausal/ postmenopausal)	Liver cell adenoma, endometrium	Sufficient limited
Methoxypsoralen with UV light	Skin	Sufficient
Oxymetholone	Liver	Limited
Phenacetin	Renal pelvis (carcinoma)	Sufficient
Phenytoin (diphenylhydantoin)	Lymphoma, neuroblastoma	Limited
Thorotrast	Liver (angiosarcoma)	Sufficient
Nitrofurazone	Rat breast	Sufficient
Metronidazole	Mouse lung and lymphomas	Sufficient
Chloroform	Mouse liver	Sufficient
Cyclophosphamide	Mouse lung, human bladder	Sufficient

Table 13.4 : Some of the plants or plant products having carcinogenic risks

Toxin	Natural sources	Affected site / organ
Alkylbenzenes (e.g. safrole, sesamol)	Spices, herbs and vegetables (essential oils of plants)	Liver
Allylisothiocyanate	*Brassica* genus (broccoli, brussels sprouts, cabbage)	Urinary bladder
Quercetin, shikimic acid and ptaquiloside	*Pteridium aquilinum* (Bracken fern)	Stomach, intestine, urinary bladder
Cinnamic acid derivatives (caffeic acid, chlorogenic acid etc.)	Vegetables, fruits and seasonings (coffee beans, apple, basil, lavender)	Forestomach, lung, kidney
Flavonoids (quercetin, kaempferol)	Fruits, berries, leaf and root vegetables, cereal grains, tea, coffee and cocoa	Liver
Furocoumarins (psoralen, angelicin)	Members of *Rutaceae* and *Umbelliferae* families (parsley, carrots, celery etc.)	Skin, kidney
Hydrazines and their derivatives (agaritine, cycasin)	Mushrooms (*Agaricus bisporus*) and cycad seeds	Intestines, brain, lungs, kidney, and mammary glands, liver
Mycotoxins-aflatoxins (B_1, B_2, G_1, G_2, M and aflatoxicol) fumonisin B_1, citrinin	*Aspergillus* sp., *Fusarium moniliforme*, *Penicillium* sp.	Liver, lungs and forestomach
N-nitroso compounds	Vegetables	Stomach
Pyrrolizidine alkaloids (clivorine, isatidine, symphytine, lasiocarpine etc.)	Members of *Compositae, Leguminosae* and *Boraginaceae*	Liver, skin, lung, ileum
Terpenes (monoterpene e.g. limonene)	Volatile oil obtained from plant materials e.g. oils of citrus fruit peel	Kidney
Betel quid (arecoline, dimethylsulfoxide extract)	Leaf of Piper betle, aqueous extract of *Acacia catechu*	Buccal pouch, skin, oropharynx, oesophagus
Tannins	Oak (*Quercus* sp.) and *Rhus semilata*, tea (*Camellia sinensis*)	Liver, oesophagus

The sources of chemical exposure are often unknown but majority of them are found to be associated with industry or pharmaceutical preparations. Asbestos represents a good example of the interactions between the employees and industry in relation to hazards and disease. It is found to be associated with lung cancer, stomach and laryngeal cancer. Besides, aromatic amines and polycyclic aromatic hydrocarbons used in chemical and dye industries are well known cancer inducers in humans.

Several drugs including immunosuppressants and chemotherapeutic agents are capable of posing an increased risk of malignant diseases e.g. cyclophosphamide and chlornaphazine, both can produce bladder cancer. In addition, a few specific agents such as phenacetin are capable of inducing renal tract tumours. Epidemiological studies showed that renal transplant patients on immunosuppressants therapy have a 60 fold excess of non-Hodgkin lymphomas (NHL) and other malignancies, especially of the skin. Another similar study has shown that cases with chronic renal failure on dialysis who do not receive immunosuppressive drugs also have an excess risk of NHL.

Epidemiological studies have demonstrated that aflatoxin, especially aflatoxin B_1, which is produced by some strains of *Aspergillus flavus*, are potent hepatocarcinogens in animals and humans. Other naturally occurring chemicals in plants such as the pyrrolizidine alkaloids, and products of the metabolism of contaminating molds which include carcinogenic nitroso compounds have been found to be carcinogenic in experimental systems.

Mechanisms of carcinogenesis: A large number of compounds have been found to induce carcinogenesis by a variety of mechanisms. Since, carcinogenesis is a complex and multistage process of conversion of normal cells into transformed cells. The detailed mechanism of initiation, progression and promotion of cancer is not completely understood. Some of the known basic events involved in chemical carcinogenesis are:

(i)   Metabolic activation of chemical ;
(ii)   DNA adduct formation ;
(iii)   Activation of proto-oncogenes ; and
(iv)   Inactivation of tumor suppressor genes.

Cancer Prevention : Cancer cells are produced by multiple carcinogenic agents and have multiple genetic alterations, therefore, there are multiple approaches to cancer prevention. Traditonal methods of treatment of cancer include surgery, radiotherapy and chemotherapy. The limitation of surgery is the excision of a visible tumour. Similarly, for radiotherapy, tumour must be visible in order to aim the ionizing radiations at the tumour. However, treatment with ionizing radiations decreases the incidence of visible recurrence of the tumour in those sites. The ultimate cure from cancer is chemotherapy because of the ability of drugs to seek out tumour cells that are not visible and those which remain unaffected from radiotherapy.

Several anticancer drugs are known today. Anticancer analogues such as 5-fluorouracil, 6-thioguanine, 6-mercaptopurine, methotrexate and others owe their anticancer effects to the fact that synthetic processes for nucleic acids proceed rapidly in tumorous cells compared to that in normal cells. Interference with the biosynthesis of intermediates or the polymerization reactions that lead to final products serve to slow the growth of cancer. These effects are shared by a number of antibiotics, which also affect processes of high orders of activity in cancer cells. Actinomycin D and adriamycin affect vital nucleolar biosynthetic reactions. Hormone therapy has been useful but not curative in breast cancer and prostrate cancer. Such therapy relates to the ability of the endocrine receptors to be influenced by the circulating hormone.

A newer approach for the treatment of cancer is the therapeutic targeting of cellular defects in oncogenes and tumour suppressor genes.

There are three approaches for the targeting of oncogenes and their products antisense oligodeoxynucleotide or RNA procedures, neutralizing monoclonal antibodies and novel drug development based on altered oncoprotein structure or function. Manipulation of tumour suppressor genes for successful cancer therapy currently seems significantly more distant than potential oncogene based therapies. However, the insertion of wild type tumour suppressor genes into malignant cells may restore normal growth regulation. Suppression of osteosarcoma and prostate carcinoma cell growth *in vivo* is observed after retroviral mediated transfer of RB gene into these cells *in vitro*.

13.2 GENETIC TOXICOLOGY

Neeraj Sinha

Science has improved the modern life of mankind by its achievements and contributions. Science has helped the human beings in many basic problems such as food, clothes, shelter, transport, comfort, communications etc. Not only this, during the past few decades a great stride has occurred in the field of humans as well as animal medicine. It is really a matter of great achievement that the discovery of modern medicines and vaccines has given the answer for various epidemics like small pox, plague etc. which were quite prevalent in the olden times. Not only this, in modern industrial society, new drugs and chemicals (synthetic and natural) have a major place in our daily life and role in achieving the quality and life expectancy that we are enjoying today. However, with the progressive advent of new chemicals and medicines, synthetic or natural, there is always the potential for new problems relating to humans and other biological systems.

Most of the chemicals or chemical products consumed by humans contain a wide variety of both beneficial and toxic substances. Apart from the general toxic effects, exposure of the embryo and the foetus to various chemicals and compounds is a matter of increasing concern. The malformed infants whose mothers got exposed to thalidomide during the early 1960s is one of the most painful reminder of the fact. It is also an established fact that certain chemical agents can alter DNA and fundamentally change gene product or expression. Gene mutations are changes in the sequence of deoxyribonucleic acid (DNA) within a gene. Gene mutations due to exposure to chemicals, drugs, natural plant products are rare in veterinary medicine. Changes are commonly associated with the electrophilic agents that covalently bind to DNA. Specific toxic agents may locate at specific positions on DNA. The type of mutation is determined by where the mutagen binds to the DNA, as well as by the alterations in DNA that the mutagen produces. Lot of drugs/chemicals have been shown to be genotoxic based on *in vitro* gene mutation tests on mouse lymphoma e.g. formaldehyde, acrylonitrile, vincristine, methotrexate, mitomycin C, aliphatic nitrosamines, dimethylhydrazine etc.

Gene mutation, cell transformation and chromosomal aberrations represent the ultimate consequences of deleterious chemical/DNA interactions. As a result of interaction, the sequence of events eventually determine whether the initial DNA lesions result in heritable genetic damage or in just cell lethality. Mutation in specific genes in germinal cells has been shown to be the cause of some stable inherited metabolic disorders expressed in the next gene generation.

Chromosomal aberrations are gross alterations of chromosome structure. Chromosomal abnormalities have been found to be common in conceptus loss and birth defects. It has also been reported that there is a genetic element in atleast 10% of all human pathological conditions, thus genetic changes, if produced in living beings, could be of serious consequence. Extensive epidemiological investigations reveal the importance and role of mutations and chromosomal alterations in human health and genetic disorders. Some of the heritable diseases, still births and spontaneous abortions etc. have been shown to be related to changes in DNA molecules and to chromosomal aberrations due to exposure to environmental pollutants.

Aneuploidy is a deviation from an exact multiple of the haploid number of chromosomes e.g. if the normal deploid chromosome number is 46 (haploid : 23), then 45 or 47 chromosomes would be an aneuploid condition. Aneuploidy is the condition of karyotype which is responsible for various diseases viz. Klinefelters syndrome and Down's syndrome which is due to trisomies where as Turner's syndrome is due to monosomy. All these are resultant of aneuploidy in early embryonic somatic cells or germ cells. Familial polyposis, neurofibromatiosis, reticoblastoma, hepaticporphyria etc. are the examples of dominant mutations whereas xeroderma pigmentosa, haemophilia, sickle cell anaemia, Fanconi's syndrome and albinism, cystic fibrosis, phenylketonuria etc. are typical examples of diseases caused by expression of recessive mutations.

Polyploidy is an increase in a complete set of chromosomes e.g. a change from a diploid to a triploid number of chromosomes.

Besides causing diseases that exhibit Mendelian inheritance, gene mutations undoubtedly contribute to diseases through the genetic component of disorders with a complex etiology. About 6 per cent infants have congenital abnormalities and if one includes multifactorial disorders, that often have a late onset, such as heart disease, hypertension and diabetes, the proportion of population affected may increase to more than 60 per cent. Most of the chromosomal aberrations cause foetal death and others cause serious abnormalities. It has been estimated that 5 per cent of the all recognized pregnancies involve chromosomal abnormalities, as do about 6 per cent of infant deaths and 30 per cent of all spontaneous embryonic and foetal deaths. Out of such ailments, aneuploidy is the most common followed by polyploidy and structural aberrations constitute about 5 per cent of the total. Polygenic type of DNA alterations have been found to be associated with birth defects such as congenital heart defects, clubfoot, idiopathic epilepsy, cleft lip or hare lip, spinabifida and anencephaly.

Association between mutation and cancer has long been evident on indirect grounds. An association of human chromosome instability syndromes and DNA repair defects with increased incidence of cancer risk has been reported. This association has been further strengthened by finding the involvement of specific chromosomal alterations in many leukemias and solid tumours. A normal cell proliferation requires balance between the factors that promote the growth and the factors that restrict the growth. The proto-oncogenes which are responsible for the normal cellular growth when get genetically altered, these become oncogenes which actually stimulate the transformation of normal cells into cancer cells. On the other side tumour suppresser genes, which normally restrain cellular proliferation, if get genetically altered, free the cells from their inhibitory influence. Protooncogenes can be converted into active oncogenes by point mutation or chromosomal alterations and can give rise to many types of mammalian tumours, for instance Burkitt's lymphoma and various other haematopoietic cancers.

Why conduct genetic toxicity tests : With the above on going discussion it is clear that most of the products whether it is man made (synthetic) or natural (of marine or plant origin etc.) may be mutagen and the mutation so induced can be detrimental and could be of serious consequence to the health of future generations if the extent of exposure is sufficient to permit expression of genotoxic properties in somatic or germ cells. Over the last few decades many substances which have been shown to be mutagenic have also been shown to be carcinogenic. Numerous sequential mutations, especially in gene(s) controlling cell growth, are not recognised in common human tumours. Therefore, it is understandable, that tests for genetic changes in germ and somatic cells receive considerable attention when predicting the safety of new materials such as drugs, pesticides and industrial chemical etc. when they are being introduced into commercial use. Moreover correlation between mutagenicity and carcinogenicity stimulated a new interest in the somatic theory of cancer of the 1950s and has resulted in intensive study of the genetic effects of chemical substances.

CHAPTER 14

REPRODUCTIVE TOXICITY, TERATOGENICITY AND EMBRYOTOXICITY

S.P. Singh

14.1 REPRODUCTIVE TOXICITY

Reproductive toxicity can be defined as the adverse effect of an agent, natural or synthetic chemicals including polychlorinated biphenyls (PCBs), dioxins, insecticides, herbicides, drugs and other xenobiotics on reproductive processes in animals. Any stage of reproductive cycle including gametogenesis, copulation, fertilization, preimplantation, implantation, embryonic period, foetal period, maternal-placental-foetal relationship, parturition, suckling and postnatal development to puberty may be vulnerable to toxic agents. This could be manifested by interference with libido, oestrus, oogenesis, spermatogenesis, abnormal mating behaviour, embryocidal effect, still births, teratogenicity and other birth defects.

Toxic effects on reproductive system : Toxic agents may produce deleterious effects on the development and maturation of male and female organs as well as the reproductive phenomena.

Male reproductive system : In males, toxicants may act directly on a particular stage of spermatogenesis i.e. on spermatogonia, spermatocytes, spermatids or spermatozoa, thus resulting in absence of mature fertile sperms. The toxicants may also affect the accessory sex organs. Injury of interstitial cells, leydig cells or adenohypophysis may result in sterile sperms.

Female reproductive system : In females, toxic agents can interfere with various prenatal processes during development and maturation of ovum, ovulation, fertilization, transplantation, gestation or parturition. Many of the xenobiotics, except a few large molecules, diffuse through placenta and/or firmly bind to maternal tissues and produce embryotoxic effects. In addition, they activate one or more embryopathic mechanisms - the interdependent processes of differentiation, growth and morphogenesis. This may result in death either directly or secondary to malformation. The nature of these effects depend on the specificity and toxicity of agent and on the gestational age or period at the time of exposure.

To achieve the objectives of studying the potential of an agent to cause reproductive toxicity and to extrapolate the results to man and animals, basic toxicodynamic and toxicokinetic principles regarding absorption, distribution, biological half-life, binding with intercellular and intracellular macromolecules such as proteins and nucleic acids etc., bioactivation and detoxication, interaction with germ cells, affinity for specific cells or tissues, accumulation in body fluids and tissues and elimination from the body need to be thoroughly understood.

Both acute as well as chronic toxicity studies should be undertaken for screening the toxic effects of test materials on the reproductive performance of the animals. In general, toxicity to male and female reproductive organs and functions can result from direct and indirect action of the chemicals on germ cells with or without any effect on endocrine glands, accessory sex organs and interference with hormone mediated mechanisms.

Agents causing reproductive toxicity : On the basis of mechanism of reproductive toxicity, the toxicants can be divided into various groups :

a) Agents affecting chromosomes, genes and DNA :
(i)   Oligopeptides- nitropsin and distamycin A inhibit RNA and DNA synthesis.
(ii) Polycyclic compounds - acridine and actinomycin D disrupt DNA.
(iii) Anticancerous agents - niridazole suppresses meiotic division.
(iv) Heavy metals - mercury, lead and cadmium interfere with vital enzymatic activity of the cells.

b) Agents affecting spermatogenesis :
(i)   Chlorinated hydrocarbons - hexachlorophene, DDT, heptachlor
(ii) Organophosphates - dichlorovos
(iii) Nitrobenezene - nitrobenzamide
(iv) Alkylating agents - cyclophosphamide, vinblastin.

c) Agents affecting accessory sex glands : Hexachlorophene, imidazole, ketoconazole.

d) Agents affecting ovary : Vinblastin, cyclophosphamide, procarbazine.

Some of the common agents/groups of drugs/chemicals causing reproductive toxicity are descibed below:

Steroids : Both male and female reproductive cycles are affected by various male and female hormonal steroids such as diethylstilbestrol, testosterone, progestogens etc. These compounds interfere with the normal process of reproduction. Any variation in the normal level of these steroids may affect the reproductive processes. In female reproductive cycle, the steroidal compounds such as contraceptive drugs produce toxic effects on fallopian tubes, ovaries and the cervix. Hormonal imbalance also disturbs the functions of endometrium, pituitary gland and hypothalamus. They may have either inhibitory or stimulatory effect on ovulation, morphology of the ovaries, corpora lutea and other follicle developmental processes. Morphologic changes such as hyperplasia of cervical glands increase vascularity and stromal oedema observed in cervix. The changes in endometrium include rapid progression from a proliferative to a secretory pattern leading to glandular atrophy. Estrogen may effect the secretion of leutinising hormone (LH) during estrous cycle. Mestronol in small doses stimulate follicle stimulating hormone (FSH) secretion while higher doses inhibit FSH secretion. The combination of estrogens and progestogens suppress FSH secretions. Daily administration of mestronol in women prevents the maturation of follicles and corpus luteum. It is thus apparent that many of the steroid formulations inhibit various compounds of reproductive cycle in man and animals. The functions of male reproductive system are also affected by these compounds. Diethylstilbestrol may inhibit the maturation of sperm by interfering with the development of spermatocytes. Methallibuse at 20-80 mg/kg/day for 28 days in dogs has been found to inhibit spermatic metamorphosis. Alphosone acetophenide at dose of 1.65 mg/kg and 8.25 mg/kg daily for 48 days intramuscularly suppresses spermatogenesis and sexual activities in cats. Clomiphene affects spermatogenesis at the primary spermatocyte level. Testosterone enanthate suppresses spermatogenesis in man. Steroids also produce adverse effects on the male reproductive system and other organs that secrete essential hormones required for its development. In females, as well, estrogen and progesterone may be the cause of the irregular oestrus cycle, reduced conception rate and infertility.

CNS depressants : Compounds that depress the central nervous system such as phenothiazine tranquilizers- chlorprormazine, perphenazine and promazine and monoamine oxidase inhibitors interfere with the ovulatory process in mice as a result of their inhibitory effect on the release of gonadotropic hormones such as LH-RF or FSH-RF.

Triphenylethylene compounds : These compounds having estrogenic properties interfere with the estrogen-dependent implantation without any toxic effect on the blastocyst in rat, mouse and dog. In addition, they also affect the sperm capacitation, ovum fertilization, cleavage, growth and development.

Non-narcotic analgesics : Prostaglandins synthesis inhibitors such as aspirin and phenacetin following prolonged administration have been found to affect spermatogenesis in rats. They may be harmful to reproductive phenomena in man and animals.

Pesticides : Various insecticides such as DDT, endrin and dieldrin- the organochlorines; malathion, dichlorovos and monocrotophos- the organophosphates, and synthetic pyrethroids- cypermethrin and alphamethrin produce deleterious effects on the reproductive processes including degenerative changes in testicular epithelium.

Metals and trace elements : Aluminium, boron, cadmium, cobalt, zinc, selenium, mercury, molybdenum, nickel, silver, lithium, mercury, lead and other heavy metals and trace elements cause testicular damage to semeniferous epithelium and leydig cells in several species of animals.

Alkylating agents : Antineoplastic alkylating agents such as hydrazines, nitrogen mustard, procarbazine and cyclophosphamide produce reproductive toxicity in man and animals. Actinomycin D, adriamycin and bleomycin also have deleterious effect on development and maturation of germ cells. Aziridine compounds cause toxic effects to spermatogonia and spermatids by causing chromosomal abnormalities.

Esters of methane sulphonic acid like ethyl methane sulphonate at an oral dose of 5 mg/kg produce adverse effects on spermatids and mature sperms. Diminished fertility of males was also reported in rats treated with busulphan. Busulphan and dimethyl busulphan act primarily on spermatogonia.

Anticonvulsants and tranquilizers : Repeated and prolonged use of phenothiazines, reserpine, MAO inhibitors and anticonvulsants like phenytoin have potential to produce toxic effects on reproductive cells and organs in male and females.

Miscellaneous therapeutic agents : Various therapeutic agents including diuretics e.g. thiazides, antimetabolites e.g. amino acid analogues; trimethotrexate, folic acid antagonists; caffeine and theobromine- xanthine derivatives and guanethidine-an antiadrenergic agent may also cause toxic effects on the reproductive system in males and females.

Environmental pollutants : A large number of environmental contaminants such as herbicides - 2,4-D, 2,4,5-T, diquat and paraquat; rodenticide-fluoroacetate; fungicides and fumigants-alpholate, captan and thiocarbamate; polycyclic aromatic hydrocarbons (PAHs)- dimethylbenzathracene etc.; solvents - benzene, carbon disulphide etc. and phthalate plasticizers can cause damage to reproductive processes in animals.

Radiation : Physical factors like heat, light, hypoxia and atmospheric pressure may cause toxicity to reproductive system directly or may act as predisposing factors to other agents. Environmental radiation with *alpha, beta, gamma,* and X-rays cause detrimental effect on the development and maturation of the germ cells in addition to tumorous growth in genital organs in man and animals.

14.2 TERATOGENICITY AND EMBRYOTOXICITY

Extensive use of thalidomide as sedative and hypnotic agent by pregnant women during 1960-61 in Germany, Japan and other parts of the world resulted in birth of about 10,000 malformed children with phocomelia, a teratogenic condition characterized by reduction of limbs. Clinical observations revealed that women who had taken thalidomide between days 35 and 50 of pregnancy gave birth to malformed children. Thalidomide tragedy gave a new dimension to the toxicological evaluation of the drugs, chemicals and pesticides before their introduction for the welfare of man and animals.

Teratological changes may occur as a result of physiological and biochemical alterations in the formation of cells, tissues and organs. Thus teratogenic changes can only occur during the period of formation of cells, tissues and organs i.e. during the period of organogenesis but not after. Teratogenic changes affect the structure and function of the developing cells, tissues and organs and are observed in developing organisms. These are manifested as external anomalies- anatomic defects such as skeletal and visceral abnormalities, loss of viability and still birth at the time of parturition. Mental disorders, however, may be apparent during infantile growth.

Teratology : The term teratology has been derived from the word '*terata*' means 'monster' and is thus the study of malformed foetuses. Teratology focuses on the study of malformation of the foetuses as a result of toxic effects of physical, nutritional, hereditary or chemical nature. Thus, teratology can be defined as the branch of toxicology which embraces the study of adverse effects of physical, chemical and nutritional agents or hereditary factors on embryonic development during organogenesis phase of gestation period. In brief teratology is the study of congenital malformations. The agent having teratogenic potential is called teratogen and the phenomenon is teratogenesis.

It deals with persistent development defects that are manifested subsequent to parturition or hatching . Persistent defects involving behaviour, reproductive capacity, metabolic function, motor coordination, neoplasia, immune function and other malformations associated with specific organs or organ systems like heart, kidney, liver etc. can occur as a result of morphological defects.

Exposure of embryo *in utero* in man and animals to drugs and other chemicals may result not only in transient and reversible toxicity but also more serious and permanent or semipermanent detrimental defects for entire life.

Physical teratogenesis is produced by physical agents such as atmospheric pressure, temperature and radiations. Chemical teratogenesis is produced by chemicals such as drugs, pesticides and environmental pollutants. Congenital malformations caused by inheritance of genes and chromosomal combinations is called hereditary teratogenesis. Maternal overfeeding or malnutrition leading to foetal malformation is the subject of study under nutritional teratogenesis.

Embryotoxicity vs teratogenicity : Both terms are closely related and explain the toxic effects of aetiological agents on the embryonic development. The only difference is the period of exposure of the pregnant animals. Any adverse effect on the embryonic development at any time during the whole gestation period by any causative agent resulting in the birth of malformed foetus is referred as embryotoxicity. Teratogenicity is referred as the adverse effect resulting into abnormal development of the organs or systems due to exposure to a causative agents only during organogenesis phase of gestation. These abnormalities are, however, manifested in offspring only after parturition or hatching. Most of the embryotoxic agents may be teratogenic but some of the agents are specific for producing teratogenicity.

Basic principles of teratogenicity

Prenatal Development : Prenatal development period may be divided into three different periods :

a) Predifferentiation period : This period consists of interval between the fertilization of the oocyte and its implantation in the endometrium. During this period, the ovum while remaining little undifferentiated, undergoes a series of mitotic divisions changing from unicellular zygote to a multicellular blastocyte. Chemicals producing malformation during the later phases of development are thought to be without teratogenic potential as this period is not susceptible for teratogenesis.

b) Embryonic period : Completion of major organ formation or organogenesis is referred as embryonic period. This phase starts soon after implantation. During organogenesis, the embryo is maximally susceptible to gross structural changes if exposed to teratogenic agents. The nature of defects will depend on embryonic cells and the gestational stage at the time of exposure.

Teratogenic agents initiate a series of pathological effects, beginning at cellular level. An agent may have a very specific toxic action at cellular level. It might affect cellular physiology such as protein synthesis or enzyme activity, ionic imbalance, cellular permeability or any other change making cells more vulnerable to teratogenic effects. The net outcome may be the abnormal development and malformed growth. The nature of defects depends on the gestation stage, type of the toxicant, the cellular mechanism, organ involved, duration of exposure to the toxicant and nutritional factors. Toxicokinetic factors such as absorption, distribution, metabolism, diffusion through placenta and half-life of the toxicants also play an important role in determining the teratogenic potential of the toxicants.

c) Period of foetus : Distribution of the toxicants and their metabolites within the foetus greatly determine their effects on the foetus. Foetal exposure is largely dependent on the concentration of the toxicants in foetal circulation, protein binding and distribution in the foetal tissues. Foetotoxic agents may cause growth retardation, functional disruption and foetal death. A number of agents including drugs, pesticides and other environmental pollutants have been found to produce foetal injury (Table 14.1).

Table 14.1 : Agents causing foetal injury

Growth retardation	Foetal death	Functional disruption
Aspirin	Aspirin	Antidepressants
Caffeine	Cocaine	Benzodiazepines
Marijuana	Narcotics	Carbamazepine
Narcotics	Nicotine	Methadone
Propranolol	Propranolol	
Steroids		
Thiazides		

Epidemiological information is a valuable tool in the early identification of possible teratogens. Identification of thalidomide embryopathy exemplifies the usefulness of epidemiological data. The incidence of limb reductions was steadily increased with the increased use of thalidomide as sedative in pregnant women in certain geographical areas. Finally thalidomide was identified as a teratogen. Five parameters have been identified that may influence the detection of a new teratogen by "birth defect monitoring system" :

i)   The frequency of exposure to suspected teratogen during critical stage of development.

ii) The strength with which the suspected agent exerts its teratogenic effects.

iii) The proportion of exposed foetuses that will be affected by the suspected teratogens.

iv) The incidence of malformation within the general population.

v)   The size of population to be monitored.

The association between *in utero* exposure to valproic acid and the development of spina bifida was first recognised by the birth defects monitoring system. In another example, the teratogenic potential of anticonvulsants was also established in epileptic women, characterized by higher incidence of facial clefts in offsprings.

Epidemiological data depicting malformation associated with *in utero* exposure to a xenobiotic needs to be confirmed by controlled animal studies. A teratogen is an agent that causes a physical malformation when maternal administration results in significant foetal exposure during the period of organogenesis. Malformations as morphological defects may result from an intrinsically abnormal developmental process. These defects would thus be described as disruptions and not malformations. The following basic principles may be stressed upon to consider the teratogenic potential of therapeutic agents :

i)   Variation in the individual susceptibility to teratogens.

ii) Phenotypic expression depends on exposure of the susceptible individuals during a critical period of development.

iii) A teratogen exerts its effect through specific target sites at cellular level.

iv) Teratogenic potential is manifested in the form of foetal death, malformation, growth retardation or the anomalies in the development of a function systems.

v)   Exposure of the foetus to a teratogen is dependent on the physicochemical properties of the agent.

vi) A dose- response relationship exists with increasing doses resulting in foetal compromise.

vii) Pharmacokinetic factors of a teratogen influence its teratogenic potential in an individual.

The pharmacokinetic profile of xenobiotics may influence their apparent teratogenicity because it largely determines the foetal drug exposure by the function of maternal placental - foetal system. The pharmacokinetic patterns of the toxicant i.e. absorption, distribution and elimination influence the availability of toxicant for foetal exposure. Rate and extent to which these agents cross the placenta are determined by their physico-chemical characteristics such as lipid solubility, polarity and molecular weight. This has further been established that the teratogenic potential is dose-dependent. The dose-dependency of the embryotoxic effects may be an important difference between teratogenesis, carcinogenesis and mutagenesis. In the latter, structural abnormality results from the alteration of a single cell. Therefore, any dose capable of affecting a single cell may result in carcinogenesis or mutagenesis. On the other hand, in foetal malformation, a critical number of cells must be injured and / or cellular regeneration must be impaired indicating a dose-effect threshold. Many of the environmental and industrial chemicals including heavy metals, pesticides etc. have been found to produce embryotoxicity and teratogenicity in animals.

Histological examination of neonatal tissues during neonatal period has been somewhat neglected. When functional defects are observed in motor or sensory systems, histological investigation should be undertaken to determine, if any, visible pathological and biochemical changes in various areas of peripheral and central nervous system. Similarly other functional changes should be confirmed by histopathological and biochemical examination.

Mechanisms of teratogenic action : Most of the teratogens exert their cellular effects through specific mechanism of action. These cellular effects result in abnormal development of the specific organ systems, without damaging the other organ systems at the same time.

Various mechanisms may be involved in teratogenicity. Some of these mechanisms have been confirmed experimentally, however, others are also well documented. Following cellular mechanisms may be responsible for teratogenesis :

i)   Mutation;
ii)   Chromosomal non-disjunction and break ;
iii)   Mitotic interference ;
iv)   Altered nucleic acid integrity and function ;
v)   Lack of precursors and substrate ;
vi)   Osmolar imbalance ;
vii)   Altered membrane characteristics ;
viii)   Altered energy sources ;
ix)   Altered metabolic processes ; and/or
x)   Enzyme inhibition.

Agents causing teratogenicity and embryotoxicity : Most of the physical and chemical agents causing reproductive toxicity in animals may also produce teratogenic and embryotoxic effects in susceptible animals if exposed during gestation period in appropriate dose levels. The etiological agents for teratogenicity may be classified into different categories as described below and the suspected teratogens are listed in Table 14.2.

Physical agents : Hyper- or hypothermia in dams and even atmospheric temperature and pressure may induce teratogenic effects in animals. Hyperthermia causes abortion and eye and teeth deformities in rats, mice and guinea pigs. Irradiation with *alpha, beta, gamma* and X-rays is also known to induce congenital malformations in man and animals.

Hormones : Estrogen and their analogues produce adverse effects on the embryonic development. Diethylstilbestrol produces vaginal adenosis in females whereas androgens cause masculinization. Clomiphene, an ovulation inducing agent causes cataract in rats. Corticosteroids have been found to produce cleft palate and other malformations in laboratory animals and premature parturition in goats and sheep.

Table 14.2 : Some of the suspected teratogens

Drugs/agents	Suspected abnormalities
Alcohol	Growth retardation, cardiac, limb and facial abnormalities
Alkylating agents	Multiple defects
Amphetamine	Learning and motor deficits
Androgens	Cleft lip palate, muscular and skeletal
Aminoglycosides	8th cranial nerve damage
Barbiturates	Cleft lip palate
beta- cis-retenoic acid	Hydrocephalus
Benzodiazepines	Craniofacial
Cigarette smoke	Intrauterine retarded growth
Cocaine	Cardiac, skeletal and renal
Chlorinated biphenyles	Multiple defects
Chloroquine	Deafness and haemolysis
Cytarabine	Limb and ear defects
Dicoumarol	Bleeding and foetal death
Diethylstilbestrol	Vaginal adenosis
Diphenyl hydantoin	Facial, cardiac and limb
Ethanol	Foetal alcohol syndrome
Folate antagonist	Multiple defects
Griseofulvin	Cleft palate
Heparin	Facial, optic and skeletal malformation
Heroin	Foetal death
Inorganic iodides	Congenital goiter
Lithium	Hypothyroidism
Methyl mercury	CNS lesions
Penicillin	Mental allergic response
Penicillamines	Loose skin
Propranolol	Growth retardation
Steroids	Skeletal and muscle
Tetracycline	Skeletal
Thalidomide	Phocomelia
Thiazides	Neonatal jaundice
Tolbutamide	Foetal death
Valproic acid	Neural tube defects
Warfarin	Blindness, skeletal

Nutritional agents : Deficiency and excess of various vitamins and other nutrients including amino acids, minerals, carbohydrates and lipids may cause congenital malformations. Deficiency of vitamins A, D and E and minerals like iron, copper, selenium, zinc, iodine and other trace elements during organogenesis may result in teratogenicity in man and animals e.g. deficiency of manganese causes arthrogryphosis, excess of selenium results in hoof abnormalities etc.

Therapeutic agents : Various therapeutic agents including CNS depressants, nonsteroidal antiinflammatory drugs, antineoplastic agents, antibiotics etc. have been suspected to cause teratogenicity and embryotoxicity in man and animals.

Environmental pollutants : Extensive use of chemicals such as insecticides, fungicides, herbicides etc. in agriculture and animal husbandry practices has made man and animals more vulnerable to their toxic effects including teratogenic and embryotoxic. Organochlorines e.g., DDT, aldrin, dieldrin, endrin etc., organophosphates, e.g. monocrotophos, dichlorovos etc., carbamates e.g. aldicarb, carbaryl etc. and synthetic pyrethroids e.g. alphamethrin, cypermethrin etc. have been shown to cause congenital malformations.

Heavy metals : Mercury, cadmium, arsenic and lead are most potent heavy metals and cause reproductive disorders and congenital malformation in man and animals. Methyl mercury causes lesions in CNS leading to nervous anomalies.

Miscellaenous agents : Alkylating agents including nitroso compound such as dimethyl nitrosamine, nitrosomethyl urea and nitrosoethyl urea cause embryotoxic and teratogenic effects in rats. Polychlorinated biphenyls like arochlor cause reproductive toxicity and teratogenicity in experimental animals. Aflatoxins produced by genus *Aspergillus* are potent teratogenic toxins to man and animals.

Teratogenic plants : Ingestion of a number of vegetative alkaloids and glycosides cause teratogenicity in animals. *Vincristine alba* and *Veratrum californicum* are well known teratogenic plants. Some other plants known or suspected to produce teratogenic and embryotoxic effects in animals are listed in Table 14.3.

Table 14.3 : Suspected teratogenic plants

Name of the plant	Suspected malformation
Astragalus sp.	Skeletal
Conium maculatum (Hemlock)	Skeletal
Gossypol	Spermatogenesis and oogenesis
Lathyrus sp.	Skeletal (Osteolathyrism)
Locoweed	Foetal oedema, abortion and death, enlargement of heart
Lupinus sp.	Skeletal (crooked calf disease)
Nicotiana sp.	Skeletal, cleft palate
Pinus ponderosa	Retarded growth and abortion
Veratrum californicum	Skeletal and facial defect (cyclopian eye), abortion and foetal death

REFERENCES AND FURTHER READINGS

1. *ACS Monograph* No. 182 (1984). American Chemical Society, Washington DC.

2. Adams, H.R. (1995). Adrenergic agonists and antagonists. In: *Veterinary Pharmacology and Therapeutics* (Adams, H.R. , editor). 7th edition. Iowa State University Press, Ames. pp 87-113.

3. Adam, S.E.I., Tartour, G., Obeid, H.M. and Idris,O.F. (1973). Effect of *Ipomoea carnea* on the liver and on some serum enzymes in young ruminants. *Journal of Comparative Pathology* **83**: 531-534.

4. Ajl, S.J., Kadis, S. and Montic, T.C. (1971). *Microbial Toxins.* Vol. I-VIII. Academic Press, New York.

5. Alam, M. (1980). Marine biotoxins. In: *Environment and Health.* (Trieff, N.M.,editor). Ann Arbor Science Publishers Inc., Ann Arbor, Michigan. pp 479-498.

6. Along, J.E., Fehrenbach, F.J., Freer, J.H. and Jeljaszwicz, J. (1984). *Bacterial Protein Toxins.* Academic Press Inc., London.

7. Aplin, T.E.H.; Steele, P. and Nottle, M.C. (1983). *Toxic ferns of Western Australia,* South Perth 6151, Western Australia, Department of Agriculture, Jurral Road *Technical Bulletin,* No. 63, p 15.

8. Aposhian, H.V. and Aposhian, M.M. (1990). Meso-2,3-dimercaptosuccinic acid: chemical, pharmacological and toxicological properties of orally effective chelating agents. *Annual Review of Pharmacology* **39**: 279-306.

9. Ayub Shah, M.A., Garg, Satish Kumar, Garg, K.M., Farooqui, M.M., Anwar, A. and Sabir, M. (1994). Toxicological evaluation of Stomp 30 EC (*Pendimethalin*) with particular reference to haematological and haemo-biochemical changes in rats. *Indian Journal Toxicology* **1**: 17-24.

10. Ayub Shah, M.A., Garg, Satish Kumar and Garg, K.M. (1997). Subacute toxicity studies on pendimethalin in rats. *Indian Journal of Pharmacology* **29**: 322-324.

11. Ayub Shah, M.A., Garg, Satish Kumar, Garg, K.M., Farooqui, M.M. and Sabir, M. (1998). Chronic toxicity studies on pendimethalin in rats. *Indian Journal of Toxicology* **5** (1): 41-44.

12. Ayub Shah, M.A. and Gupta, P.K. (1997). Biochemico-toxicological study on permethrin- a synthetic pyrethroid insecticide in rats. *Indian Journal of Toxicology* **4** (1): 57-60.

13. Ayub Shah, M.A. and Gupta, P.K. (1998). Influence of permethrin- a synthetic pyrethroid insecticide on immune responses of mice. *Indian Journal of Toxicology* 5 (2): 13-19.

14. Ayub Shah, M.A., Gupta, P.K. and Tandon, H.K.L. (1996). Effect of permethrin- a synthetic pyrethroid on pentobarbital-induced sleeping time and hepatic microsomal constituents in mice. *Indian Journal of Toxicology.* 2 (2): 19-23.

15. Bapat, B.N., Godbole, S.H., Vartak, V.D. and Wagle, P.M. (1978) Allergenicity of *Parthenium hysterophorus* Linn. *Indian Journal of Medical Research* 68 : 1007.

16. Barrass, N. and Anderson, D. (1990). Molecular genetic toxicology. In : *Experimetal Toxicology-The basic Issues* (Anderson, D and Conning, D.M., editors). Royal Society of Chemistry, Cambridge. pp 287-304.

17. Bast, A. (1996). Biotransformation: species differences and determining factors. In: *Toxicology: Principles and Applications* (Niesink, R.J.M., de Vries, J. and Hollinger, M.A., editors). CRC Press, New York, London, Tokyo. pp 67-91.

18. Beier, R.C. (1990). Natural pesticides and bioactive components in foods. *Review of Environmental Contamination and Toxicology* 113: 47-137.

19. Beier, L.C., Norman, J.O., Irvin, T.R. and Witzel, D.A. (1987). Microsomal activation by constituents of white snakeroot (*Eupatorium rugosum* Houtt) to form toxic products. *American Journal of Veterinary Research* 48: 583-585.

20. Blaauboer, B.M. (1996). Biotransformation: detoxification and bioactivation. In: *Toxicology: Principles and Applications* (Niesink, R.J.M., de Vries, J. and Hollinger, M.A., editors). CRC Press, New York, London, Tokyo. pp 41-65.

21. Blackwell, W.H. (1990). *Poisonous Plants.* Prentice-Hall Inc., New Jersey.

22. Blood, D.C. and Radostits, O.M. (1989). *Veterinary Medicine.* 7 th edition, ELBS, London.

23. Booth, D.M. (1995). The analgesic-antipyretic-antiinflammatory drugs. In: *Veterinary Pharmacology and Therapeutics* (Adams, H.R., editor). 7th edition. Iowa State University Press, Ames. pp 432-450.

24. Booth, N.H. and McDonald, L.E. (1988). *Veterinary Pharmacology and Therapeutics.* 6th edition. Panima Publishing Corporation, New Delhi.

25.   Brown, S.A. (1996). Fluoroquinolones in animal health. *Journal of Veterinary Pharmacology and Therapeutics* **19**:1-14.

26.   Buck, W.B., Osweiler, G.D. and van Gelder, G.A. (1976). *Clinical and Diagnostic Veterinary Toxicology.* Dubuque, Iowa, Kendall/Hunt.

27.   Carlson, G.P. and Schoeing, G.P. (1980). Induction of liver microsomal NADPH-cytochrome reductase and cytochrome P450 by some pyrethroids. *Toxicology and Applied Pharmacology* **52**: 507-512.

28.   Carolyn, H.M. and Anthony, T.T. (1988). Bacterial toxins (*Handbook of Natural Toxins*, Vol. IV). Marcel Dekker Inc., New York.

29.   Casida, J.E., Gammon, D.W., Gockman, A.H. and Lawrence, L.J. (1983). Mechanisms of selective action of pyrethroid insecticides. *Annual Review of Pharmacology and Toxicology* **23**:413-438.

30.   Casteel, S.W. and Barley, E.M.Jr. (1986). A review of zinc phosphide poisoning. *Veterinary and Human Toxicology* **28**:151-154.

31.   Cheeke, P.R. (1989).Toxicants of plants origin. In: *Phenolics.* Volume 1-4, C.R.C. Press, Boca Raton, FL.

32.   Cheeke, P.R. (1996). *Natural Toxicants in Feeds, Forages and Poisonous Plants.* 2nd ediion, Interscience Publishers Inc., Illinois.

33.   Cheng, L.W. (1977). Neurobiological effects of mercury- a review. *Environmental Research* **14**: 329-373.

34.   Chick, B.F.; Quinn, C.; McCleary, B.V. (1985) Pteridophyte intoxication of livestock in Australia. In: *Plant Toxicology, Proceedings of the Australia-USA Posionous Plants Symposium*, Brisbane, 1984,(Seawright, A.A., Heagarty, M.P., James, L.F., Keeler, R.F., eds.) Queensland Poisonous Plants Committee Queensland, Department of Primary Industries Animal Research Institute, Yeerogpilly, Queensland, pp. 453-464.

35.   Chopra, R.N., Chopra, I.C., Handa, K.L. and Kapoor, L.D. (1958). *Indigenous Drugs of India.* U.N. Dhar and Sons Pvt. Ltd., Calcutta. pp 260-262.

36.   Chopra, R.N., Nayar, S.L. and Chopra, I.C. (1956). *Glossary of Indian Medicinal Plants.* Council of Scientific and Industrial Research, New Delhi.

37.   Clarke, E.G.C. and Clarke, M.L.C. (1967). *Garner's Veterinary Toxicology*, 3rd edition, Bailliere Tindall & Cassel, London.

38.   Clark, I.A. and Dimmock, C.K. (1971). The toxicity of *Chelianthus sieberi* to cattle and sheep. *Australian Veterinary Journal* **47** : 149.

39.   Clarke, M.Y., Harvey, D.G. and Humphreys, D.J. (1981). *Veterinary Toxicology*. 2nd edition. ELBS and Bailliere Tindall, London.

40.   Clark, J.M. and Matsumura, F. (1982). Two different types of inhibitory effects of pyrethroids on nerve Ca^{2+} and $Ca-Mg^{2+}$ ATPase activity in squid (*Loligo pealei*). *Pesticide Biochemistry and Physiology* **18**: 180-190.

41.   Cockerham, L.G., Mickley, G.A., Walden Jr., T.L. and Stuart, B.O. (1994). Ionizing radiation. In: *Principles and Methods of Toxicology*, 3rd edition. (Hayes, A.W., editor). Raven Press Ltd., New York. pp 447-496.

42.   Colegate, S.M. and Dorling, P.R. (1994). *Plant Associated Toxins-agricultural, phytochemical and ecological aspects*. CAB International, Wallingford, U.K.

43.   Costa, D.L. and Amdur, M.O. (1996). Air pollution. In: *Casarett and Doull's Toxicology: The Basic Science of poisons*. (Klassen, C.D., editor). 5th edition. Mc Graw Hill, New York. pp 857-882.

44.   David. M. and Palmer, S.J. (1980). Water and health. In: *Environment and Health* (Trieff, N.M., editor). Ann Arbos Science Publishers Inc., Ann Arbor, Michigan. pp 163-214.

45.   Deshpande, S.S. and Sathe, S.K. (1991). Toxicants in plants. In: *Mycotoxins and Phytotoxins* (Sharma, R.P. and Salunkhe, D.K., editors). C.R.C. Press, Boca Raton, FL. pp 671-730.

46.   Devegowda, G., Raju, M.V.L.N., Afzali, N. and Swamy, H.V.N.L. (1998). Mycotoxin picture worldwide: novel solution for counteraction. *Pashudhan* **13**: 1.

47.   Diniz, M.F.; Basile, J.R.; Camargo, N.J. De.(1984) Natural poisoning of mules with *Pteridium aquilinum* in Brazil. *Arquivo Brasilero de Medicina Veterinaria e Zootecnia* **36** : 515-520.

48.   Dixon, R.L. and Hall, J.L. (1982). Reproductive toxicology. In: *Principles and Methods in Toxicology* (Hayes, A.W., editor). Raven Press, New York.

49.   Dobberstein, R.H., Tin-Wa, M., Fong, H.H.S., Crane, F.A. and Farnsworth, N.R. (1977). Flavonoid constituents from *Eupatorium altissimum* L. *Journal of Pharmaceutical Sciences* **66**: 600-602.

50.   Dorner, F. and Drews, J. (1986). Pharmacology of bacterial toxins (*International Encyclopedia of Pharmacology and Therapeutics*, Section 19) Pergamon Press, Oxford/London.

51. Dowling, R.M. and McKenzie, R.A. (1993). *Poisonous Plants- a field guide*. Information Series Q192035, Department of Primary Industries, Queensland.

52. Echobichon, D.J. (1979). Hydrolytic mechanisms of pesticide degradation. In: *Advances in Pesticide Science* (Geissbuhler, H., editor). Pergamon, New York. pp 516-524.

53. Echobichon, D.J. (1996). Toxic effects of pesticides. In: *Casarett and Doull's Toxicology: The Basic Science of poisons*. (Klassen, C.D., editor). 5th edition Mc Graw Hill, New York. pp 643-690.

54. Edwards, B.L.(1983) Poisoning by *Pteridium aquilinum* in pregnant sows. *Veterinary Record* **112** : 459-460.

55. Elliot, M., Janes, N.F., Kimmel, E.C. and Casida, J.E. (1972). Metabolic fate of pyrethrin-I, pyrethrin-II and allethrin administered orally to rats. *Journal of Agriculture and Food Chemistry* **20**: 300-313.

56. Engel, R.A. and Smith, A.H. (1994). Arsenic in drinking water and mortality from vascular disease-an ecological analysis in 30 countries in the USA. *Archieves of Environmental Health* **49** : 418-428.

57. EPA *Special Report on Ingested Inorganic Arsenic*: Skin cancer and nutritional essentiality. Risk Assessment Form US Protection Agency, Washington (1987).

58. Essen, G.E.M., Rendy, J. and Neims, A. (1983). Methyl mercury exposure in Northern Quebec II. Neurological fluids in children. *American Journal of Epidemiology* **118**: 470-478.

59. Evans, I.A. (1968). The radiomimetic nature of bracken toxin. *Cancer Research* **28** : 2252-2261.

60. Evans, W.C. and Evans, E.T.R. (1949) The effect of inclusion of the bracken (*Pteris aquilina*) in the diet of rats and the problem of bracken poisoning in farm animals. *British Veterinary Journal* **105** : 175-186.

61. Evans, W.C.; Evans, E.T.R. and Hughes, L.E. (1954) Studies on bracken poisoning in cattle-Part I. *British Veterinary Journal* **110** : 295-306.

62. Evans, E.T.R., Evans, W.C. and Roberts, H.E. (1951a) Studies on bracken poisoning in the horse. *British Veterinary Journal* **107** : 364-371.

63. Evans, E.T.R., Evans, W.C. and Roberts, H.E. (1951b) Studies on bracken poisoning in the horse. *British Veterinary Journal* **107** : 399-411.

64. Evans, I.A.; Jones, R.S. and Mainwaring-Burton, R. (1972) Passage of bracken toxicity into milk. *Nature* (London) **237**: 107-108.

65. Fenwick, G.R. (1988). Bracken (*Pteridium aquilinum*)- toxic effects and toxic consituents. *Journal of the Science of Food and Agriculture* **46** : 147-173.

66. Forysth, A.A. (1968). *British Poisonous Plants*, Maff Bulletin 161, HMSO, London, pp 28-30.

67. Fowler, M.E. (1992). *Veterinary Zootoxicology,* C.R.C. Press, Boca Raton.

68. Fox, M.W. (1961). Castor seed residue poisoning in dairy cattle. *Veterinary Record* **73**: 885-86.

69. Galli, C.L. , Murphy , S.D. and Paoletti, R. (1980*). The Principles and Methods in Modern Toxicology*. Elsevier Press. pp 125-159.

70. Gallo, M.A. (1996). History and scope of toxicology. In: *Casarett and Doull's Toxicology: The Basic Science of poisons*. 5th edition. Mc Graw Hill, New York. pp 1-11.

71. Gangolli, S.D. and Phillips, J.C. (1990). The metabolism and disposition of xenobiotics. In: *Experimental Toxicology-The Basic Issues*. (Anderson, D. and Conning, D.M. editors). Royal Society of Chemistry, Cambridge. pp 130-169.

72. Gardener, C.A. and Bennetts, H.W. (1956). *The Toxic Plants of Western Australia*. West Australian News Papers Ltd., Periodicals Division, Perth.

73. Garg, Satish K., Gupta, R.P., Verma, S.P. and Garg, B.D. (1989). Adverse effects of daily administration of gentamicin for a week in buffalo calves. *Indian Journal of Animal Sciences* **59** : 1101-1103.

74. Garg, Satish Kumar, Ayub Shah, M.A., Garg, K.M., Farooqui, M.M. and Sabir, M. (1997). Biochemical and physiological alterations following short term exposure to fluvalinate- a synthetic pyrethroid. *Indian Journal of Pharmacology* **29**: 250-254.

75. Garg, Satish Kumar, Ayub Shah, M.A., Garg, K.M., Farooqui, M.M. and Sabir, M. (1997). Antilymphocytic and immunosuppressive effects of *Lantana camara* leaves in rats. *Indian Journal of Experimental Biology* **35**:1315-1318.

76. Garg, Satish K., Makkar, H.P.S., Nagal, K.B., Wadhwa, D.R., Sharma, S.K. and Singh, B. (1992). Oak poisoning in cattle. *Veterinary* and *Human Toxicology* **34**:161-164.

77. Garg, Satish K., Rastogi, S.K., Varshneya, C. and Gupta, V.K. (1992) Toxicological profile of fluvalinate- a pyrethroid. *Indian Journal of Pharmacology* **24**:154-157.

78. Garg, Satish K., Thaker, A.M., Verma, S.P., Uppal, R.P. and Garg, B.D. (1987). Neurotoxic effects of pendimethalin- a weedicide. *Indian Journal of Experimental Biology* **25** : 463-466.

79. Garg, Satish K., Thaker, A.M. , Verma, S.P., Uppal, R.P. and Garg, B.D. (1996). Biochemical and histopathological alterations and tissue residues of gentamicin in rabbits following repeated parenteral administration. *Indian Journal of Toxicology* **3** : 45-50.

80. Gerken, D.F. (1995). *Kirk's Current Veterinary Therapy* XI, *Small Animal Practice.* W.B. Saunders, Philadelphia.

81. Ghafoor, M.A.,Sabnis, M.G., Vadlamudi, V.P. and Karalgikar, B.N. (1977) A note on *Ipomoea carnea* poisoning in goats. *Research Bulletin of Marathwada Agricultural University,* Parbhani I : 107.

82. Gibson, J.A. and O'Sullivan, B.M. (1984). Lung lesions in horses fed mist flower (*Eupatorium riparium*). *Australian Veterinary Journal* **61**: 271.

83. Gill, J.M. (1993). Selenium poisoning in dairy cattle. *Newzeland Veterinary Journal* **41**:46-52.

84. Glickman, A.H. and Casida, J.E. (1982). Species and structural variations affecting pyrethroid neurotoxicty. *Neurobehavioural Toxicology and Teratology* **4** : 793-799.

85. Gockerham, L.G., Mickley, G.A. Walden, T.L. Jr. and Staurt, B.O. (1994). Ionizing radiation. In: *Principles and Methods in Toxicology* (Hayes, A.W., editor). 3rd edition. Raven Press, New York. pp 447-496.

86. Goyer, R.A. (1996). Toxic effects of metals. In: *Casarett and Doull's Toxicology: The Basic Science of Poisons.* (Klassen, C.D., editor). 5th edition. Mc Graw Hill, New York. pp 691-736.

87. Granth, D.L., Phillips, W.L. Jr. and Hatiana, G.V. (1977). Effects of hexachlorobenzene on reproduction in the rats. *Archieves of Environmental Contamination* **5**: 207-216.

88. Grasso, P. (1990). Testing for carcinogenicity. In: *Experimental Toxicology-The Basic Issues.* (Anderson, D. and Conning, D.M. editors). Royal Society of Chemistry, Cambridge. pp 309-330.

89. Gray, A.P. (1984). Design and structure- activity relationships of antidotes to organophosphorus anticholinesterase agents. *Drug Metabolism Review* **15**: 557-589.

90. Gregg, K. (1995). Engineering gut flora of ruminant livestcok to reduce forage toxicity: progress and problems. *Trends in Biotechnology* **13**: 418.

91.   Grollman, A. and Slaughter, D. (1947). *Cushny's Pharmacology and Therapeutics*. 13th edition. Lea and Febiger, Philadelphia.

92.   Gupta, P.K. (1986). *Pesticides in The Indian Environment*. Interprint. New Delhi.

93.   Gupta, I. and Nauriyal, M.M.(1966) *Acacia leucophloea* Wild (*Raunja*) poisoning in livestock. *Indian Veterinary Journal* **43**: 538-540.

94.   Harding, J.D.J. (1972). Bracken poisoning in pigs. *Agriculture* **9**: 313-314.

95.   Harley, N.H. (1996). Toxic effects of radiations and radioactive materials. (1996). In: *Casarett and Doull's Toxicology: The Basic Science of Poisons*. (Klassen, C.D., editor). 5 th edition. Mc Graw Hill, New York. pp 773-800.

96.   Harvey, R.B., Kubena, L.F., Phillips, T.D., Corrier, D.E., Elissalde, M.H. and Huff, W.E. (1991). Dimunition of aflatoxin toxicity to growing lambs by dietary supplementation with sodium calcium aluminosilicate. *American Journal of Veterinary Research* **52**: 152.

97.   Hatch, R.C. (1988). Poisons causing respiratory insufficiency. In: *Veterinary Pharmacology and Therapeutics* (Booth, N.H. and McDonald, L.E., editors). 6th edition. Panima Publishing Corporation, New Delhi. pp 1007-1052.

98.   Hatch, R.C. (1988). Poisons causing nervous stimulation or depression. In: *Veterinary Pharmacology and Therapeutics* (Booth, N.H. and McDonald, L.E., editors). 6th edition. Panima Publishing Corporation, New Delhi. pp 1053-1101.

99.   Hatch, R.C. (1988). Poisons causing abdominal distress or liver or kidney damage. In: *Veterinary Pharmacology and Therapeutics* (Booth, N.H. and McDonald, L.E., editors). 6th edition. Panima Publishing Corporation, New Delhi. pp 1102-1125.

100.   Hatch, R.C. (1988). Poisons causing lameness or visible disfigurement. In: *Veterinary Pharmacology and Therapeutics* (Booth, N.H. and McDonald, L.E., editors). 6th edition. Panima Publishing Corporation, New Delhi. pp 1126-1131.

101.   Hatch, R.C. (1988). Poisons having unique effects. In: *Veterinary Pharmacology and Therapeutics* (Booth, N.H. and McDonald, L.E., editors). 6th edition. Panima Publishing Corporation, New Delhi. pp 1132-1142.

102.   Hayes, A.W. (1994). *Principles and Methods of Toxicology*. 3rd edition. Raven Press Ltd., New York.

103.   Hayes, J.A. (1989). Metal toxicity. In: *A Guide to General Toxicology*. (Marquis, J.K. and Mass, B., editors), 2nd edition. Karger. pp 179-189.

104. Herz, W., Watanabe, H., Miyazaki, M. and Kishida,Y. (1962) The structures of parthenin and ambrosin. *Journal of American Chemical Society* **84**: 2601.

105. Hirono I., Kono, Y., Takahishi, K., Yamada, K., Niwa, H., Ojika, M., Kigoshi, H., Niiyama, K. and Uosaki, Y. (1984) Reproduction of acute bracken poisoning in a calf with ptaquiloside, a bracken constituent. *Veterinary Record* **115**: 375-378.

106. Hindmarsh, W.L. (1930) The lethal dose of hydrocyanic acid for ruminants. *Journal of Council of Scientific and Industrial Research* **3**: 12.

107. Hobbs, W.R., Rall, T.W. and Verdoorn, T.A. (1996). Hypnotics and sedatives: ethanol. In: *Goodman and Gilman's The Pharmacological Basis of Therapeutics* (Hardman, J.G., Limbird, L.E., Molinoff, P.B., Ruddon, R.W. and Gilman, A.G., editors). 9th edition. McGraw-Hill, New York. pp 361-398.

108. Hoffmann, G.R. (1996). Genetic toxicology. In: *Casarett and Doull's Toxicology: The Basic Science of poisons* (Klaassen, C.D., editor). 5th edition. Mc Graw Hill, New York. pp 269-300.

109. Hungerford, T.G. (1990). *Hungerford's Diseases of Livestock*. 9th edition. McGraw-Hill Book Company, Sydney.

110. Hurst, E. (1942). *The Poisonous Plants of New South Wales*, New Southwales Poisonous Plants Committee, Sydney.

111. IARC *Scientific Publication* No. 78 (1986). International Agency for Research on Cancer, Lyon, France.

112. Idris, O.F., Tartpur, G., Adam, S.E.I. and Obeid, H.M. (1973). Toxicity to goats by *Ipomoea carnea*. *Tropical Animal Health and Production* **5**: 119-123.

113. Insel, P.A. (1996). Analgesic-antipyretic and antiinflammatory agents and drugs employed in the treatment of gout. In: *Goodman and Gilman's The Pharmacological Basis of Therapeutics* (Hardman, J.G., Limbird, L.E., Molinoff, P.B., Ruddon, R.W. and Gilman, A.G., editors). 9th edition. McGraw-Hill, New York. pp 617-658.

114. Jelinek, C.F., Pohlan, A.E. and Wood, G.E. (1989). Worldwide occurrence of mycotoxins in foods and feeds- an update. *Journal of Association of Official Analytical Chemists* **72**: 223-230.

115. Jones, M.M. (1991). New development in therapeutic-chelating agents as antidotes for metal poisoning. *Critical Review of Toxicology* **21**: 209-233.

116. Jones, H.W., Hoerr, N.L. and Osol, A. (1949). *Blakinston's New Gould Medical Dictionary*. Blakinston, Philadelphia.

117. Joy, R.M. (1994). Pyrethrins and pyrethroid insecticides. In: *Pesticides and Neurological Diseases* (Echobichon, D.J. and Joy, R.M., editors). 2nd edition. Boca Raton, FL:CRC. pp 291-312.

118. Jubb, K.V.F., Kennedy, P.C. and Palmer, N. (1985). Nervous system -Anoxia and anoxic poisons, malacia and malacic diseases. In: *Pathology of Domestic Animals- Volume I.* 3rd edition. Academic Press, London. pp. 249-274.

119. Jubb, K.V.F., Kennedy, P.C. and Palmer, N. (1985). Skin and appendages - Physico-chemical diseases of skin. In: *Pathology of Domestic Animals- Volume I.* 3rd edition. Academic Press, London. pp 433-443.

120. Jubb, K.V.F., Kennedy, P.C. and Palmer, N. (1985).Haematopoietic system- disorders of haemoglobin. In: *Pathology of Domestic Animals- Volume III.* 3rd edition. Academic Press, London. pp 84-216.

121. Kalant, H. and Orrego, H. (1989). Drugs, alcohol and the liver. In: *Principles of Medical Pharmacology.* 5th edition. (Kalant, H. and Roschlau, W.H.E., editors). B.C. Dekker Inc., Toronto. pp 620-631.

122. Kaldrumidou, E., Poluzopoulou, T. and Pepasteriadas, A. (1994). Subclinical lead poisoning in sheep: ultrastructural study of lesions in the liver and kidneys. *Bulletin of Hellenic Medical Society* **45**: 283-290.

123. Kaur, H, Srivastava, A.K., Garg, Satish K. and Prakash, D. (1997). Chlorpyriphos-induced pathomorphological changes in secondary lymphoid organs in kids. *Indian Journal of Veterinary Pathology* **21**: 145-147.

124. Kaur, H, Srivastava, A.K., Garg, Satish K. and Prakash, D.(1998). Studies on pathophysiological aspects of acute chorpyriphos toxicity in goats. *Indian Journal of Toxicology* **5**: 53-58.

125. Kaur, H, Srivastava, A.K., Garg, Satish K. and Prakash, D.(2000). Acute chlorpyriphos toxicity in goats- a pathomorphological study. *Indian Journal of Veterinary Pathology* (in press).

126. Kaur, H., Srivastava, A.K., Garg, Satish K. and Prakash, D.(2000). Subacute oral toxicity of chlorpyriphos in goats with particular reference to blood biochemical and pathomorphological alterations. *Indian Journal of Toxicology,* **7**: (In press).

127. Kaur, H., Srivastava, A.K., Garg, Satish K. and Prakash, D.(2000). Subacute dermal toxicity of chlorpyriphos in goats with particular reference to clinicopathological aspects. *Indian Journal of Veterinary Research,* **8** (in press).

128. Klaassen, C.D. (1996). *Casarett and Doull's Toxicology: The Basic Science of poisons.* 5th edition. Mc Graw Hill, New York.

129. Klaassen, C.D. (1996). Heavy metals and heavy metal antgonists. In: *Goodman and Gilman's The Pharmacological Basis of Therapeutics* (Hardman, J.G., Limbird, L.E., Molinoff, P.B., Ruddon, R.W. and Gilman, A.G., editors). 9th edition. Mc Graw-Hill, New York. pp 1649-1672.

130. Klaassen, C.D. (1996). Non-metallic environmental toxicants: air pollutants, solvents, vapours and pesticides. In: *Goodman and Gilman's The Pharmacological Basis of Therapeutics* (Hardman, J.G., Limbird, L.E., Molinoff, P.B., Ruddon, R.W. and Gilman, A.G., editors). 9th edition. McGraw-Hill, New York. pp 1673-1696.

131. Keeler, R.F., Van Kampen, K.R. and James, L.F. (1978). *Effects of Poisonous Plants on Livestock*. Academic Press, New York/San Francisco/London.

132. Keeler, R.F., Van Kampen, K.R. and James, L.F. (1978). Oxalate poisoning in livestock. In: *Effects of Poisonous Plants on Livestock* (Keeler, R.F., Van Kampen, K.R. and James, L.F., editors) Academic Press, New York/San Francisco/London. pp 139-146.

133. Koeman, J.H. (1996). Toxicology, history and scope of the field. In: *Toxicology: Principles and Applications* (Niesink, R.J.M., de Vries, J. and Hollinger, M.A., editors). CRC Press, New York, London, Tokyo. pp 3-16.

134. Konshi, T.; Ichijo, S. (1984) Experimentally-induced equine bracken poisoning by thermostable antithiamine factor (SF factor) extracted from dried bracken. *Journal of Japan Veterinary Medical Association* **37**: 730-734.

135. Kotsonis, F.N., Burdock, G.A. and Flamm, W.G. (1996). Food toxicology. In : *Casarett and Doull's Toxicology: The Basic Science of poisons* (Klaassen, C.D., editor). 5th edition. Mc Graw Hill, New York. pp 909-950.

136. Leinweber, F.J. (1987). Possible physiologiacl roles of carboxylic ester hydrolases. *Drug Metabolism Review* **18** (4): 379-439.

137. Loomis, T.A. (1978). *Essentials of Toxicology*, 3rd edition. Lea and Febiger, Philadelphia.

138. Lorgue, G., Lechnet, J. and Riviere, A. (1996). *Clinical Veterinary Toxicology*. Blackwell Science Inc., Cambridge.

139. Lu, L.P., Zhuo, Ma, WenFen, Li and Chang, XueFeng (1997). Studies on lead-Cd poisoning in sheep. *Chinese Journal of Veterinary Sciences* **17**: 166-169.

140. MacPharson, A., Milne, E.m. and MacPharson, A.J. (1997). Copper poisoning in ewes. *Veterinary Record* **141**: 631.

141. Malzia, E.L., Sarceinelli and Andreucci (1977). *Ricinus* poisoning : a familiar epidemic. *Acta Pharmacologie et Toxicologie* **41** (*Supplement* 1): 351-361.

142. Mannering, G.J. (1976). Microsomal enzyme systems which catalyze drug metabolism. In : *Fundamentals of Drug Metabolism and Drug Disposition* (Bret, N., Mandel, H.G. and Way, E.L., editors). Williams and Wilkins Co., Baltimore, USA. pp 206-252.

143. Marletta, M.A. (1989). Chemical basis of toxicology. In: *A Guide to General Toxicology.* (Marquis, J.K. and Mass, B., editors), 2nd edition. Karger. pp 1-23.

144. Marquis, J.K. (1989). General toxicology of pesticides. In: *A Guide to General Toxicology* (Marquis, J.K. and Mass, B, editors), 2nd edition. Karger. pp 157-178.

145. Martin,J.H., Couch,J.F.and Briese,P.R. (1938). Hydrocyanic acid content of different parts of the *Sorghum* plant. *Journal of American Society of Agronomy* **30** : 725.

146. McCera, C.T.; Head, K.W. (1981). Sheep tumours in Northeast Yorkshire II: prevalence on seven Moorland farms. *British Veterinary Journal* **137**: 21-30.

147. McCleod, N.S.M.; Greig, A.; Bonn, J.M.; Angus, K.W. (1978). Poisoning of cattle associated with *Dryopteris flix-max. Veterinary Record* **102** : 239-240.

148. McGuigan, M.A. (1989). Clinical Toxicology. In: *A Guide to General Toxicology.* (Marquis, J.K. and Mass, B, editors), 2nd edition. Karger. pp 216-262.

149. McGuigan, M.A. (1989). Principles of toxicology and poisons and antidotes. In: *Principles of Medical Pharmacology.* 5th edition. (Kalant, H. and Roschlau, W.H.E., editors). B.C. Dekker Inc., Toronto. pp 712-722.

150. McKellar, Q.A. (1996). Clinical relevance of the pharmacologic properties of fluoroquinolnes. *Compendium on Continuing Education for Practicing Veterinarians* **18** *(2 Supplement)* :14-21.

151. Menzler, A.D. and Schreiner, A.W. (1970). Copper-induced acute haemolytic anaemia: a new complication of haemodialysis. *Annale of Internal Medicine* **73**: 409-412.

152. Miller, W.C. and West, G.P. (1953). *Black's Veterinary Dictionary.* 3rd edition. Adam and Charles Black, London.

153. Mishra,A.and Misra,S.N. (1965) Study on the toxic effect of *Ipomoea carnea* in sheep, goats and cattle. *Indian Veterinary Journal* **42** : 703.

154. Montgomery, R.D. (1980). Cyanogens. In: *Toxic Constituents of Foodstuffs* (Liener, I.E., editor) Academic Press, New York. pp 143-160.

155. More, P.R., Vadlamudi, V.P., Deshpande, B.B. and Qureshi, M.I. (1981). A note on pathological changes in *Parthenium hysterophorus* L. toxicity in buffalo calves.*Indian Journal of Veterinary Pathology* **5** : 21.

156. More, P.R., Vadlamudi, V.P. and Qureshi, M.I. (1982) Note on the toxicity *of Parthenium hysterophorus* in livestock. *Indian Journal of Animal Sciences* **52** : 456.

157. Moslen, M.T. (1996). Toxic responses of the liver. In: *Casarett and Doull's Toxicology: The Basic Science of poisons (*Klaassen, C.D., editor). 5th edition. Mc Graw Hill, New York. pp 403-416.

158. Mount, M. (1993). Toxicology. In: *Textbook of Veterinary Internal Medicine* (Ettinger, S.J., editor). 4th edition. W.B. Saunders, Philadelphia.

159. Murphy, M.J. (1996). *A Field Guide to Common Animal Poisons.* Iowa State University Press, Ames.

160. Murphy, S.D. (1980). Pesticides. In :*Casarett and Doull's Toxicology: The Basic Science of Poisons* (Doull, J., Klassen, CD. and Amdur, M.O., editors). 2nd edition. McMillan Publishing Co. Inc., New York. pp 357-408.

161. Musch, A. (1996). Exposure: qualitative and quantitative aspects. . In: *Toxicology: Principles and Applications* (Niesink, R.J.M., de Vries, J. and Hollinger, M.A., editors). CRC Press, New York, London, Tokyo. pp 17-28.

162. Nadkarni, A.K. (1954). *Indian Materia Medica.* Popular Prakashan, Bombay.

163. Nagy, B. and Fekete, P. Z. (1999). Enterotoxigenic *Eschercia coli* (ETEC) in farm animals. *Veterinary Research* **30**: 259-284.

164. Narahashi, T. (1992). Nerve membrane Na^+-channels as targets of insecticides. *Trends in Pharmacological Sciences* **13**: 236-241.

165. Narasimhan, T.R., Ananth, M., Naryanaswamy, M., Rajendra Babu, M., Mangala, A. and Subba Rao, P.V. (1977). Toxicity of *Parthenium hysterophorus. Current Science* **46** : 15.

166. Narasimhan, T.R., Ananth, M., Narayana Swamy, M., Rajendra Babu M., Mangala, A. and Subba Rao, P.V. (1977) Toxicity of *Parthenium hysterophorus* L to cattle and buffaloes. *Experentia*, **33**: 1358.

167. Niesink, R.J.M (1996). Absorption, distribution and elimination of xenobiotics. In: *Toxicology: Principles and Applications* (Niesink, R.J.M., de Vries, J. and Hollinger, M.A., editors). CRC Press, New York, London, Tokyo. pp 93-135.

168. Norton, S. (1996). Toxic effects of plants. In: *Casarett and Doull's Toxicology: The Basic Science of poisons.* (Klassen, C.D., editor). 5th edition. Mc Graw Hill, New York. pp 841-854.

169. O'Brien, R.D. (1967). Carbamates. In: *Insecticides: Action and Metabolism* (O'Brien, R.D., editor). Academic Press Inc., New York. pp 86-107.

170. Oehme, F.W. (1987). Plant toxicities. In: *Current Therapy in Equine Medicine* II. W.B. Saunders, Philadelphia.

171. Olsnes, S. and Pihl, A. (1978), Abrin and ricin: two toxic lectins. *Trends in Biochemical Science* 3: 7-10.

172. O'Sullivan, B.M. (1985). Investigations into Crofton weed (*Eupatorium adenophorum*) toxicity in horses. *Australian Veterinary Journal* **62**: 30-32.

173. Osweiler, G.D. (1996). *Toxicology.* 1st edition. Williams and Wilkins. Philadelphia.

174. Osweiler, G.D., Carson, T.L., Buck, W.B. and van Gelder, G.A. (1985). *Clinical and Diagnostic Veterinary Toxicology.* 3rd edition. Debuque: Kendall/Hunt, Iowa.

175. O' Toole, D. and Reibeck, M.F. (1995). Pathology of experimentally-induced chronic selenosis (alkali disease) in yearling cattle. *Journal of Veterinary Diagnostic Investigations* 7: 364-373.

176. Palmer, S.J. (1980). Food, nutrition and human health. In: *Environment and Health* (Trieff, N.M.,editor). Ann Arbor Science Publishers Inc., Ann Arbor, Michigan. pp. 411-448.

177. Parkinson, A. (1996). Biotransformation of xenobiotics. In: *Casarett and Doull's Toxicology: The Basic Science of poisons* (Klaassen, C.D., editor). 5th edition. Mc Graw Hill, New York. pp 113-186.

178. Pass, M.A. and Stewart, C. (1984). Administration of activated charcoal for the treatment of *Lantana* poisoning of sheep and cattle. *Journal of Applied Toxicology* 4: 267-269.

179. Pass. M.A. (1986). Current ideas on the pathophysiology and treatment of *Lantana* poisoning of ruminants. *Australian Veterinary Journal* **63**: 169-171.

180. Pass, M.A. (1991). Poisoning of livestock by *Lantana* plants. In: *Toxicology of Plant and Animal Compound* (Keeler, R. and Anthony, T., editors). Tu Marcel Dekker Inc., New York.

181. Pavlic, M.R. and Zorko, M. (1983). Role of dehydration of the esteratic site in acetylation of cholinesterase. In: *Cholinesterases: Fundamentals and Applied Aspects* (Burzin, N., Bernard, A. and Sket, D., editors). *Proceedings of 2nd International Meeting on Cholinesterases.* Yugoslavia. pp 21-36.

182. Penberthy, J. (1893) Vegetable poisoning (?) simulating antharax in cattle. *Journal of Comparative Pathology and Therapeutics* **6**: 266-275.

183. Phillips, T.D., Kubena, L.F., Harvey, R.B., Yayler, D.R. and Heidelbaugh, H.D. (1988). Hydrated sodium calcium aluminosilicate: a highly sorbent for aflatoxin. *Poultry Science* **67**: 243.

184. Pitot III, HC and Dragan, Y.P. (1996). Chemical carcinogenesis. In: *Casarett and Doull's Toxicology: The Basic Science of poisons* (Klaassen, C.D., editor). 5th edition. Mc Graw Hill, New York. pp 201-268.

185. Poulton, J.E. (1983). Cyanogenic compounds in higher plants and their toxic effects. In: *Handbook of Natural Toxins, Volume I, Plant and Fungal Toxins* (Keeler, R.F. and Tu, A.T., editors). Marcel Dekker, New York. pp 117-160.

186. Parke, D.V. (1974). Pesticides. In: *The Biochemistry of Foreign Compounds* (Parke, D.V., editor). Pergamon Press Ltd. New York. pp 206-252.

187. Prasad, B., Joshi, H.C., Choudhari, P.C. (1977). Role of fern *Diplazium esculentum* in chronic bovine haematuria; some clinical, biochemical and pathological studies. *Pantnagar Journal of Research* **2**: 69-73.

188. Prober, C. (1989). Antimicrobial agents that affect synthesis of cellular proteins. In: *Principles of Medical Pharmacology.* 5th edition. (Kalant, H. and Roschlau, W.H.E., editors). B.C. Dekker Inc., Toronto. pp 557-568.

189. Prober, C. and Uetrecht, J. (1989). Drugs affecting cellular nucleic acid synthesis. In: *Principles of Medical Pharmacology.* 5th edition. (Kalant, H. and Roschlau, W.H.E., editors). B.C. Dekker Inc., Toronto. pp 569-578.

190. Qureshi, M.I., Vadlamudi, V.P and Wagh, K.R. (1980). A study on subacute toxicity of *Parthenium hysterophorus* L. in goats. *Livestock Advisor* **V**: 39.

191. Radeleff, R.D. (1970). *Veterinary Toxicology.* 2nd edition. Lea and Febiger, Philadelphia.

192. Radostits, O.M., Blood, D.C. and Gay, C.C. (1994). *Vetrinary Medicine*. A Text book of the Diseases Cattle, Sheep, Pigs, Goats and Horses. 8th edition. ELBS and Bailliere Tindall, London.

193. Rahman, A. and Mia, A.S. (1972). *Abrus precatorius* poisoning in cattle. *Indian Veterinary Journal* **49**:1045-47.

194. Rang, H.P., Dale, M.M. and Ritter, J.M. (1999). *Pharmacology*. 4th edition. Churchill Livingstone, London.

195. Reddy, C.S. and Hayes, A.W. (1994). Food borne toxicants. In: *Principles and Methods in Toxicology* (Hayes, A.W., editor). 3rd edition. Raven Press, New York. pp 317-360.

196. Rhodes, M.L. (1974). Hypoxic protection in paraquat poisoning. A model for respiratory distress syndrome. *Chest* **66**: 341-342.

197. Riviere, J.E. and Spoo, J.W. (1995). Aminoglycoside antibiotics. In: *Veterinary Pharmacology and Therapeutics* (Adams, H.R., editor). 7th edition. Iowa State University Press, Ames. pp 797-819.

198. Roberson, E.L. (1988). Antinematodal drugs. In: *Veterinary Pharmacology and Therapeutics* (Booth, N.H. and McDonald, L.E., editors). Panima Publishing Corporation, New Delhi. pp 882-927.

199. Robertson, J.B. (1989). Toxicology of ionizing radiation. In: *A Guide to General Toxicology* (Marquis, J.K. and Mass, B, editors), 2nd edition. Karger. pp 141-156.

200. Rozman, K.K. and Klaassen, C.D. (1996). In: *Casarett and Doull's Toxicology: The Basic Science of poisons* (Klaassen, C.D., editor). 5th edition. Mc Graw Hill, New York. pp 91-112.

201. Runnells, R.A., Monlux, W.S. and Monlux, A.W. (1976). *Principles of Veterinary Pathology*. 2nd edition. Scientific Book Agency, Calcutta.

202. Russell, F.E. (1996). Toxic effects of animal toxins. In: *Casarett and Doull's Toxicology: The Basic Science of poisons.* (Klassen, C.D., editor). 5th edition Mc Graw Hill, New York. pp 801-840.

203. Saunders, D.S. and Harper, C. (1994). Pesticides. In: *Principles and Methods of Toxicology* (Hayes, W., editor). 3rd edition. Raven Press Ltd., New York. pp 389-415.

204. Schwarting, A.E. (1961). *Toxicology, Mechanisms and Analytical Methods*, Vol. II (Stewart, C.P. and Stolman, A.P., editors). Academic Press, London.

205. Severs, R.K. (1980). Air pollution and health. In: *Environment and Health* (Trieff, N.M.,editor). Ann Arbor Science Publishers Inc., Ann Arbor, Michigan. pp 123-142.

206. Sharma O.P. (1984). Review of the biochemical effects of *Lantana camara* toxicity. *Veterinary* and *Human Toxicology* **26**: 488-493.

207. Sharma, O.P., Makkar, H.P.S. and Dawra, R.K. (1988). A review of the noxious plant *Lantana camara. Toxicon* **26**: 975-987.

208. Smith, H.A., Jones, T.C. and Hunt, R.D. (1972). *Veterinary Pathology*. Lea and Febiger, Philadelphia.

209. Smith, L.L. (1987). The mechanism of paraquat toxicity in the lungs. *Review of Biochemical Toxicology* **8**: 37-71.

210. Spoo, J.W. and Riviere, J.E. (1995). Sulfonamides. In: *Veterinary Pharmacology and Therapeutics* (Adams, H.R., editor*)*. 7th edition. Iowa State University Press, Ames. pp 753-773.

211. Spoo, J.W. and Riviere, J.E. (1995). Chloramphenicol, macrolides, lincosamides, fluoroquinolones and miscellaneous antibiotics. In: *Veterinary Pharmacology and Therapeutics* (Adams, H.R., editor*)*. 7th edition. Iowa State University Press, Ames. pp 820-854.

212. Springfield, A.C., Carlson, G.P. and De Feo, J.J. (1973). Liver elargement and modification of hepatic microsaomal drug metabolism in rtas by pyrethrum. *Toxicology and Applied Pharmacology* **24**: 298-308.

213. Stewart, C., Lamberton, J.A., Fairclough, R.J. and Pass, M.A. (1988). Vaccination as a possible means of preventing lantana poisoning. *Australian Veterinary Journal* **65**: 349-352.

214. Storrar, D.M. (1893). Cases of vegetable poisoning in cattle. *Journal of Comparative Pathololgy and. Therapeutics* **6**: 276-279.

215. Stryczek, J. (1984) Male fern causing intestinal changes in cattle. *Medycyna Weterynaryjna* **40** : 488-489.

216. Subba Rao, P.V., Mangala, A., Subba Rao,B.S. and Prakash, K.M. (1977). Clinical and immunological studies on persons exposed to *Parthenium hysterophorus* L. *Experentia* **33** : 1387.

217. Sunderman, F.M. (1987) Bracken poisoning in sheep. *Australian Veterinary Journal* **64**: 25-26.

218. Swaminathan, M. and Daniel, V.A. (1970). Toxicants naturally occuring in feeds. *Indian Journal of Nutrition and Dietetics* **7**: 105.

219. Tahirov, T.H.O., Lu, T.H. and Liaw, Y.C. (1994). A new crystal form of abrin-a from the seeds of *Abrus precatorius*. *Journal of Molecular Biology* **235**: 1152-1153.

220. Tartour, G., Adam, S.E.I., Obeid, H.M. and Idris, O.F. (1974). Development of anaemia in goats fed with *Ipomoea carnea*. *British Veterinary Journal* **130** : 271.

221. Tartour, G., Obeid, H.M., Adam, S.E.I. and Idris, O.F. (1973). Haematological changes in sheep and calves following prolonged oral administration of *Ipomoea carnea*. *Tropical Animal Health and Production* **5** : 284.

222. Tondon, S.K. and Magos, L. (1980). Effect of kidney damage on the mobilisation of mercury by thiol complexing agents. *British Journal of Industrial Medicine* **37**: 128-132.

223. Towers, G.H.N.,Mitchell,J.C.,Rodriguez,E.,Bennett,F.D. and Subba Rao, P.V. (1977). Biology and chemistry of *Parthenium hysterophorus* L., a problem weed in India. *Journal of Scientific and Industrial Research* **36** : 672.

224. Trieff, N.M. (1980). *Environment and Health* . Ann Arbor Science Publishers Inc., Ann Arbor, Michigan.

225. Vancutsem, P.M., Babish, J.G. and Schwark, W.S. (1990). The fluoroquinolone antimicrobials: structure, antimicrobial activity, pharmacokinetics, clinical use in domestic animals and toxicities. *Cornell Veterinary* **80**:173-186.

226. van den Bercken, J. and Vijverberg, H.P.M. (1983). Interaction of pyrethroids and DDT-like compounds with the sodium channels in the nerve membrane. In: *Pesticide Chemistry. Human Welfare and Environment. Mode of Action, Metabolism and Toxicology* (Miyamito, J. and Kearney, P.C., editors). Pergamon, Oxford, London. pp 115-121.

227. Wallis, T.E. (1967). *Text Book of Pharmacognosy*. 5th edition. J.and A., Churchill Ltd., London.

228. Wang, Y.D.; Xu, L.R.; Wen, L.J.; Ma, W.L.; You, S.Z.; He, K.R. and Yang, R.X. (1984). Studies on experimental bovine bracken poisoning. *Acta Veterinaria et Zootechnica Sinica* **15**: 235-239.

229. Weswig, P.H.; Freed, A.M.; Haag, J.R. (1946). Antithiamine activity of plant materials. *Journal of Biological Chemistry* **165**: 737-738.

230. Wiber, C.G. (1980). Toxicology of selenium: a review. *Clinical Toxicology* **7**: 171-230.

231. Williams, P.D., Holohan, P.D. and Ross, C.R. (1981). Gentamicin nephrotoxicity: acute biochemical correlates in rats. *Toxicology and Applied Pharmacoloy* **61**: 234-242.

232. Wilson, J.G. (1965). *Teratology: Principles and Techniques*. University of Chicago Press, Chicago. pp 262-277.

233. Wilson, J.G. (1977). Current status of teratology. In: *Handbook of Teratology* (Wilson, J.G. and Fraser, F.L. editors). Pleneram, New York. pp 476.

234. Winge, D.R. and Mehra, R.K. (1990). Host defenses agents in copper toxicity. *International Review of Experimental Pathology* **31**: 47-83.

235. Wintzer, H.J. (1986) *Equine Diseases*, Verlag Paul Parley, Berlin. pp 415-416.

236. WHO (1990). *Permethrin*. Environmental Health Criteria No. 94. World health Organization, Geneva.

237. Zakir, M., Vadlamudi, V.P. and More, P.R. (1986). Haemolytic property of *Ipomoea carnea* leaves in goats. *Journal of Maharashtra Veterinary Asssociation* **I** : 15.

238. Zakir, M., Vadlamudi, V.P., Qureshi, M.I. and Degloorkar, M.N. (1987). Haematological changes in *Ipomoea carnea* (*Besharam*) toxicity in goats. *Livestock Advisor* **XII** : 41.

239. Zakrzewski, S.F. (1991). *Principles of Environmental Toxicology*. American Chemical Society, Wasington DC.

Index